MEDICAL INTELLIGENCE UNIT

DNA Methylation and Cancer Therapy

Moshe Szyf, Ph.D.

Department of Pharmacology and Therapeutics
McGill University
Montreal, Quebec, Canada

LANDES BIOSCIENCE / EUREKAH.COM
GEORGETOWN, TEXAS
U.S.A.

KLUWER ACADEMIC / PLENUM PUBLISHERS
NEW YORK, NEW YORK
U.S.A.

DNA METHYLATION AND CANCER THERAPY

Medical Intelligence Unit

Landes Bioscience / Eurekah.com
Kluwer Academic / Plenum Publishers

Printed in the U.S.A.

Kluwer Academic / Plenum Publishers, 233 Spring Street, New York, New York, U.S.A. 10013
http://www.wkap.nl/

Please address all inquiries to the Publishers:
Landes Bioscience / Eurekah.com, 810 South Church Street
Georgetown, Texas, U.S.A. 78626
Phone: 512/ 863 7762; FAX: 512/ 863 0081
www.Eurekah.com
www.landesbioscience.com

DNA Methylation and Cancer Therapy, edited by Moshe Szyf, Landes / Kluwer dual imprint / Landes series: Medical Intelligence Unit

ISBN: 0-306-47848-X

While the authors, editors and publisher believe that drug selection and dosage and the specifications and usage of equipment and devices, as set forth in this book, are in accord with current recommendations and practice at the time of publication, they make no warranty, expressed or implied, with respect to material described in this book. In view of the ongoing research, equipment development, changes in governmental regulations and the rapid accumulation of information relating to the biomedical sciences, the reader is urged to carefully review and evaluate the information provided herein.

Library of Congress Cataloging-in-Publication Data

DNA methylation and cancer therapy / [edited by] Moshe Szyf.
    p. ; cm. -- (Medical intelligence unit)
  Includes bibliographical references and index.
  ISBN 0-306-47848-X
  1. DNA--Methylation. 2. Cancer--Molecular aspects. I. Szyf, Moshe. II. Series: Medical intelligence unit (Unnumbered : 2003)
    [DNLM: 1. Neoplasms--genetics. 2. DNA Methylation. 3. Neoplasms--therapy. QZ 200 D629 2004]
QP624.5.M46D625 2004
616.99'406--dc22

                    2004022708

This book is dedicated to my parents
and Vicky whose support has enabled me to delve
into the secrets of the epigeome.

CONTENTS

EDITOR

Moshe Szyf, Ph.D.
Department of Pharmacology and Therapeutics
McGill University
Montreal, Quebec, Canada
Chapters 10, 12, Epilogue

CONTRIBUTORS

Pascal Bigey
Unite de Pharmacologie Chimique
   et Genetique
Universite Ren Descartes
Paris, France
Chapter 16

Veronica Bovenzi
Department of Pharmacology
   and Therapeutics
McGill University
Montreal, Quebec, Canada
Chapter 12

Paola Caiafa
Department of Cellular Biotechnologies
   and Haematology
Universite di Roma "La Sapienza"
Rome, Italy
Chapter 11

Paul M. Campbell
Department of Pharmacology
   and Therapeutics
McGill University
Montreal, Quebec, Canada
Chapter 12

Chuan-Mu Chen
Department of Zoology
National Chung Hsing University
Taiwan, Republic of China
Chapter 8

Adriana de Capoa
Dipartimento di Genetica e Biologia
   Molecolare
Universita di Roma
Roma, Italy
Chapter 7

Nancy Detich
Department of Pharmacology
   and Therapeutics
McGill University
Montreal, Quebec, Canada
Chapters 10, 12

Emmanuel Drouet
Laboratoire de Virologie Moleculaire
   et Structurale
Faculte de M decine et Pharmacie
Universit Joseph Fourier de Grenoble
La Tronche, France
Chapter 7

Jean-Marc Dumollard
Service d'Anatomie et de Cytologie
   Pathologiques
Hopital Bellevue
Saint-Etienne, France
Chapter 7

Melanie Ehrlich
Program in Human Genetics
   and Tulane Cancer Center
Tulane Medical School
New Orleans, Louisiana, U.S.A.
Chapter 3

Manel Esteller
Cancer Epigenetics Laboratory
Spanish National Cancer Center (CNIO)
Madrid, Spain
Chapter 6

Lucien Frappart
Unit INSERM 403
Facult de M decine Laennec
Lyon, France
Chapter 7

Claudio Grappelli
Dipartimento di Genetica e Biologia
 Molecolare
Universita di Roma 1 La Sapienza
Rome, Italy
Chapter 7

Carolina Haefliger
Epigenomics AG
Berlin, Germany
Chapter 9

Stefan Hamm
Department of Pharmacology
 and Therapeutics
McGill University
Montreal, Quebec, Canada
Chapter 12

Michel Herranz
Cancer Epigenetics Laboratory
Spanish National Cancer Center (CNIO)
Madrid, Spain
Chapter 6

Tim H.-M. Huang
Department of Pathology
 and Anatomical Sciences
Ellis Fischel Cancer Center
University of Missouri School
 of Medicine
Columbia, Missouri, U.S.A.
Chapter 8

Jeremy R. Jass
Department of Pathology
McGill University
Montreal, Quebec, Canada
Chapter 5

Guanchao Jiang
Program in Human Genetics
 and Tulane Cancer Center
Tulane Medical School
New Orleans, Louisiana, U.S.A.
Chapter 3

Barbara A. Leggett
Conjoint Gastroenterology Laboratory
Bancroft Centre
Brisbane, Queensland, Australia
Chapter 5

A. Robert MacLeod
Department of Molecular Biology
MethylGene Inc.
Montreal, Quebec, Canada
Chapter 14

Sabine Maier
Epigenomics AG
Berlin, Germany
Chapter 9

Maria Malanga
Department of Biological Chemistry
University "Federico II" of Naples
Naples, Italy
Chapter 11

Richard L. Momparler
Department de Pharmacologie
University de Montreal
Centre de Recherche
Hosptial Sainte-Justine
Montreal, Quebec, Canada
Chapter 15

Sylvester L. Mosley
School of Chemistry and Biochemistry
Georgia Institute of Technology
Atlanta, Georgia, U.S.A.
Chapter 13

Alain Niveleau
Laboratoire de Virologie Structurale
  et Moleculaire
Faculte de Medecine
Universite Joseph Fourier de Grenoble
La Tronche, France
Chapter 7

Alexander Olek
Epigenomics AG
Berlin, Germany
Chapter 9

Jing Ni Ou
Department of Pharmacology
  and Therapeutics
McGill University
Montreal, Quebec, Canada
Chapter 12

Chandrika Piyathilake
Department of Nutrition Sciences
Division of Nutritional Biochemistry
  and Molecular Biology
University of Alabama at Birmingham
Birmingham, Alabama, U.S.A.
Chapter 7

Aharon Razin
Department of Cellular Biochemistry
  and Human Genetics
The Hebrew University - Hadassah
  Medical School
Jerusalem, Israel
Chapter 1

Anna Reale
Department of Cellular Biotechnologies
  and Haematology
University of Rome "La Sapienza"
Rome, Italy
Chapter 11

Gregory K. Reid
Department of Clinical Research
MethylGene Inc.
Montreal, Quebec, Canada
Chapter 14

Keith D. Robertson
Epigenetic Gene Regulation and Cancer
  Section
National Cancer Institute
National Institutes of Health
Bethesda, Maryland
Chapter 2

Daniel Scherman
Unite de Pharmacologie Chimique
  et Genetique
Universite Ren Descartes
Paris, France
Chapter 16

Wolfgang A. Schulz
Department of Urology
Heinrich-Heine-University
Dusseldorf, Germany
Chapter 4

Hans-Helge Seifert
Department of Urology
Heinrich-Heine-University
Dusseldorf, Germany
Chapter 4

Katherine L. Seley
School of Chemistry and Biochemistry
Georgia Institute of Technology
Atlanta, Georgia, U.S.A.
Chapter 13

Susan H. Wei
Department of Pathology
  and Anatomical Sciences
Ellis Fischel Cancer Center
University of Missouri School
  of Medicine
Columbia, Missouri, U.S.A.
Chapter 8

Vicky L.J. Whitehall
Conjoint Gastroenterology Laboratory
Bancroft Centre
Brisbane, Queensland, Australia
Chapter 5

Timothy T.-C. Yip
Department of Clinical Oncology
Queen Elizabeth Hospital
Kowloon, Hong Kong
Chapter 8

Joanne Young
Conjoint Gastroenterology Laboratory
Bancroft Centre
Brisbane, Queensland, Australia
Chapter 5

Giuseppe Zardo
Department of Cellular Biotechnologies
  and Haematology
University of Rome "La Sapienza"
Rome, Italy
Chapter 11

Jordanka Zlatanova
Department of Chemistry and Chemical
  Engineering
Polytechnic University
Brooklyn, New York, U.S.A.
Chapter 11

PREFACE

DNA methylation has bewildered molecular biologists since Hotchkiss discovered it almost six decades ago (Hotchkiss RDJ. Biol Cem 1948; 175:315-332). The fact that the chemical structure of our genome consists of two components that are covalently bound, the genetic information that is replicated by the DNA replication machinery and DNA methylation that is maintained by independent enzymatic machinery, has predictably stimulated the imagination and curiosity of generations of molecular biologists. An obvious question was whether DNA methylation was a bearer of additional information to the genetic information and what was the nature of this information? It was tempting to speculate that DNA methylation applied some form of control over programming of the genome's expression profile. Once techniques to probe the methylation profile of whole genomes as well as specific genes became available, it became clear that DNA methylation patterns are gene and tissue specific and that patterns of gene expression correlate with patterns of methylation. DNA methylation patterns emerged as the only component of the chemical structure of DNA that exhibited tissue and cell specificity. This data seemingly provided an attractively simple explanation for the longstanding dilemma of how could one identical genome manifest itself in so many different forms in multicellular organisms? The DNA methylation pattern has thus become the only known factor to confer upon DNA a unique cellular identity. This important set of data provided strong support for the hypothesis that DNA methylation played an important role in controlling tissue specific gene expression. However, the naïve early models that predicted that DNA methylation would ultimately explain cell specific gene expression programs were later replaced by confusing and complex sets of data, cynicism and sarcasm. The fact that lower organisms such as flies and nematodes developed elaborate gene expression programs in the absence of any detectable DNA methylation has further shaken the belief that DNA methylation played any role in gene regulation. It became clear that proteins such as transacting and trans repressing factors interacting with *cis* acting factors in DNA as well as chromatin, chromatin modifications and the proteins that modify histones were principally responsible for cell specific gene expression profiles. However, recent advances in the field of chromatin modification and the discovery of methylated binding proteins are starting to clarify how DNA methylation is integrated with other epigenomic factors in regulating programmed gene expression. This clearer picture of the factors involved in regulating gene expression has brought DNA methylation back to the forefront of molecular biology. The first two chapters of the book by Razin and by Robertson will provide a review of our current understanding of how DNA methylation is integrated with the complex machinery, which controls gene expression in vertebrates.

An additional issue that is obviously coupled with the question of the functional role of DNA methylation is to understand the mechanisms responsible for generating and maintaining the DNA methylation pattern. The fact that gene expression and DNA methylation patterns correlate does not necessarily imply that there is a causal relation between DNA methylation and gene expression. The reverse possibility that DNA methylation is directed by gene expression or that a common factor determines both DNA methylation patterns and gene expression is also consistent with such a correlation. To address this issue we need to understand what defines DNA methylation patterns. This obviously has important implications for our understanding of the changes in DNA methylation seen in cancer as will be discussed below.

The first unresolved question is the enzymology of DNA methylation. A number of DNA methyltransferases were discovered and cloned as will be discussed by Robertson. The presence of an enzyme that reverses the DNA methylation reaction has been extremely controversial. It has been long believed that DNA methylation is an irreversible reaction and that an enzyme that truly demethylates DNA and reverses the methylation reaction does not exist. The reason behind this strong dogma is that the bond between the methyl moiety and the cytosine ring is considered to be a strong bond that could not be broken by an enzymatic process. Our entire understanding of DNA methylation is based on this assumption. I will discuss recent data suggesting that DNA methylation is a reversible reaction and that the steady state DNA methylation pattern is an equilibrium of DNA methylation and demethylation reactions. This clearly changes our conception of DNA methylation patterns and how they are formed and maintained.

It is clear that none of the enzymes that catalyze either DNA methylation or demethylation show distinct specificity. This raises the question of what determines the specificity of the DNA methylation reactions. New data suggests an important relation between the chromatin modifying proteins and enzymes that catalyze either DNA methylation or demethylation, which could also explain the correlation between DNA methylation and chromatin structure. This will be discussed in Robertson's and my chapter.

The tight correlation between DNA methylation and programmed gene expression begs the question whether DNA methylation aberrations play a role in cancer. There is now overwhelming data supporting the conclusion that alterations in DNA methylation are a hallmark of cancer. This has both diagnostic and therapeutic applications, which will be discussed in this book. Four chapters in this book will focus on discussing the nature of DNA methylation alterations in cancer. The paradox of DNA methylation patterns in cancer is the coexistence of global hypomethylation and regional hypermethylation. The chapter by Ehrlich will focus on this issue. Three other chapters by Schulz et al, by Jass et al, and by Esteller et al will discuss specific methylation aberrations observed in cancer.

Another aberration of the DNA methylation machinery observed in cancer is the deregulation of expression of DNA methyltransferases, which will be discussed by Detich. Demethylases and their potential clinical application will be discussed. Caiafa et al will discuss the possible role of poly ADP ribosylation in hypermethylation.

The changes in DNA methylation observed in cancer have potentially important implications in therapeutics as well as diagnostics in addition to challenging us scientifically. The diagnostic applications of DNA methylation in cancer will be discussed as well as novel methods to measure global and regional hypermethylation in cancer. Niveleau et al will discuss immunochemistry approaches whereas Wei et al and Haefliger et al will discuss the use of new microarray technology and bioinformatics to unravel profiles of DNA methylation that can potentially serve as diagnostic tools for cancer, cancer stages and predictors of clinical progression of the disease.

The last part of the book will focus on preclinical and clinical attempts to target the DNA methylation machinery in cancer therapy. Seley et al will discuss the synthesis of novel DNA methylase inhibitors. Momparler will discuss the lessons derived from preclinical and clinical trials with the DNA methyltransferase inhibitor 5-aza-CdR and Scherman and Bigey will discuss the use of electrotransfer for knockdown of methylated DNA binding protein 2.

DNA methylation patterns and their relation to cancer have confused and bewildered us on one hand and stimulated our curiosity and enchanted us on the other hand. Recent advances in DNA methylation enzymology, methylated DNA binding proteins and chromatin have begun to clarify the role of DNA methylation in gene expression and cancer. They also raised the attractive possibility that enzymes of the DNA methylation machinery might serve as targets for anticancer drugs. More work needs to be done and future trials will determine whether the pioneering work with DNA methylation modulators will indeed translate to first-rate anticancer therapeutics. We hope that this book unravels some of these advances and therapeutic potential of DNA methylation as well as inspires the reader to further understanding of this emerging field in cancer biology and therapeutics.

Moshe Szyf, Ph.D.

CHAPTER 1

DNA Methylation:
Three Decades in Search of Function

Aharon Razin

D NA methylation is an epigenetic mark that is involved in control mechanisms of a variety of biological processes. Being symmetrically positioned on the two complementary DNA strands the methyl groups represent a clonally inheritable feature of the DNA. Once established during embryogenesis, methylation patterns are maintained for many cell generations by a maintenance methyltransferase. These methylation patterns are interpreted by proteins that interact with the DNA depending on its state of methylation. Since methylation patterns provide a universal code for DNA-protein interactions, it is not surprising that methylation takes part in many biological processes such as: control of gene expression, DNA replication and cell cycle, DNA repair, imprinting, inactivation of the X-chromosome in eutherian females and much more. In essence, DNA methylation patterns fulfill their task by guiding specific proteins to target sites on the DNA.

DNA Methylation in Prokaryotes

My interest in DNA methylation began with the discovery of a single residue of 5-methylcytosine (5-metCyt) in the bacteriophage φX174 genome.[1] My interest grew further when it became apparent that this methylation takes place on the viral replicating DNA[2] by the cell dcm methyltransferase which is induced by the virus.[3] Subsequent experiments revealed that the 5-metCyt residue on the phage DNA plays a critical role in the excision of one genome long φX174 DNA stretch from the rolling circle replicating DNA.[4,5] It has been suggested that the methyl moiety serves as a recognition site for the viral gene A product, the endonuclease responsible for this excision process.[6] This demonstration that a specific protein can interact with methylated DNA to carry out a biological function came at a time when the restriction-modification phenomenon was under extensive investigation. The observation that methylation at specific sequences protects DNA from being cleaved by the restriction endonuclease counterpart[7] revealed that DNA methylation can promote DNA-protein interaction in many cases but could sometimes prevent these interactions.

It became gradually evident that DNA methylation in bacteria is involved in biological processes other than restriction-modification. DNA methylation is clearly involved in post replication mismatch repair as a device for strand discrimination to distinguish the mutated newly synthesized DNA strand from the "wild type" parental strand.[8] Strand discrimination obviously takes place within the replication fork since only at that stage DNA is hemimethylated. In this regard it should be noted that the bacterial genome is generally methylated symmetrically on both strands[9] and methylation of the newly synthesized strand lags behind the fork for not more than 30 seconds (equivalent to an Okazaki fragment).[10]

Only very few sequences escape the maintenance methylation at the fork. Examples of sequences that are methylated late include the origin of replication and the promoter of the dnaA gene. These two adjacent sequences that are required for initiating a new round of replication stay hemimethylated for 13 min and, as a result, reinitiation of replication does not take

DNA Methylation and Cancer Therapy, edited by Moshe Szyf. ©2005 Eurekah.com and Kluwer Academic/Plenum Publishers.

place more than once in each cell cycle.[11,12] The hemimethylated state of the origin allows its sequestration within the outer cell membrane and the hemimethylated state of the promoter of the dnaA gene renders the gene inactive, thus transiently depleting the cell from the dnaA protein. This protein is required in substantial amounts, 20-40 molecules per origin, to trigger initiation of replication. Another important function of hemimethylated DNA had been implicated in activation of transposable elements. Methylation of these sequences also lag behind the movement of the replication fork.[13]

DNA Methylation in Eukaryotes

In the very early days of DNA methylation research in eukaryotes it was hypothesized that DNA methylation may play a role in cell differentiation.[14-16] Although initially these hypotheses were based on very limited experimental data, they gained strong support by a series of observations that have been made in the early eighties. These years have seen rapid progress in the isolation and cloning of an increasing number of genes. The new analytical approach based on digestion of genomic DNA by methylation sensitive restriction enzymes followed by electorphoretic separation of the restriction fragments and Southern blotting revealed gene specific methylation patterns that inversely correlate with gene expression.[17-20] It became clear that once methylation patterns have been established during embryogenesis, they must be maintained and serve as a memory device.[21] Transfection experiments revealed that gene specific methylation patterns are indeed clonally inherited,[22-24] possibly by the maintenance methyltransferase activity that had been shown to prevail in mammalian cell extracts[25,26] and to methylate replicating DNA within the fork.[27]

The involvement of methylation in the control of gene expression was suggested by transfection experiments with in vitro methylated genes[28-30] on one hand, and activation of silent endogenous genes by demethylation, on the other hand.[31] For example, genes on the inactive X chromosome could be activated by treating cells with the potent methyltransferase inhibitor, 5aza-cytidine.[32,33] From this point, the DNA methylation field of research developed very quickly. This is reflected in the fact that while all the existing data by the end of the seventies could be reviewed in one article,[21] it required an entire book in 1984.[34]

Methylation and Embryogenesis

Although methylation patterns correlated clearly with gene activity and in vivo experiments corroborated the possibility that promoter methylation is involved in gene silencing, the role played by methylation in the regulation of gene expression still demanded direct proof. Progress towards understanding the function of DNA methylation in regulation of gene activity had been made when the process of establishing gene specific methylation patterns during embryogenesis began to clarify.[35]

Several important principles were discovered in these studies. First, genome wide demethylation was found to erase the methylation patterns inherited from the gametes. This demethylation takes place post fertilization resulting in very low genomic methylation levels in the blastocyst[36] and in complete demethylation of all gene sequences studied.[35] Subsequently, global de novo methylation appeared to take place at the time of implantation before gastrulation. These global methylation changes in the early embryo result in a bimodal pattern of methylation in which the entire genome is methylated by two de novo methyltransferases Dnmt3a and Dnmt3b,[37,38] and CpG islands[39] that are characteristic of promoter/exon 1 regions of housekeeping genes[18] remain unmethylated.[35] The protection of CpG islands from becoming methylated had been studied in embryonic cells[40] and shown to involve Sp1 elements.[41,42] It was suggested that once the bimodal pattern of methylation is established, gene and tissue specific demethylations take place to activate tissue specific genes in the appropriate tissues as exemplified by the developmental activation of the phospoenol pyruvate carboxykinase (PEPCK) gene.[43,44] The genome-wide demethylation that is observed postfertilization and the gene specific demethylations preceding differentiation raised questions concerning the mechanistic aspects of these processes.

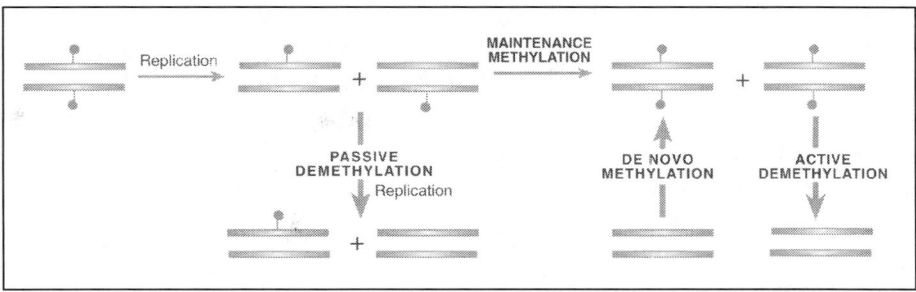

Figure 1. Establishment, maintenance and erasure of methylation patterns. Methylation (lollipops) is of DNA undergoing dynamic changes in living cells. Replication produces hemimethylated DNA (at the replication fork). Maintenance methylation performed by a methyltransferase that acts specifically on hemimethylated DNA restores the symmetric methylation on the two DNA strands. A second round of replication, if not accompanied by maintenance methylation, produces completely unmethylated DNA. This process, when taking place in vivo, is called passive demethylation. Unmethylated DNA can also be obtained by an active demethylase that works in an as yet uncharacterized reaction. Unmethylated DNA is a substrate for de novo methyltransferases so that methylation patterns can be regenerated. While de novo methylation and demethylation are processes that occur predominantly during embryogenesis, maintenance methylation is more specific to somatic cells.

In principle, demethylation could be achieved by either a passive mechanism that involves several rounds of replication in the absence of concomitant maintenance methylation, or rather by an active mechanism based on an enzymatic demethylation activity (Fig. 1). The genome-wide demethylation that was observed in the very early embryo[35,36] had been shown to take place also on the parental DNA strand during one round of replication, implying an active demethylation mechanism.[45] Previous observations also suggested the existence of an active mechanism to demethylate DNA. Induction of the lytic cycle of the Epstein Barr virus (EBV) is associated with viral DNA hypomethylation preceding viral DNA replication.[46] Similarly, no replication was required for the demethylation of the chicken δ crystalin,[47] the mouse myoblast α actin,[48] chick vitellogenin[49] and other genes.[50] All these observations suggested an active demethylation mechanism and prompted investigators to search for a demethylation enzymatic activity.

An early study revealed a repair type mode of demethylation that results in replacement of 5 methylcytosine by cytosine residues in cells undergoing demethylation during differentiation.[51] In fact, demethylase activity that involves excision repair of 5 methylcytosine was subsequently observed in chick embryos.[52] This observation and the discovery of a human DNA glycosylase involved in 5 methylcytosine removal[53] supported the suggested repair type mechanism.[51] These observations combined, encouraged investigators to attempt identification, purification and characterization of enzymatic demethylase activities.

A 5 metCyt-DNA glycosylase activity that causes demethylation of DNA was purified from chick embryos and shown to require both protein and RNA.[54] A similar activity was then shown to participate in the genome-wide loss of DNA methylation during mouse myoblast differentiation.[55] The possibility that RNA is required for the catalytic activity of a demethylase was also suggested in another attempt to study demethylase activity in myoblast cell extracts.[56] An amazing demethylase activity that transforms methylated cytosine bases in the DNA to cytosine residues and methanol had recently been cloned and characterized.[57,58] Obviously, these observations do not allow as yet to draw a universal mode of demethylase activity. Moreover, none of the demethylase activities described above have been proven as yet to function in vivo.

A milestone in the search of a biological role for DNA methylation was laid by the cloning of the mouse maintenance methyltransferase gene (Dnmt1)[59] and its subsequent targeted mutation.[60] Mice deficient in Dnmt1 activity showed embryonic defects that caused their death at day 8-9 post coitum. This lethality suggested a critical role of DNA methylation in normal mammalian embryonic development, but the developmental processes that are critically dependent on DNA methylation were not known. However, DNA methylation had been implicated in X-chromosome inactivation[61] and genomic imprinting.[62]

DNA Methylation and Imprinting

Many genes which are involved in mammalian developmental processes do not obey Mendel's rules and are expressed monoallelically in a parent of origin fashion. DNA methylation turned out to be a critical element in the imprinting process by marking the alleles and establishing a monoallelic expression pattern of the imprinted genes. Much had been learned about the importance of DNA methylation in imprinting from the Dnmt1 deficient mouse. Loss of methylation of the imprinted Xist gene in Dnmt1 deficient cells did not activate Xist.[61] This was surprising in light of the fact that the RNA product of the Xist gene is thought to spread the inactive state along one of the X-chromosomes in eutherian females. In contrast, the monoallelic expression of the imprinted genes H19, Igf2 and Igf2r was disrupted in the Dnmt1 mutant cells. This observation established for the first time a causal link between DNA methylation and gene activity.[63]

Differentially methylated regions (DMRs) in imprinted genes were found to serve as imprinting boxes that control the imprinting of the Igf2r gene[64] and the Prader Willi/Angelman syndrome domain.[65] DMRs are also key elements in regulating reciprocal monoallelic expression of the maternal regulation of the imprinted H19 and Igf2 genes.[66] Experiments with Dnmt1 deficient mice revealed that the establishment of DMRs in imprinted genes required transmission through the germ line. Rescue of the Dnmt1 genotype in ES cells did not restore imprinted methylation of the above mentioned genes while it reestablished imprinting in the whole animal.[67] This is in accord with the fact that methylation of the Snrpn gene on the maternal allele is established during oogenesis and maintained thereafter.[68]

DNA Methylation in Gene Silencing

In all studies on methylation and gene expression which were described above, chromatin was ignored. This was in spite of the fact that nucleosomal DNA was found very early to be richer in methylation than internucleosomal DNA.[69] It is even more surprising in light of the fact that correlation between undermethylation and DNaseI sensitivity, a general property of transcriptionally active regions of chromatin, was demonstrated almost three decades ago.[70] For many years it was therefore difficult to explain how methylation in higher eukaryotes affects such divergent phenomena as genome stability and gene silencing.

Deciphering the mechanism by which methylation affects gene expression was of special importance since it became clear that methylation is involved in parent of origin specific monoallelic expression of imprinted genes,[71] inactivation of the X-chromosome in eutherian female cells[72] and silencing of tissue specific genes[20] and sequences of foreign origin.[73] A hint that the effect of methylation on gene expression may involve a chromatin protein was given when histone H1 had been shown to mediate inhibition of transcription initiation of methylated templates in vitro.[74] A recent study in *Ascobolus immersus* shows, however, that histone H1 is dispensable for methylation-associated gene silencing in fungi.[75]

Over the last decade much progress had been made towards understanding the molecular mechanisms that underlie methylation-dependent processes that result in gene silencing. Although a correlative interrelationship between DNA methylation and chromatin structure had been suggested, as mentioned above, a long time ago, it is only recently that these two epigenetic marks were connected mechanistically.[76] The turning point in our understanding of this important biological issue is attributed to a critical discovery made almost a decade ago by Adrian Bird and colleagues of two methyl binding proteins MeCP1[77] and MeCP2.[78] MeCP1 was implicated in methylation dependent transcriptional repression,[79-82] but turned out to be too complex to purify and clone. However, MeCP2 was successfully purified, cloned and characterized.[83] MeCP2 had been shown to localize to methylation CpG-rich heterochromatin[84] and was later shown to contain a methyl binding domain (MBD) and a transcriptional repressory domain (TRD).[85]

Further progress had been made when it was found that MeCP2 binds to the methylated DNA by its MBD and recruits the corepressor Sin3A through its TRD.[86,87] MeCP2 can therefore be considered as an anchor for binding to the DNA a multiprotein repressory complex that causes histone deacetylation and chromatin remodeling. Two histone deacetylase activities, HDAC1 and HDAC2 are components of this repressory complex. Transcriptional inactivation caused by this deacetylation could be alleviated by the HDAC inhibitor trichostatin A (TSA).[86-88] These observations finally solved the long suspected three-way connection between DNA methylation, condensed chromatin structure and gene silencing.[76]

It is now clear that MeCP2 is not the sole methyl binding protein involved in transcriptional repression. A search of EST database revealed four additional methyl binding proteins, MBD1-4.[89] Three of them are components of histone deacetylase chromatin remodeling and repressory complexes. MeCP1, that was discovered thirteen years ago[77] and its composition remained obscure for the entire last decade, turns out to be a histone deacetylase multiprotein complex composed of ten components that include MBD2.[90,91] MeCP1 was found to share most of its components with the Mi2/NuRD histone deacetylase-chromatin remodeling complex that was deciphered earlier.[92] One of these components, MBD3, had been shown to play a distinctive role in mouse development.[93]

Interestingly, DNA methyltransferases were also found to be components of histone deacetylase repressory complexes. The de novo DNA methyltransferase Dnmt3a binds deacetylases and is recruited by the sequence specific repressory DNA binding protein RP58 to silence transcription.[94] The human maintenance DNA methyltransferase (DNMT1) forms a complex with Rb, E2F1 and HDAC1 and represses transcription from E2F-responsive promoters.[95] In parallel, DNMT1 binds HDAC2 and the corepressory DMAP1 to form a complex at replication foci.[96] Although MBD1 has been shown to cause methylation mediated transcription silencing in euchromatin,[97,98] it is not known to participate in any of the known histone deacetylase multiprotein repressory complexes. Nevertheless, its repressory effect can be alleviated by TSA, suggesting that MBD1 is part of a yet unknown histone deacetylase repressory complex. For a comprehensive discussion of all methylation associated repressory complexes, see review by Kantor and Razin.[99]

It has recently become clear that the flow of epigenetic information may be bidirectional. DNA methylation affects histone modification which in turn can affect DNA methylation (Fig. 2). The interaction between these covalent modifications of chromatin may shed light on the yet unsolved mechanisms concerning the establishment of heterochromatin, its spreading along large domains of the genome and its stable inheritance. The DNA methylation-heterochromatin cycle described in Figure 2 includes the three epigenetic marks, DNA methylation, histone acetylation and histone methylation and their interrelations.

DNA Methylation and Disease

The progress made in our understanding of the methylation machinery, its role in important developmental processes such as genomic imprinting and X-chromosome inactivation, the involvement of DNA methylation in the control of the cell cycle, in the process of cell differentiation, in host defense and gene silencing can now be used in deciphering the molecular events that lead to disease. Since DNA methylation in cancer is the theme of this book, I will not dwell on this important involvement of DNA methylation in disease.

Interestingly, when biological processes that involve DNA methylation go awry, neurodevelopmental disorders occur. This is true for ATR-X, ICF, Rett and Fragile X syndromes and for Prader-Willi, Angelman and Beckwith-Wiedemann syndromes that are associated with disruption of imprinting processes. Common to all these diseases is mental impairment, suggesting that DNA methylation-associated gene control is particularly important in

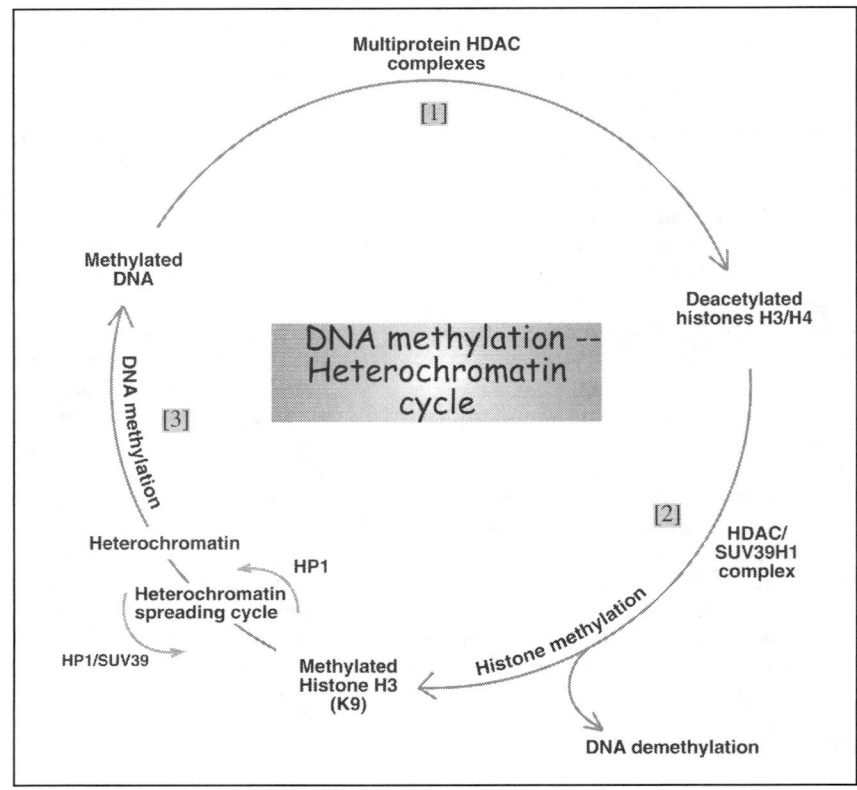

Figure 2. DNA methylation patterns which are clonally inherited by a maintenance methyltransferase (Dnmt1) create DNA binding sites used to assemble HDAC repressory multiprotein complexes that deacetylate histones H3/H4. The level of histone acetylation is determined by a balance between the activities of several histone acetylases (HATs) and histone deacetylases (HDACs).[123] Obviously, DNA methylation affects specific local alterations in this balance (reaction [1] in Fig. 2). DNA hypomethylation at specific sequences both in Neurospora[124] and in mammalian cells[125] have been reported to be associated with histone deacetylation followed by methylation of the lysine residue in position 9 (K9) of histone H3. The replacement of K9 in histone H3 results in canceling the loss in DNA methylation in vivo in Neurospora.[126] The complex HDAC/SUV39H1 deacetylates histone H3 and methylates K9 at histone H3 (reaction [2] in Fig. 2). The methylated K9 of histone H3 is recognized by heterochromatin assembly protein 1 (HP1) which can recruit histone methylase to methylate K9 on an adjacent nucleosome. The newly methylated nucleosomes further recruit HP1, thus spreading the heterochromatization of the domain until a CTCF boundary is encountered.[127] This can shed light on the heterochromatin spreading mechanism (Heterochromatin spreading cycle. Heterochromatin, in turn, can recruit a DNA methyltransferase to restore and maintain the methylated status of the DNA (reaction [3]).

brain development and function. ATR-X (α-Thalassemia, mental retardation, X-linked) patients show severe mental retardation and are known to have methylation defects with high methyltransferase activity in neurons.[100] In mice this high activity of methyltransferase contributes to delayed ischemic brain damage in mice.[101] Methylation pattern defects of hypo or hypermethylation of repetitive sequences characterize this disease.[102]

ICF (immunodeficiency, centromeric instability and facial anomaly) syndrome is linked to mutations in the de novo methyltransferase gene DNMT3B (mapped to chromosome 20q) affecting its carboxy terminal catalytic domain.[103] Deficiency of DNMT3B in ICF cells affects chromosome stability[104] almost exclusively on chromosomes 1, 9 and 16 which contain satellite repeat arrays.[105,106] Satellite sequences, repetitive elements elsewhere in the genome[107] and single copy sequences on the inactive X-chromosomes[108,109] are normally heavily methylated in the human genome but are hypomethylated in ICF cells. Mice knocked out in Dnmt3b show similar demethylation and could therefore serve as an experimental ICF model.[38] How DNMT3B deficiency affects brain development remains to be elucidated.

Another syndrome that is manifested in mental retardation and is associated with methylation dependent gene silencing is the X-linked Rett syndrome. Only recently has it been discovered that the disease results from mutations in the MeCP2 gene.[110] Mutations that cause the disease disrupt the integrity of the methyl binding domain (MBD) or the transcription repressory domain (TRD) of MeCP2 whose function in gene repression had been discussed above in the section "DNA methylation and gene silencing". How MeCP2 mutations lead to developmental defects in the brain remains to be found.

The most common form of inherited mental retardation after Down syndrome is the Fragile-X syndrome. The X-linked gene that is associated with the disease, Fragile-X mental retardation 1 (FMR1), contains highly polymorphic CGG repeats with an average length of 29 repeats in normal individuals and 200-600 repeats in Fragile-X patients. In addition, the CpG island at the 5' end of the gene in patients is abnormally methylated and histone deacetylated, causing silencing of the gene.[111,112] The reasons for this de novo methylation and the mechanisms driving this de novo methylation are, as yet, unclear.

As mentioned above, the imprinted genes are characterized by differentially methylated regions (DMRs) that are critical for the regulation of the imprinting process.[62] A number of neurobehavioral disorders are caused by loss of function of imprinted genes. Such epigenetic defects within a 2 Mb domain on human chromosome 15q11-q13 cause two different syndromes. Prader-Willi syndrome is caused by loss of function of a large number of paternally expressed genes[113] while silencing of the maternally expressed genes within the domain causes Angelman Syndrome.[114] The imprinting of this entire domain is regulated by an imprinting center that constitutes a DMR within the 5' region of the imprinted SNRPN gene. Individuals with deletions of this region on the paternal allele have Prader-Willi while another sequence located 35 kb upstream to SNRPN confers methylation of the SNRPN DMR on the maternal allele thereby inactivating the paternally expressed genes on the maternal allele. When this upstream region is deleted on the maternal allele, the SNRPN DMR does not become methylated.[115] Consequently, the entire domain on the maternal allele remains unmethylated and all paternally expressed genes on the maternal allele are activated while maternally expressed genes are silenced, thus causing Angelman Syndrome. A model had been proposed suggesting that the upstream sequence, together with the SNRPN DMR constitute a complex imprinting box responsible for both the establishment and maintenance of the imprinting state at PWS/AS domain on both alleles.[65]

Altered allelic methylation and expression patterns of the imprinted gene IGF2 have been found in Beckwith-Wiedemann Syndrome (BWS) patients. BWS is a pre and post natal growth syndrome associated with predisposition for childhood tumors. Translocation breakpoints in a number of BWS patients map to the imprinted gene KCNQ1 which is located in the center of the 800 kb BWS region on human chromosome 11p15.5. The translocations in BWS are associated with loss of imprinting of IGF2 but not H19.[116] It appears that this impairment in imprinting involves the differentially methylated intronic CpG island in KCNQ1. In a small number of BWS patients, hypomethylation of the KCNQ1 CpG island correlated with biallelic expression of IGF2.[117,118] Deletion of this CpG island on the paternal chromosome 11 leads to silencing of KCNQ1 antisense transcript and activation of KCNQ1, p57^{KIP2} and SMS4 that are located downstream on the normally repressed paternal allele. It is therefore possible that this CpG island is at least part of an imprinting center on human chromosome 11p15.5 and its orthologous region on mouse chromosome 7.

Reflections

DNA methylation may have evolved as a luxury device to promote central biological processes in a wide variety of living organisms. Methylation may even be dispensable in single cell organisms such as bacteria[119] or yeast[120] but critical in complex multicellular organisms, with the exception of *Drosophila*[121] and *C. elegans.*[122]

The increasing complexity of the mammalian genome with a multi-level hierarchy of gene expression control may have required the introduction of methylation to drive evolution. As a result, the mammalian genome became permissive to the invasion of foreign genes that could now be silenced by methylation.[73] Nevertheless, DNA methylation must have been advantageous to mammals, being conserved in evolution in spite of the fact that organisms that use methylation pay a price in the form of mutations in genes playing central roles in the well being of the cell.

Acknowledgements

My apology to colleagues whose work has not been cited. These will certainly be cited in other chapters of this book. I am thankful to my students and research associates that worked with me over the last three decades. I am especially thankful to my close friend and collaborator, Dr. Howard Cedar. My laboratory had been continually supported by the National Institutes of Health grant GM20483 and by grants from the Israel Science Foundation, Israel-U.S. Binational Foundation, the Council for Tobacco Research, U.S.A., Inc., German-Israeli Foundation for Scientific Research and Development, Israel Ministry of Science, Israel Ministry of Health, Rett Syndrome Research Foundation and Conquer Fragile X foundation, Inc.

References

1. Razin A, Sedat JW, Sinsheimer RL. Structure of the DNA of bacteriophage φX174. VII. Methylation. J Mol Biol 1970; 53:251-259.
2. Razin A, Sedat JW, Sinsheimer RL. In vivo methylation of replicating bacteriophage φX174 DNA. J Mol Biol 1973; 78:417-425.
3. Razin A. DNA methylase induced by bacteriophage φX174. Proc Natl Acad Sci USA 1973; 70:3773-3775.
4. Friedman J, Razin A. Studies on the biological role of DNA methylation; II. Role of φX174 DNA methylation in the process of viral progeny DNA synthesis. Nucleic Acids Res 1976; 3:2665-2675.
5. Friedman J, Friedmann A, Razin A. Studies on the biological role of DNA methylation: III Role in excision of one-genome long single-stranded φX174 DNA. Nucleic Acids Res 1977; 4:3483-3496.
6. Fujisawa H, Hayashi M. Gene A product of φX174 is required for site-specific endonucleolytic cleavage during single-stranded DNA synthesis in vivo. J Virol 1976; 19:416-424.
7. Kuhnlein U, Arber W. Host specificity of DNA produced by Escherichia coli. XV. The role of nucleotide methylation in in vitro B-specific modification. J Mol Biol 1972; 63:9-19.
8. Radman M, Wagner R. Mismatch repair in Escherichia coli. Annu Rev Genet 1986; 20:523-538.
9. Razin A, Urieli S, Pollack Y et al. Studies on the biological role of DNA methylation; IV. Mode of methylation of DNA in E. coli cells. Nucleic Acids Res 1980; 8:1783-1792.
10. Szyf M, Gruenbaum Y, Urieli-Shoval S et al. Studies on the biological role of DNA methylation: V. The pattern of E. coli DNA methylation. Nucleic Acids Res 1982; 10:7247-7259.
11. Ogden GB, Pratt MJ, Schaechter M. The replicative origin of the E. coli chromosome binds to cell membranes only when hemimethylated. Cell 1988; 54:127-135.
12. Campbell JL, Kleckner NE. coli oriC and the dnaA gene promoter are sequestered from dam methyltransferase following the passage of the chromosomal replication fork. Cell 1990; 62:967-979.
13. Roberts D, Hoopes BC, McClure WR et al. IS10 transposition is regulated by DNA adenine methylation. Cell 1985; 43:117-130.
14. Scarano E. The control of gene function in cell differentiation and in embryogenesis. Adv Cytopharmacol 1971; 1:13-24.
15. Holliday R, Pugh JE. DNA modification mechanisms and gene activity during development. Science 1975; 187:226-232.
16. Riggs AD. X inactivation, differentiation, and DNA methylation. Cytogenet. Cell Genet 1975; 14:9-25.
17. Naveh-Many T, Cedar H. Active gene sequences are undermethylated. Proc Natl Acad Sci USA 1981; 78:4246-4250.

18. Stein R, Sciaky-Gallili N, Razin A et al. Pattern of methylation of two genes coding for housekeeping functions. Proc Natl Acad Sci USA 1983; 80:2422-2426.

19. Yisraeli J, Szyf M. Gene methylation patterns and expression. In: Razin A, Cedar H, Riggs AD, eds. DNA methylation: Biochemistry and Biological Significance. New York: Springer-Verlag, 1984:352-370.

20. Yeivin A, Razin A. Gene methylation patterns and expression. In: Jost JP, Saluz HP, eds. DNA Methylation: Molecular Biology and Biological Significance. Basel: Birkhauser Verlag, 1993:523-568.

21. Razin A, Riggs AD. DNA methylation and gene function. Science 1980; 210:604-610.

22. Pollack Y, Stein R, Razin A et al. Methylation of foreign DNA sequences in eukaryotic cells. Proc Natl Acad Sci USA 1980; 77:6463-6467.

23. Wigler M, Levy D, Perucho M. The somatic replication of DNA methylation. Cell 1981; 24:33-40.

24. Stein R, Gruenbaum Y, Pollack Y et al. Clonal inheritance of the pattern DNA methylation in mouse cells. Proc Natl Acad Sci USA 1982; 79:61-65.

25. Gruenbaum Y, Cedar H, Razin A. Substrate and sequence specificity of a eukaryotic DNA methylase. Nature 1982; 292:620-622.

26. Bestor TH, Ingram VM. Two DNA methyltransferases from murine erythroleukemia cells: Purification, sequence specificity, and mode of interaction with DNA. Proc Natl Acad Sci USA 1983; 80:5559-5563.

27. Gruenbaum Y, Szyf M, Cedar H et al. Methylation of replicating and post-replicated mouse L-cell DNA. Proc Natl Acad Sci USA 1983; 80:4919-4921.

28. Stein R, Razin A, Cedar H. In vitro methylation of the hamster APRT gene inhibits its expression in mouse L-cells. Proc Natl Acad Sci USA 1982; 79:3418-3422.

29. Cedar H, Stein R, Gruenbaum Y et al. Effect of DNA Methylation on Gene Expression. Cold Spring Harbor Symposia on Quantitative Biology, 1983:605-609.

30. Stewart CL, Stuhlmann H, Jahner D et al. De novo methylation, expression, and infectivity of retroviral genomes introduced into embryonal carcinoma cells. Proc Natl Acad Sci USA 1982; 79:4098-4102.

31. Jones PA. Gene activation by 5-azacytidine. In: Razin A, Cedar H, Riggs AD, eds. DNA Methylation: Biochemistry and Biological Significance. New York: Springer-Verlag, 1984:165-188.

32. Mohandas T, Sparker RS, Shapiro LJ. Reactivation of an inactive human X chromosome: Evidence for X inactivation by DNA methylation. Science 1981; 211:393-396.

33. Jones PA, Taylor SM, Mohandas T et al. Cell cycle-specific reactivation of an inactive X-chromosome locus by 5- azadeoxycytidine. Proc Natl Acad Sci USA 1982; 79:1215-1219.

34. Razin A, Cedar H, Riggs AD, eds. DNA methylation: Biochemistry and Biological Significance. New York: Springer Verlag, 1984.

35. Kafri T, Ariel M, Brandeis M et al. Developmental pattern of gene-specific DNA methylation in the mouse embryo and germline. Genes Dev 1992; 6:705-714.

36. Monk M, Boubelik M, Lehnert S. Temporal and regional changes in DNA methylation in the embryonic, extraembryonic and germ cell lineages during mouse embryo development. Development 1987; 99:371-382.

37. Okano M, Xie S, Li E. Cloning and characterization of a family of novel mammalian DNA (cytosine-5) methyltransferases. Nature Genet 1998; 19:219-20.

38. Okano M, Bell DW, Haber DA et al. DNA methyltransferases Dnmt3a and Dnmt3b are essential for de novo methylation and mammalian development. Cell 1999; 99:247-57.

39. Bird AP. CpG-rich islands and the function of DNA methylation. Nature 1986; 321:209-213.

40. Frank D, Keshet I, Shani M et al. Demethylation of CpG islands in embryonic cells. Nature 1991; 351:239-241.

41. Brandeis M, Frank D, Keshet I et al. Sp1 elements protect a CpG island from de novo methylation. Nature 1994; 371:435-438.

42. Macleod D, Charlton J, Mullins J et al. Sp1 sites in the mouse Aprt gene promoter are required to prevent methylation of the CpG island. Genes Dev 1994; 8:2282-2292.

43. Benvenisty N, Szyf M, Mencher D et al. Tissue-specific hypomethylation and expression of rat phosphoenolpyruvate carboxykinase gene induced by in vivo treatment of fetuses and neonates with 5-azacytidine. Biochemistry 1985; 24:5015-5019.

44. Benvenisty N, Mencher D, Meyuchas O et al. Sequential changes in DNA methylation patterns of the rat phosphoenolpyruvate carboxykinase gene during development. Proc Natl Acad Sci USA 1985; 82:267-271.

45. Kafri T, Gao X, Razin A. Mechanistic aspects of genome-wide demethylation in the preimplantation mouse embryo. Proc Natl Acad Sci USA 1993; 90:10558-10562.

46. Szyf M, Eliasson L, Mann V et al. Cellular and viral DNA hypomethylation associated with induction of Epstein-Barr virus lytic cycle. Proc Natl Acad Sci USA 1985; 82:8090-8094.

47. Sullivan CH, Grainger RM. δ-Crystallin genes become hypomethylated in postmitotic lens cells during chicken development. Proc Natl Acad Sci USA 1986; 83:329-333.
48. Paroush Z, Keshet I, Yisraeli J et al. Dynamics of demethylation and activation of the α actin gene in myoblasts. Cell 1990; 63:1229-1237.
49. Jost JP, Seldran M. Association of transcriptionally active vitellogenin II gene with the nuclear matrix of chicken liver. EMBO J 1984; 3:2005-2008.
50. Frank D, Mintzer-Lichenstein M, Paroush Z et al. Demethylation of genes in animal cells. Philo Trans Royal Soc 1990; 326:241-251.
51. Razin A, Szyf M, Kafri T et al. Replacement or 5-methylcytosine by cytosine: A possible mechanism for transient DNA demethylation during differentiation. Proc Natl Acad Sci USA 1986; 83:2827-2831.
52. Jost JP. Nuclear extracts of chicken embryos promote an active demethylation of DNA by excision repair of 5-methyldeoxycytidine. Proc Natl Acad Sci USA 1993; 90:4684-4688.
53. Vairapandi M, Duker N. Enzymic removal of 5-methylcytosine from DNA by a human DNA -glycosylase. Nucleic Acids Res 1993; 21:5323-5327.
54. Fremont M, Siegmann M, Gaulis S et al. Demethylation of DNA by purified chick embryo 5-methylcytosine-DNA glycosylase requires both protein and RNA. Nucleic Acids Res 1997; 25:2375-2380.
55. Jost JP, Oakeley EJ, Zhu B et al. 5-Methylcytosine DNA glycosylase participates in the genome-wide loss of DNA methylation occurring during mouse myoblast differentiation. Nucleic Acids Res 2001; 29:4452-4461.
56. Weiss A, Keshet I, Razin A et al. DNA demethylation in vitro: Involvement of RNA. Cell 1996; 86:709-718.
57. Bhattacharya SK, Ramchandani S, Cervoni N et al. A mammalian protein with specific demethylase activity for mCpG DNA. Nature 1999; 397:579-583.
58. Cervoni N, Bhattacharya S, Szyf M. DNA demethylase is a processive enzyme. J Biol Chem 1999; 274:8363-8366.
59. Bestor TH. Cloning of a mammalian DNA methyltransferase. Gene 1988; 74:9-12.
60. Li E, Bestor TH, Jaenisch R. Targeted mutation of the DNA methyltransferase gene results in embryonic lethality. Cell 1992; 69:915-926.
61. Beard C, Li E, Jaenisch R. Loss of methylation activates Xist in somatic but not in embryonic cells. Genes Dev 1995; 9:2325-2334.
62. Razin A, Cedar H. DNA methylation and genomic imprinting. Cell 1994; 77:473-476.
63. Li E, Beard C, Jaenisch R. Role for DNA methylation in genomic imprinting. Nature 1993; 366:362-365.
64. Birger Y, Shemer R, Perk J et al. The imprinting box of the mouse Igf2r gene. Nature 1999; 397:84-88.
65. Shemer R, Hershko AY, Perk J et al. The imprinting box of the Prader-Willi/Angelman Syndrome domain. Nature Genet 2000; 26:440-443.
66. Bartolomei MS, Webber AL, Brunkow ME et al. Epigenetic mechanisms underlying the imprinting of the mouse H19 gene. Genes Dev 1993; 7:1663-1673.
67. Tucker KL, Beard C, Dausman J et al. Germ-line passage is required for establishment of methylation and expression patterns of imprinted but not non imprinted genes. Genes Dev 1996; 10:1008-1020.
68. Shemer R, Birger Y, Riggs AD et al. Structure of the imprinted mouse Snrpn gene and establishment of its parental-specific methylation pattern. Proc Natl Acad Sci USA 1997; 94:10267-10272.
69. Razin A, Cedar H. Distribution of 5-methylcytosine in chromatin. Proc Natl Acad Sci USA 1977; 74:2725-2728.
70. Weintraub H, Groudine M. Chromosomal subunits in active genes have an altered conformation. Science 1976; 193:848-856.
71. Shemer R, Razin A. Establishment of imprinted methylation patterns during development. In: Russo VEA, Martienssen RA, Riggs AD, eds. Epigenetic Mechanisms of Gene Regulation: Cold Spring Harbor Laboratory Press, 1996:215-229.
72. Migeon BR. X-chromosome inactivation: Molecular mechanisms and genetic consequences. Trends Genet 1994; 10:230-235.
73. Yoder JA, Walsh CP, Bestor TH. Cytosine methylation and the ecology of intragenomic parasites. Trends Genet 1997; 13:335-340.
74. Levine A, Yeivin A, Ben-Asher E et al. Histone H1 mediated inhibition of transcription initiation of methylated templates in vitro. J Biol Chem 1993; 268:21754-21759.
75. Barra JL, Rhounim L, Rossignol JL et al. Histone H1 is dispensable for methylation-associated gene silencing in Ascobolus immersus and essential for long life span. Mol Cell Biol 2000; 20:61-69.

76. Razin A. CpG methylation, chromatin structure and gene silencing—a three-way connection. EMBO J 1998; 17:4905-4908.
77. Meehan RR, Lewis JD, McKay S et al. Identification of a mammalian protein that binds specifically to DNA containing methylated CpGs. Cell 1989; 58:499-507.
78. Lewis JD, Meehan RR, Henzel WJ, et al. Purification, sequence, and cellular localization of a novel chromosomal protein that binds to methylated DNA. Cell 1992; 69:905-14.
79. Boyes J, Bird A. DNA methylation inhibits transcription indirectly via a methyl-CpG biding protein. Cell 1991; 64:1123-1134.
80. Boyes J, Bird A. Repression of genes by DNA methylation depends on CpG density and promoter strength; evidence for involvement of a methyl-CpG binding protein. EMBO J 1992; 11:327-333.
81. Levine A, Cantoni GL, Razin A. Inhibition of promoter activity by methylation: Possible involvement of protein mediators. Proc Natl Acad Sci USA 1991; 88:6515-6518.
82. Levine A, Cantoni GL, Razin A. Methylation in the preinitiation domain suppresses gene transcription by an indirect mechanism. Proc Natl Acad Sci USA 1992; 89:10119-10123.
83. Nan X, Meehan RR, Bird A. Dissection of the methyl-CpG binding domain from the chromosomal protein MeCP2. Nucleic Acids Res 1993; 21:4886-4892.
84. Nan X, Tate P, Li E et al. DNA methylation specifies chromosomal localization of MeCP2. Mol Cell Biol 1996; 16:414-421.
85. Nan X, Campoy FJ, Bird A. MeCP2 is a transcriptional repressor with abundant binding sites in genomic chromatin. Cell 1997; 88:471-481.
86. Nan X, Ng H-H, Johnson CA et al. Transcriptional repression by the methyl-CpG-binding protein MeCP2 involves a histone deacetylase complex. Nature 1998; 393:386-389.
87. Jones PL, Veenstra GJC, Wade PA et al. Methylated DNA and MeCP2 recruit histone deacetylase to repress transcription. Nature Genet 1998; 19:187-191.
88. Eden S, Hashimshony T, Keshet I et al. DNA methylation models histone acetylation. Nature 1998; 394:842-843.
89. Hendrich B, Bird A. Identification and characterization of a family of mammalian methyl-CpG binding proteins. Mol Cell Biol 1998; 18:6538-6547.
90. Ng HH, Zhang Y, Hendrich B et al. MBD2 is a transcriptional repressor belonging to the MeCP1 histone deacetylase complex. Nature Genet 1999; 23:58-61.
91. Feng Q, Zhang Y. The MeCP1 complex represses transcription through preferential binding, remodeling, and deacetylating methylated nucleosomes. Genes Dev 2001; 15:827-832.
92. Zhang Y, Ng HH, Erdjument-Bromage H et al. Analysis of the NuRD subunits reveals a histone deacetylase core complex and a connection with DNA methylation. Genes Dev 1999; 13:1924-1935.
93. Hendrich B, Guy J, Ramsahoye B et al. Closely related proteins MBD2 and MBD3 play distinctive but interacting roles in mouse development. Genes Dev 2001; 15:710-723.
94. Fuks F, Burgers WA, Godin N et al. Dnmt3a binds deacetylases and is recruited by a sequence-specific repressor to silence transcription. EMBO J 2001; 20:2536-2544.
95. Robertson KD, Ait-Si-Ali S, Yokochi T et al. DNMT1 forms a complex with Rb, E2F1 and HDAC1 and represses transcription from E2F-responsive promoters. Nature Genet 2000; 25:338-342.
96. Rountree MR, Bachman KE, Baylin SB. DNMT1 binds HDAC2 and a new corepressor, DMAP1, to form a complex at replication foci. Nature Genet 2000; 25:269-277.
97. Fujita N, Takebayashi S, Okumura K et al. Methylation-mediated transcriptional silencing in euchromatin by methyl- CpG binding protein MBD1 isoforms. Mol Cell Biol 1999; 19:6415-6426.
98. Ng HH, Jeppesen P, Bird A. Active repression of methylated genes by the chromosomal protein MBD1. Mol Cell Biol 2000; 20:1394-406.
99. Kantor B, Razin A. DNA methylation, histone deacetylase repressory complexes and development. Gene Func Dis 2001; 2:69-75.
100. Goto K, Numata M, Komura JI et al. Expression of DNA methyltransferase gene in mature and immature neurons as well as proliferating cells in mice. Differentiation 1994; 56:39-44.
101. Endres M, Meisel A, Biniszkiewicz D et al. DNA methyltransferase contributes to delayed ischemic brain injury. J Neurosci 2000; 20:3175-3181.
102. Gibbons RJ, McDowell TL, Raman S et al. Mutations in ATRX, encoding a SWI/SNF-like protein, cause diverse changes in the pattern of DNA methylation. Nature Genet 2000; 24:368-371.
103. Hansen RS, Wijmenga C, Luo P et al. The DNMT3B DNA methyltransferase gene is mutated in the ICF immunodeficiency syndrome. Proc Natl Acad Sci USA 1999; 96:14412-14417.
104. Xu GL, Bestor TH, Bourc'his D et al. Chromosome instability and immunodeficiency syndrome caused by mutations in a DNA methyltransferase gene. Nature 1999; 402:187-91.
105. Tagarro I, Fernandez-Peralta AM, Gonzalez-Aguilera JJ. Chromosomal localization of human satellites 2 and 3 by a FISH method using oligonucleotides as probes. Hum Genet 1994; 93:383-388.

106. Jeanpierre M, Turleau C, Aurias A et al. An embryonic-like methylation pattern of classical satellite DNA is observed in ICF syndrome. Hum Mol Genet 1993; 2:731-735.
107. Kondo T, Bobek MP, Kuick R et al. Whole-genome methylation scan in ICF syndrome: Hypomethylation of non satellite DNA repeats D4Z4 and NBL2. Hum Mol Genet 2000; 9:597-604.
108. Bourc'his D, Miniou P, Jeanpierre M et al. Abnormal methylation does not prevent X inactivation in ICF patients. Cytogenet Cell Genet 1999; 84:245-252.
109. Miniou P, Jeanpierre M, Blanquet V et al. Abnormal methylation pattern in constitutive and facultative (X inactive chromosome) heterochromatin of ICF patients. Hum Mol Genet 1994; 3:2093-2102.
110. Amir RE, Van den Veyver IB, Wan M et al. Rett syndrome is caused by mutations in X-linked MECP2, encoding methyl- CpG-binding protein 2. Nature Genet 1999; 23:185-188.
111. Oberle I, Rousseau F, Heitz D et al. Instability of a 550-base pair DNA segment and abnormal methylation in fragile X syndrome. Science 1991; 252:1097-1102.
112. Coffee B, Zhang F, Warren ST et al. Acetylated histones are associated with FMR1 in normal but not fragile X-syndrome cells. Nature Genet 1999; 22:98-101.
113. Buiting K, Saitoh S, Gross S et al. Inherited microdeletions in the Angelman and Prader-Willi syndromes define an imprinting centre on human chromosome 15. Nature Genet 1995; 9:395-400.
114. Reis A, Dittrich B, Greger V et al. Imprinting mutations suggested by abnormal DNA methylation patterns in familial Angelman and Prader-Willi syndromes. Am J Hum Genet 1994; 54:741-747.
115. Perk J, Makedonsky K, Lande L et al. On the imprinting mechanism of the Prader Willi/Angelman regional control center. EMBO J 2002; 21:5807-5814.
116. Brown KW, Villar AJ, Bickmore W et al. Imprinting mutation in the Beckwith-Wiedemann syndrome leads to biallelic IGF2 expression through an H19-independent pathway. Hum Mol Genet 1996; 5:2027-32.
117. Smilinich NJ, Day CD, Fitzpatrick GV et al. A maternally methylated CpG island in KvLQT1 is associated with an antisense paternal transcript and loss of imprinting in Beckwith- Wiedemann syndrome. Proc Natl Acad Sci USA 1999; 96:8064-8069.
118. Paulsen M, El-Maarri O, Engemann S et al. Sequence conservation and variability of imprinting in the Beckwith- Wiedemann syndrome gene cluster in human and mouse. Hum Mol Genet 2000; 9:1829-1841.
119. Marinus MG. Methylation of prokaryotic DNA. In: Razin A, Cedar H, Riggs AD, eds. DNA methylation: Biochemistry and Biological Significance. New York: Springer Verlag, 1984:81-109.
120. Proffitt JH, Davie JR, Swinton D et al. 5-Methylcytosine is not detectable in Saccharomyces cerevisiae DNA. Mol Cell Biol 1984; 4:985-988.
121. Urieli-Shoval S, Gruenbaum Y, Sedat J et al. The absence of detectable methylated bases in Drosophila melanogaster DNA. FEBS Lett 1982; 146:148-52.
122. Simpson VJ, Johnson TE, Hammen RF. Caenorhabditis elegans DNA does not contain 5-methylcytosine at any time during development or aging. Nucleic Acids Res 1986; 14:6711-6719.
123. Vogelauer M, Wu J, Suka N et al. Global histone acetylation and deacetylation in yeast. Nature 2000; 408:495-498.
124. Selker EU. Trichostatin A causes selective loss of DNA methylation in Neurospora. Proc Natl Acad Sci USA 1998; 95:9430-9435.
125. Cervoni N, Szyf M. Demethylase activity is directed by histone acetylation. J Biol Chem 2001; 27:27.
126. Tamaru H, Selker EU. A histone H3 methyltransferase controls DNA methylation in Neurospora crassa. Nature 2001; 414:277-283.
127. Berger SL, Felsenfeld G. Chromatin goes global. Mol Cell 2001; 8:263-268.

CHAPTER 2

Epigenetic Mechanisms of Gene Regulation:
Relationships between DNA Methylation, Histone Modification, and Chromatin Structure

Keith D. Robertson

Abstract

DNA methylation is a post-replicative, or epigenetic, modification of the genome that is critical for proper mammalian embryonic development, gene silencing, X chromosome inactivation, and imprinting. Genome-wide DNA methylation patterns are nonrandomly distributed and undergo significant remodeling events during embryogenesis. DNA methylation patterns are also frequently 'remodeled' in tumor cells in a way that directly contributes to tumor suppressor gene inactivation and genomic instability. The mechanisms for the establishment and maintenance of genomic DNA methylation patterns during development and in somatic cells remains a very important and unanswered question in the DNA methylation field. Emerging evidence suggests that protein-protein interactions between components of the DNA methylation machinery (the DNA methyltransferases) and aspects of chromatin structure such as histone tail modifications and chromatin remodeling, directly determine which regions of the genome are to be methylated. By studying these mechanisms in detail we should be able gain insights into how DNA methylation patterns become disrupted in tumor cells and how these defects may be corrected.

Introduction

Methylated DNA refers to DNA strands containing nucleotide bases modified to contain a methyl group ($-CH_3$). Early work in the DNA methylation field centered on the study of the restriction-modification system, a mechanism of bacterial genome protection. In this system, a restriction endonuclease designed to cleave invading viral DNA is coexpressed with a DNA methylase. The DNA methylase methylates the bacterial genome at the same sequence cleaved by the restriction endonuclease, which inhibits cleavage of the host genome and thus selectively destroys the invading DNA sequence.[1,2] It was not surprising that researchers suspected that similar activities might exist in higher eukaryotes. Beginning in the late 1960s, there were several reports on the purification of an activity from mammalian sources able to produce 5-methylcytosine.[3-5] Although the biochemical properties of 5-methylcytosine in mammalian cells were analyzed extensively through two decades, it was not until 1988 that the first cloning and sequencing of a murine DNA methyltransferase was reported.[6] This enzyme is now termed DNMT1, and our understanding of the players performing on the stage of mammalian DNA methylation has been growing by leaps and bounds ever since.

From bacteria to humans, DNA methyltransferases are highly conserved during evolution, and are thus regarded as important regulators of a variety of aspects of cellular function.[7] There are three types of DNA methyltransferases, classified by the nucleotide targeted for modification, N4-methyladenine, N6-methyladenine, and C5-methylcytosine DNA methyltransferases.

DNA Methylation and Cancer Therapy, edited by Moshe Szyf. ©2005 Eurekah.com and Kluwer Academic/Plenum Publishers.

Only one type of DNA methyltransferase is known in mammalian cells, 5-methylcytosine DNA methyltransferase, which transfers a methyl group to the 5-position of cytosine within the CpG dinucleotide recognition sequence.[8] The product of this methylation reaction, 5-methylcytosine, has drawn considerable attention because methylated DNA is believed to be associated with transcriptional regulation and higher order chromatin structure.[9-12]

In mammals, DNA methylation patterns are not randomly distributed throughout the genome, but rather methylated DNA is localized to discrete regions of the genome enriched in repetitive DNA and transposable elements, imprinted domains, and the inactive X chromosome in females.[11,13] In these regions, DNA methylation may serve to suppress spurious transcription, transposition, and recombination. Furthermore, DNA methylation patterns are quite dynamic during mammalian development, with genome-wide methylation remodeling events occurring following fertilization and embryo implantation.[14] DNA methylation patterns also change substantially during the process of tumorigenesis and these changes appear to be early events contributing directly to the transformed phenotype.[9] Tumor cells exhibit global losses of methylation from repetitive sequences and region-specific gains in methylation, primarily within CpG-rich gene regulatory regions known as CpG islands.[15] Promoter region CpG island methylation can silence expression of the associated gene with great efficiency. If the gene is a tumor suppressor gene then the aberrant methylation can provide the cell with a growth advantage as if the sequence had been deleted.[9,10] Thus, although DNA methylation patterns are generally very stable in somatic cells, they can undergo dramatic changes during embryogenesis and tumorigenesis and these changes have profound effects on cell growth and development.

This chapter will first review what is known about the enzymes that are directly responsible for methylated DNA modification in mammalian cells—the DNA methyltransferases (DNMTs). I will then summarize current knowledge of the proteins known to associate with the DNMTs that may alter their enzymatic activity or nuclear targeting. Finally, I will discuss exciting emerging connections between DNA methylation and histone modifications and chromatin remodeling proteins that may soon provide answers to the perplexing question of how cellular DNA methylation patterns are established during development and maintained in somatic cells. These studies may also shed light on the nature of the defect in the cellular DNA methylation machinery that contributes to cellular transformation.

The Mammalian DNA Methyltransferases (DNMTs)

Five genes encoding DNMTs (including potential DNMT-like genes that may not be enzymatically active) have been identified in mammalian cells, DNMT1, 2, 3A, 3B, and 3L.[6,16-18] Each gene is designated by the numbers 1, 2, 3, in the order in which they were identified. For the members of DNMT3 family, the additional letters A, B and L were used. The five genes can be divided into three catagories, based primarily on function: the maintenance DNA methyltransferase DNMT1, the de novo DNA methyltransferases DNMT3A and DNMT3B, and the DNMT-like proteins DNMT2 and DNMT3L (Fig. 1, Table 1). DNMT1 is referred to as the maintenance methyltransferase due to its preference for hemimethylated DNA (which exists as a by-product of DNA replication),[19,20] its targeting to replication foci during S-phase,[21,22] and its interactions with the replication foci-associated proteins proliferating cell nuclear antigen (PCNA),[23] and the retinoblastoma gene product, Rb[24] (Fig. 1, Table 4). DNMT1 also has several alternatively spliced sex-specific isoforms which appear to be involved in the establishment of DNA methylation patterns and imprinting in germ cells and the developing embryo[25] (Table 2). The de novo DNA methyltransferases, DNMT3A and DNMT3B, have been shown to be essential for the waves of de novo methylation in embryonic cells following implantation.[26] These enzymes also mediate de novo methylation of newly intregated parasitic DNA sequences, such as retroviruses, as part of a host cell 'genome defense system'.[13,26] Like DNMT1, DNMT3B has a number of isoforms resulting from alternative splicing events that are expressed in a tissue-specific fashion and which may alter catalytic activity or DNA binding (Table 2).[17,27] The DNMT-like proteins DNMT2.[16] and DNMT3L [18] possess all or some of

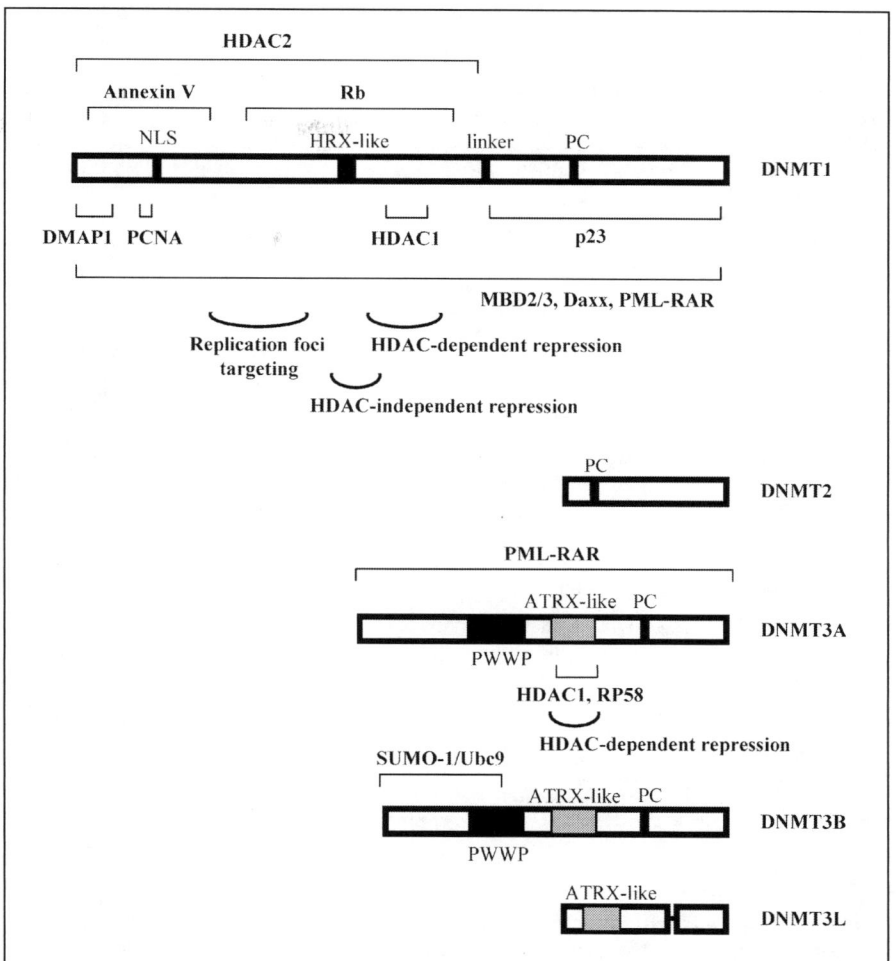

Figure 1. Schematic structures of all known mammalian DNA methyltransferases. Important domains are indicated with boxes and include the nuclear localization sequence (NLS), an HRX-like region, the lysine-glycine repeat region (linker), the catalytic active site proline-cysteine dipeptide (PC), an ATRX-like plant homeodomain region (ATRX-like),[17] and a PWWP motif[86] Proteins known to interact with each of the DNMTs are shown with brackets to denote the interacting region, where known. A bracket encompassing the entire protein indicates that the interaction domain has not been mapped. Transcriptional repression domains are indicated with rounded brackets. DNMT3L lacks the PC motif.

the highly conserved methyltransferase catalytic motifs, respectively, but have not been shown to display enzymatic activity in vitro. Thus their true roles in DNA methylation metabolism remain unclear.

Targeted inactivation of each of the DNA methyltransferase genes in murine embryonic stem (ES) cells has been performed and these studies have provided important information regarding the roles of each DNMT in the establishment and maintenance of genome-wide DNA methylation patterns. The results of all the DNMT knockout studies, summarized in Table 3, indicate that all of the DNMTs that have been shown to possess enzymatic activity in vitro are also absolutely essential for proper embryonic development in mice. DNMT1

Table 1. Properties of the mammalian DNA methyltransferases

DNA Methyl Transferase	Size of Human Protein (Amino acids)	Chromo- somal Location	mRNA Expression Profile	Subcellular Localization	Catalytic Activity in Vitro
DNMT1	1616	19p13.2	Placenta, brain, lung, heart, cell cycle dependent[87,88]	Nucleoplasm during G1 and G2 phases and replication foci throughout S phase[21,22]	Strong (preference for hemimet hylated DNA)[19]
DNMT2	391	10p12- 10p14	Ubiquitous at low levels[16,89]	Not determined	Not detected
DNMT3A	912	2p23	Abundant in ES cells, ubiquitous at very low level in embryos and adult tissues, cell cycle independent[17,27]	Discrete nuclear foci throughout the cell cycle, however replication foci during late-S phase?[44]	Weak (preference for unmethylated DNA)[90,91]
DNMT3B	853	20q11.2	Undifferentiated ES cells, embryos, and testis, cell cycle dependent[17,27]	Diffuse nuclear distribution in NIH3T3 cells, pericentromeric heterochromatin in ES cells[44]	Weak (preference not precisely determined)[92]
DNMT3L	387	21q22.3	Testis and embryos[18]	Not determined	Not determined

and DNMT3B knockout mice die very early in embryonic development, while DNMT3A knockout mice die soon after birth.[26,28] Knockout of DNMT3L, which is not likely to encode a functional DNA methyltransferase, resulted in a more subtle phenotype. Homozygous mutant animals of both sexes were viable but sterile and methylation analysis revealed that loss of Dnmt3L resulted in a lack of maternal methylation imprints in homozygous oocytes. Thus it appears that DNMT3L contributes to imprint establishment during oogenesis, but not to genome-wide methylation patterning.[29] Although DNMT2 possesses all of motifs believed to be important for catalysis, no enzymatic activity has been detected from DNMT2 and DNMT2-knockout mice appeared completely normal (Table 3).[16]

Interaction between DNMTs and Other Proteins

As was mentioned in the introductory remarks, genome-wide DNA methylation is not randomly distributed, yet DNA methyltransferases display little sequence specificity in vitro other than requiring the CpG dinucleotide recognition sequence. Therefore, recent efforts have begun to focus on identifying the protein interaction partners of each of the DNMTs. Emerging evidence suggests that protein-protein interactions dictate which regions of the genome become methylated and which will be protected from methylation.[11] This next section will discuss a number of the proteins known to interact with DNMT1, 3A, and 3B (summarized in Fig. 1 and Table 4), and describe how these interactions may mediate DNMT catalytic activity, subnuclear localization, and sequence specificity in vivo.

Table 2. Splice variants of mammalian DNA methyltransferases

DNA Methyl-transferase Gene	Splice Variants	Tissue Specificity of Expression	Size Difference Compared to the Most 'Typical' Form	Translatable	Catalytic Activity
DNMT1	DNMT1s	The most typical somatic form (usually referred to as DNMT1)	1616 aa (human DNMT1)[6,87]	Yes	Yes
	DNMT1b	Somatic tissues	+16 aa (human),[4] -2 aa (mouse), alternatively spliced at exon 4[93,9]	Yes	Yes
	DNMT1o	Oocytes	-118 aa (mouse), alternatively spliced at exon 1[25]	Yes	Yes
	DNMT1p	Pachytene sperm-atocytes and skeletal muscle	-118 aa (mouse), alternatively spliced at exon 1[25,68]	Yes in skeletal muscle, no in spermatocytes	Not determined, but most likely yes
DNMT3A	Short form	Undifferentiated ES cells and 10.5 day embryos (usually referred to as DNMT3A)	912 aa (human DNMT3A), 4.2-kb in murine cells [17]	Yes	Yes
	Long form	Most adult tissues and differentiated embryos, not in ES cells	9.5-kb as murine mRNA (possibly containing large 5' UTR?)[17]	Not determined	Not determined
DNMT3B	DNMT3B1	Undifferentiated ES cells, embryos, and testis	853 aa (human DNMT3B)	Yes	Yes
	DNMT3B2	Undifferentiated ES cells, embryos, and testis	-20 aa (human and mouse)[26]	Yes	Yes
	DNMT3B3	Undifferentiated ES cells, embryos, and testis	-63 aa (human and mouse)[26]	Yes	No
	DNMT3B4	Testis	-109 aa, alternatively spliced at C-terminal region (human)[27]	Not determined	Not determined
	DNMT3B5	Testis	-41 aa, alternatively spliced at C-terminal region (human)[27]	Not determined	Not determined

Table 3. Effects of targeted disruption of DNA methyltransferase genes in mice

DNA Methyl-transferase	Homozygous ES Cells	Heterozygous Knockout Mice	Homozygous Knockout Mice
DNMT1	Viable and normal morphology, 70% decrease in total 5-methylcytosine.	Indistinguishable from wild type.	Failed to develop beyond mid-gestation, embryonic lethality.[28]
DNMT1o	Viable, characteristics not described in detail.	Normal	Homozygous mutant males showed normal fertility. Homozygous mutant females were infertile. Heterozygous offspring of homozygous females showed demethylation at certain imprinted loci, but not over the whole genome.[95]
DNMT2	Viable and normal.	Normal	No significant phenotype.[16]
DNMT3A	Viable and normal undifferentiated morphology. The de novo methylation activity on proviral DNA was normal. Centromeric-minor satellite DNA repeats were normally methylated.	Normal and fertile.	Appeared normal at birth, showed undergrowth at 18 days and died by 4 weeks of age. Retroviral DNA was methylated at normal levels.[26]
DNMT3B	Viable and normal undifferentiated morphology. The de novo methylation activity on proviral DNA was normal. Centromeric-minor satellite DNA repeats were substantially demethylated.	Normal and fertile.	No viable homozygous mice were born. Retroviral DNA was slightly undermethylated.[26]
DNMT3A and DNMT3B	Double mutant ES cells completely lacked de novo methylation activity on proviral DNA. Centromeric-minor satellite DNA repeats were demethylated to the same level as DNMT3B-\- ES cells.	Not reported.	Double homozygous embryos showed smaller size at E8.5 and died before E11.5. Retroviral DNA was highly undermethylated.[26]
DNMT3L	Viable, characteristics not described in detail.	Normal and fertile.	Both sexes born normal but sterile. Adult testes had severe hypogonadism. Females showed a maternal-effect lethal in that heterozygous progeny of homozygous females died before midgestation. Maternal methylation imprints were markedly disrupted, while genome-wide methylation patterns were normal.

Table 4. Proteins interacting with mammalian DNA methyltransferases

DNA Methyl-transferase	Interacting Protein	Function of Interacting Protein	Possible Role in Vivo
DNMT1	HDAC1/2	Histone deacetylase	Modification of chromatin by histone deacetylation, resulting in chromosome condensation, targeting DNA methylation?[24,42,46]
	Rb	Tumor suppressor, Cell-cycle regulation	Sequester DNMT1 in non-dividing cell, target or modulate DNMT activity at replication foci?[24,37]
	DMAP1	Co-repressor	Recruiting other repressors, transcriptional repression.[46]
	PML-RAR	Oncogenic transcription factor	DNA-binding and interaction with other transcriptional co-regulators, targeting methylation.[62]
	MBD2/3	Methyl-CpG binding proteins	Transcriptional repression in methylated regions, possible targeting of DNMT1 to hemimethylated DNA at replication foci?[65]
	PCNA	"Sliding platform" that can mediate the interaction of proteins with DNA, essential for processivity of DNA polymerase	Targeting DNMT1 to replication foci.[23]
	p23	Subunit of a progesterone receptor complex	Proper protein folding, regulation of DNMT1 catalytic activity?[96]
	Daxx	Transcriptional repressor?	Connection between DNMT1 and other transcription factors, mediate PML – DNMT1 interaction?[97]
DNMT1o	Annexin V	Ca-2+ dependent phospholipid-binding protein	Cytosol-nuclear translocation of Dnmt1o during oogenesis via membrane trafficking?[67]
DNMT3A	HDAC1	Histone deacetylase	Modification of chromatin by histone deacetylation, targeting DNA methylation?[43,44]
	RP58	Transcription factor	Sequence-specific DNA binding, targeting repression, maybe methylation as well?[43]
	PML-RAR	Oncogenic transcription factor	DNA-binding and interaction with other transcriptional co-regulators, targeting methylation.[62]
DNMT3B	HDAC1	Histone deacetylase	Modification of chromatin by histone deacetylation, targeting DNA methylation?[44]
	SUMO-1/Ubc9	Sumo ligase	Modification of protein by sumoylation, altered localization or enzymatic activity?[98]

DNA Methylation and DNA Replication

PCNA

PCNA, or the polymerase processivity factor, is an essential protein in DNA replication. Its heterotrimeric ring-shaped structure allows PCNA to encircle double-stranded DNA and provide a platform for the assembly of other replication-associated proteins.[30] PCNA has been shown to bind to a number of other cellular proteins involved in DNA replication, mismatch repair, and cell cycle regulation.[31] One of these interacting factors is DNMT1 (Fig. 1).[23] DNMT1 was shown to bind to and colocalize with PCNA at early S-phase replication foci. PCNA binding to DNMT1 did not affect its methyltransferase activity, suggesting that PCNA does not regulate DNMT1 activity directly. Rather, DNMT1 appears to be recruited by PCNA to foci of newly replicated DNA to allow for remethylation of hemimethylated DNA. This result is consistent with the traditional model of DNMT1 as a maintenance methyltransferase of newly replicated DNA (Table 4).[23]

Rb

The retinoblastoma tumor suppressor protein (Rb) controls cell growth by regulating the expression of genes that promote cell cycle progression. Hypophosphorylated Rb binds to the transcription factor E2F and represses its activation function by recruiting histone deacetylases (HDACs) and histone methylases (HMTs).[32-34] When a cell is ready to divide, Rb is phosphorylated, dissociates from E2F, and transcriptional activation occurs.[35] Rb itself, or other members of the Rb regulatory pathway, are mutated in nearly all tumor cells, emphasizing the importance of Rb-mediated growth control in normal cells.[36] DNMT1 was found to copurify with Rb and E2F1 and interact directly with Rb (Fig. 1). DNMT1 could enhance Rb-mediated repression specifically at E2F-responsive promoters and the repression was partially HDAC-dependent but was independent of the DNA methyltransferase activity of DNMT1.[24] The amino terminal region of DNMT1 containing the cysteine-rich domain was found to interact with the A/B pocket domain of Rb [24] and later the B/C pocket region as well.[37] Interestingly, DNA methyltransferase activity was strongly inhibited by binding to Rb and this effect appeared to be mediated by interfering with the ability of DNMT1 to bind to DNA. This suggested that Rb may modulate DNMT1 activity in vivo, which was confirmed when Rb was over expressed in cells and a genome-wide reduction in 5-methylcytosine was observed.[37] Rb, as well as DNMT1 and PCNA, colocalizes with early S-phase perinucleolar foci that correspond to sites of active DNA synthesis.[38] Therefore, the potential roles of the Rb-DNMT1 interaction may be to sequester and repress DNMT1 enzymatic activity in nondividing cells,[11] target DNMT1 to replication foci, regulate the association of DNMT1 with PCNA, or reduce or inhibit the catalytic activity of DNMT1 specifically at early S-phase replication foci (Table 4). Interestingly, cancer-specific mutations in the A/B pocket region of Rb were shown to inhibit binding to DNMT1. Therefore mutations in Rb, or the Rb pathway, may directly lead to unscheduled or aberrant DNA methylation events in nondividing cells which would then be copied and spread with each round of cell division.[24]

Links between DNA Methylation and Histone Modification

HDAC1 and HDAC2

DNA methyltransferase and histone deacetylase (HDAC) are believed to operate along the same mechanistic pathway to silence gene expression. DNA methyltransferases establish and maintain 5-methylcytosines in the context of chromatin and methyl-CpG binding proteins of the MBD family, such as MeCP2, recognize and bind methylated DNA and recruit the corepressor/HDAC protein complex.[12] The catalytic unit HDAC can remove acetyl groups from the core histone tails, leading to assembly of tight-packed chromatin and rendering a promoter inaccessible to the transcription machinery by increasing the affinity of histones for DNA. It has long been known that transcriptionally inactive regions are hypermethylated and enriched in hypoacetylated histones. Thus the finding that proteins which bind specifically to methylated DNA interact with and recruit HDACs tied these seemingly unrelated observations together.[39-41] However, this still leaves open the question of how the region was targeted for DNA methylation to begin with. Recent results from several laboratories have revealed that DNA methylation and histone acetylation may be even more tightly linked than first thought because DNA methyltransferases and histone deacetylases directly interact.[24,42-44] In fact HDACs have now been found to interact with all of the catalytically active DNMTs and DNA methylation and histone deacetylation act synergistically to repress transcription.[45]

Initial studies with DNMT1 indicated that the amino terminal regulatory domain could act as a transcriptional repressor when fused to a heterologous DNA binding domain.[42] DNMT1-mediated transcriptional repression was shown to be comprised of both HDAC-dependent and HDAC-independent components. The HDAC-dependent component was defined using the HDAC inhibitor trichostatin A (TSA), which relieves a substantial amount of the DNMT1-mediated repression.[24,42] DNMT1 was shown to bind HDAC1 via a transcriptional repression region adjacent to the HRX-homology domain,[42] and a direct interaction between the DNMT1 amino terminal regulatory domain and HDAC2 has also been demonstrated.[46]

Both DNMT3A and DNMT3B are also capable of conferring transcriptional repression when fused to heterologous DNA binding domains via both HDAC-dependent and HDAC-independent mechanisms. Yeast two-hybrid studies have shown that DNMT3A and DNMT3B interact with HDAC1 through the PHD region (Fig. 1, Table 4),[43,44] which is not present in DNMT1. Therefore, it is possible that the repressive capabilities of DNMT1 and the DNMT3s may be caused by distinct protein-protein interactions, with histone deacetylase as a common mediator.

The functional significance of the DNMT-HDAC interaction remains unclear, although possible roles will be suggested here and in the DNA methylation and chromatin remodeling section to follow. Data from a number of sources indicates that transcriptional silencing occurs before DNA methylation and this transcriptional shutdown may be mediated by histone modifications (deacetylation and methylation) and chromatin remodeling events. For example, silencing of transcription from the X chromosome destined to be inactivated in female cells and histone deacetylation occurs before DNA methylation.[47] Silencing of transcription from newly introduced retroviral sequences also occurs before de novo methylation and can occur in the complete absence of DNMT3A and DNMT3B, the enzymes thought to mediate retroviral DNA methylation.[48] A useful analogy may be that histone tail modifications as well as chromatin remodeling, may 'close the door' on transcription while DNA methylation is the 'deadbolt lock' which ensures that the door remains closed. This model implies that histone deacetylation sets the stage, or targets, DNA methylation to a particular region by establishing a particular chromatin configuration or signature recognized by other DNMT-associated proteins or the DNMT itself and which promotes DNA methylation of the region. Recent dramatic results with studies of histone methylation strongly support this notion.

DNA Methylation and Histone Methylation

Histone tail acetylation has received much attention over the last several years as a key mediator of chromatin structure and transcriptional regulation. The histone tails can also be methylated on select residues, such as lysine and arginine, and depending on the amino acid and the histone modified, may exert a stimulatory or inhibitory effect on transcription.[49,50] Methylation of lysine 9 on histone H3 (H3K9) is associated with transcriptionally silent, heterochromatic regions of the genome.[51] Interestingly, H3K9 methylation also occurs before DNA methylation during X chromosome inactivation.[52] More direct evidence of a connection between DNA and histone methylation comes from two recent fascinating studies in *Neurospora* and *Arabidopsis*. In *Neurospora*, mutation of a H3K9 methyltransferase gene called *dim-5* eliminated all detectable cytosine methylation in the *Neurospora* genome.[53] In *Arabidopsis*, mutation of a gene homologous to *dim-5* termed *kryptonite*, results in substantial losses of methylation from the CpNpGp sequence, which, unlike mammals, is frequently methylated in plants.[54] In the latter case, it was demonstrated that the plant DNA methyltransferase responsible for CpNpGp methylation, chromomethylase 3, is targeted to DNA via interaction with the methylated lysine binding protein HP1.[55] Although DNA methylation is dispensible in *Neurospora* and mammals do not appear to have chromomethylases, a protein homologous to dim-5 and KRYPTONITE exists in mammals (SUV39H1) and is currently the subject of intense study.[56,57] We should therefore soon find out if histone methylation is a critical mediator of DNA methylation in mammals, but regardless of the existence of a homologous system in mammals, this example further reinforces the notion that chromatin modifications may set the stage for DNA methylation.

DNMTs As Transcriptional Corepressors

As we described above, DNMTs are associated with transcriptional repression in an HDAC-dependent manner. This fact, however, is not the only aspect of gene silencing mediated by DNA methyltransferases because they are also involved in transcriptional repression in an HDAC-independent (or TSA insensitive) manner. In fact all catalytically active DNMTs possess transcriptional corepression activity that is independent of histone deacetylase activity.[42,44] Interestingly, the domain responsible for the HDAC-independent repression is different between DNMT1 and the DNMT3s. The region of DNMT1 mediating this effect comprises the cysteine-rich / HRX homology domain (Fig. 1).[42] In contrast, an amino terminal region, which does not include the cysteine-rich domain, is responsible for the HDAC-independent repression capability of DNMT3A and DNMT3B (Fig. 1).[44] Although the protein-protein interactions responsible for the HDAC-independent repression are not well characterized, it is clear that mammalian DNMTs are multi-functional proteins capable of modulating transcription. In the following section, we review the DNMT-associated proteins that may contribute to the ability of the DNMTs to repress transcription and target methylation (Fig. 1, Table 4).

PML-RAR

PML-RAR is an oncogenic fusion protein resulting from the reciprocal translocation of the promyelocytic leukemia (PML) gene on chromosome 15 and the retinoic acid receptor α gene (RAR) on chromosome 17, and gives rise to acute promyelocytic leukemia (APL). Although the function of PML in normal cells remains unclear, it appears to be a critical component of discrete nuclear structures referred to as PML-oncogenic domains (PODs, ND10, or nuclear bodies).[58-61] The PML-RAR fusion protein disrupts the PODs, however they can be restored by treating cells with retinoic acid (RA) since PML-RAR retains both the DNA and ligand binding domains of RARα. RA treatment also results in differentiation of APL cells, indicating that PODs have an important role in promyelocyte differentiation.[58,62] A recent study showed that both DNMT1 and DNMT3A interacted with PML-RAR (Fig. 1, Table 4) and recruited the DNMTs to a PML-RAR target gene promoter resulting in transcriptional silencing and de novo methylation.[62] In the absence of retinoic acid, conditions where PML-RAR acts as a transcriptional repressor, HDAC-dependent transcriptional silencing of an RAR target gene occurred early on, and was then followed by promoter region de novo methylation.[62] This work represents the first example of protein-protein interactions being able to target DNA methylation to particular genomic regions and again stresses the intimate relationships between DNA methylation and histone deacetylation.

DMAP1

Yeast two hybrid screens using the amino terminal regulatory region of DNMT1 as bait identified a novel factor, DNMT1 associated protein (DMAP) 1, that interacts directly with the first 120 amino acids of DNMT1 (Fig. 1, Table 4).[46] It was assumed that DMAP1 also acted as a transcriptional repressor and subsequent two hybrid screens with DMAP1 as bait identified the potent transcriptional repressor, TSG101, as a binding partner of DMAP1. The interaction of DMAP1 and DNMT1 may be responsible for the HDAC-independent component of DNMT1-mediated transcriptional repression. DNMT1 and DMAP1 colocalized at replication foci throughout S-phase, while DNMT1 and HDAC2 colocalized only at late S-phase replication foci. These results led to the suggestion that DNMT1, DMAP1, and HDAC2 may participate in the restoration of heterochromatin structure following DNA replication. DNMT1 and DMAP1 would act to restore DNA methylation patterns following replication throughout S phase while the recruitment of HDAC2 to late replication foci, when hypoacetylated, transcriptionally silenced regions are usually replicated, may allow for rapid deactylation of newly deposited histones.[46]

RP58

DNMT3A and DNMT3B were recently shown to interact directly with a protein called RP58 via the PHD region within the amino terminal regulatory domain (Fig. 1).[43] This region of DNMT3A and DNMT3B also mediates the interaction with HDAC1. RP58 is a sequence-specific zinc finger DNA binding protein and transcriptional repressor associated with heterochromatin.[63,64] The capacity of RP58 to repress transcription was enhanced by coexpression of DNMT3A, however, this cooperative effect did not require a catalytically active form of DNMT3A.[43] This suggests that DNMT3A acts as a structural component in the RP58-mediated repression pathway. Nuclear localization studies using a DNMT3A fragment lacking the catalytic domain also support this notion. The isolated amino terminal domain of DNMT3A colocalized with heterochromatin-associated proteins like HP1α and methyl-CpG binding proteins like MeCP2.[44] The colocalization of DNMT3A with other known heterochromatin-associated proteins therefore suggests that DNMT3A may be an important component of hypermethylated, pericentromeric heterochromatin.

MBD2 and MBD3

An interaction between DNMT1 and the methyl-CpG binding proteins MBD2 and MBD3 has been reported (Fig. 1). DNMT1 coimmunoprecipitated with MBD2 and MBD3 and MBD2/MBD3 demonstrated colocalization with DNMT1 at late S-phase replication foci. Furthermore, the MBD2/MBD3 complex exhibited binding affinity for both hemimethylated and fully methylated DNA and repressed transcription in an HDAC-dependent fashion.[65] These interactions may have roles in directing DNMT1 to hemimethylated sequences following DNA replication, silencing of genes during S-phase, or deacetylation of newly deposited histones in a manner akin to the previously described DNMT1-DMAP1-HDAC2 complex (Table 4).

Daxx

The precise function of Daxx remains unclear. Roles for Daxx in apoptosis have been proposed,[66] but more recent data indicates that Daxx may be an HDAC-dependent transcriptional repressor.[60] Daxx has been found to interact with a rather diverse group of proteins in yeast two hybrid screens, including Fas, CENP-C, Pax-3, PML, and DNMT1 (Fig. 1).[66] The latter two are of interest here, especially in light of the connection between PML, RAR and DNMT1 described earlier. Daxx interacts with and colocalizes with PML in the PODs and the interaction with PML, but not the PML-RAR fusion, inhibits the repression function of Daxx. In cells lacking PML, Daxx resides in regions of condensed chromatin, consistent with a role in transcriptional repression.[60] The previous study describing the interaction between DNMT1 and PML-RAR [62] did not determine if the interaction between the two proteins was direct or indirect. Thus one potential function of the Daxx may be as a bridge between DNMT1 and PML-RAR (Table 4).

Annexin V

The amino terminal regulatory domain of DNMT1 interacts directly with annexin V (Fig. 1). It was also reported that this interaction was enhanced by calcium.[67] Annexin V is localized mainly in the cytosol, but has also been detected in the plasma membrane and the nucleus. In oocytes and four-cell embryos, both annexin V and DNMT1o, the oocyte specific isoform of DNMT1 (Table 2), colocalize in the cytoplasm.[68] Annexin V exhibits calcium-dependent binding to acidic phospholipids in the cytosol, suggesting that this protein participates in membrane-related transactions (such as membrane organization, exocytosis, and endocytosis).[69] Although the function of annexin V is poorly understood, the colocalization of annexin V and DNMT1o may help to explain the unique cytoplasmic-nuclear translocation events of DNMT1o during oogenesis and annexin V may somehow be involved in anchoring DNMT1o in the cytoplasm at this stage (Table 4).

DNA Methylation and Chromatin Remodeling

Three highly significant studies over the last few years have revealed additional links between DNA methylation and chromatin structure other than the previously described interactions between DNMTs and HDACs and histone tail modifications. While covalent modification of the core histone tails by acetylation, methylation, and phosphorylation is a major method of regulating chromatin structure, another mechanism involves ATP-dependent chromatin remodeling machines. These remodeling enzymes, or ATPases, use the energy derived from ATP hydrolysis to directly mobilize or slide nucleosomes on the DNA to permit greater access of transcription factors to DNA and therefore promote activation. Alternatively, they may reorganize nucleosomes into a more regularly spaced, closely packed format which is inhibitory to transcription.[70-72] The ATP-dependent chromatin remodeling enzymes that perform this reaction are members of the SNF2 family, so named for the first such protein identified in yeast (sucrose nonfermenter). The SNF2 superfamily can be divided into three subfamilies, SNF2-like, ISWI, and CHD, based on the presence of several conserved motifs.[72,73] SNF2 proteins are involved in transcription, DNA repair, recombination, and chromatin remodeling and have been shown to be able to assemble regularly spaced nucleosomal arrays on DNA in vitro, promote ATP-dependent disruption of a periodic nucleosomal array, stimulate factor binding, and alter nucleosome spacing.[70-72] I will next review the connections between DNA methylation and chromatin remodeling and then discuss how these processes may be connected mechanistically.

The first connection between DNA methylation and chromatin remodeling came from studies in *Arabidopsis*. Mutation of a gene called *DDM1* (decrease in DNA methylation) yielded plants with numerous growth defects and a profound loss of genomic 5-methylcytosine. The growth defects and losses of DNA methylation became progressively greater with increasing generations of inbreeding.[74] Rather than a DNA methyltransferase, DDM1 is a member of the SNF2 family of ATPases. Further evidence of a connection between DNA methylation and chromatin remodeling comes from studies of a human protein called ATRX. The *ATRX* gene is mutated in a human genetic disease called ATR-X syndrome (α-thalassemia, mental retardation, X-linked).[75] ATRX is a member of the CHD subfamily of ATPases and has been shown to associate with transcriptionally inactive heterochromatin and due to its structure, may have a role in chromatin remodeling. ATRX patients also demonstrated DNA methylation defects, although they were far more subtle than the DDM1 mutation, and included both aberrant hypomethylation and hypermethylation events occurring at several repetitive elements in the genome.[76] Lastly, a recent study made use of transgenic mice to produce a knockout of the murine homolog of DDM1 termed Lsh (lymphoid specific helicase, Hells, PASG).[77-79] Lsh-deficient mice died soon after birth with renal lesions and exhibited 50-60% reductions in genomic 5-methylcytosine levels that affected repetitive elements, single copy genes, and genomic imprinting control regions.[79] Thus inactivation of three putative ATP-dependent chromatin remodeling enzymes in plants and mammals has yielded defects in DNA methylation ranging from subtle to profound.

The three studies just described, especially the Lsh knockout study, provide compelling evidence for a connection between DNA methylation and chromatin remodeling. They also suggest, like the previously mentioned studies of histone methylation, that chromatin remodeling events can determine cellular DNA methylation patterns. This implies that chromatin remodeling takes place before DNA methylation and the DDM1, ATRX, and Lsh results indicated that both de novo and maintenance methylation, or both, could be affected. How then might histone tail modifications (such as acetylation), chromatin remodeling, and DNA methylation be linked? There can be little doubt that the linkage between DNA methyltransferases and histone deacetylases is direct and that this linkage is common to all DNMTs. It remains unclear if the ATPase(s) required for proper DNA methylation patterning is directly associated with a particular DNMT, or if the remodeling enzyme exerts its effects transiently, before the DNMT can be directed to its target DNA sites, then departs. Whether the association of the remodeling enzyme and DNMT is direct or indirect, accumulating evidence strongly suggests that modification of chromatin structure, or the establishment of a particular chromatin configuration or 'signature', must be created for the DNA methyltransferases to be directed to regions of the genome to be methylated. This system could be operational during DNA replication, when DNMT1 must access newly replicated DNA to ensure methylation patterns are faithfully copied. It could also be operational during development when genome-wide DNA methylation patterns are remodeled to first erase parental DNA methylation patterns, and then de novo methylation events establish the proper DNA methylation pattern of the developing organism.

Studies with the ISWI chromatin remodeling enzyme have shown that prior histone acetylation inhibits the ability of ISWI to bind to and remodel nucleosomes.[80] If this is a property of other remodeling enzymes like ATRX or Lsh, then the interaction of DNMTs with HDACs would be highly logical. The HDAC would first deacetylate the region to be methylated, the remodeling enzyme would create a chromatin configuration optimal for the DNMT, then the DNMT would access and methylate DNA (Fig. 2). Alternatively, the region destined for DNA methylation may first have to be methylated on histones before the remodeling enzyme and the DNMT can carry out their functions. These models will likely become testable in the very near future and may finally provide the answer to the complex and perplexing question of how DNA methylation patterns are established and maintained. The next important question to be answered will be then be how the region to be methylated is first targeted for transcriptional silencing and histone deacetylation (or methylation). This process may also involve chromatin remodeling[81] or may be the default setting in the absence of transcription or promoter-specific transcription factors.

Answers to these questions should also provide essential insights into the cause of the DNA methylation defects observed in cancer cells. DNA methylation patterns, chromatin structure, and histone tail modifications may all become disrupted in tumor cells.[82-84] ATP-dependent chromatin remodeling enzymes themselves have been found to be mutated in genetic diseases and cancer.[84,85] Assuming that DNA methylation, histone acetylation, and chromatin remodeling are directly linked as described above, it will be important to determine the relative contribution of aberrations in each of these pathways to the formation of aberrant DNA methylation patterns. If normal histone tail modifications and chromatin structure become disrupted in tumor cells, then when do these aberrations occur during the transformation process? Do they precede the DNA methylation defects? If so then DNA methylation abnormalities may be the end result of other regulatory problems and be most visible to researchers due to the extremely efficient and heritable gene silencing afforded by DNA methylation. If histone acetylation defects occur very early during tumorigenesis for example, then drugs which affect the activity of histone deacetylases or histone acetyltransferases (HATs) may be preventative and the subsequent DNA methylation changes might not occur (Fig. 2). Later stages of tumorigenesis, in which aberrations in multiple epigenetic control mechanisms have already occurred, including DNA methylation changes, may require combination therapies to reverse not only

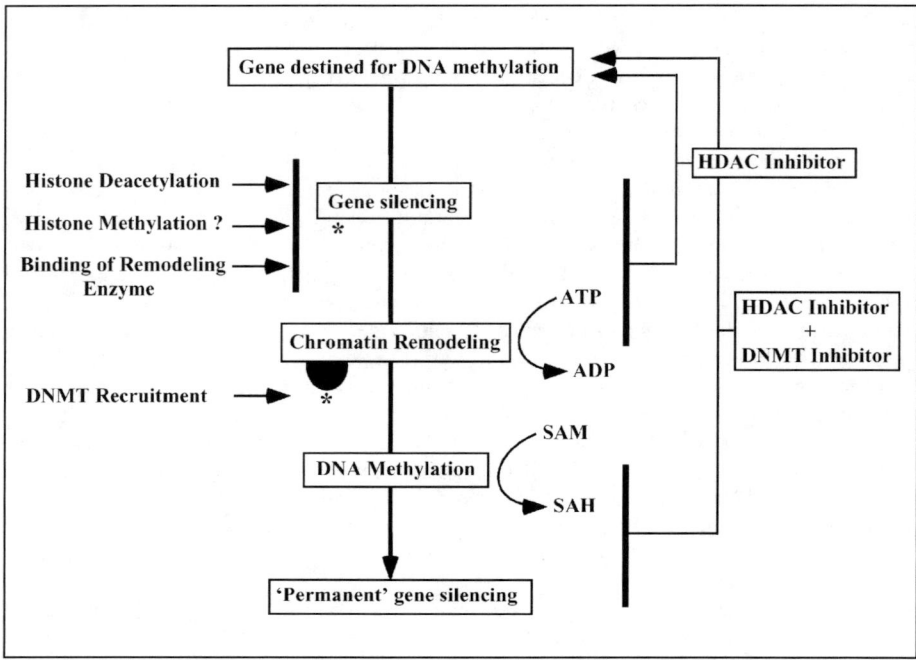

Figure 2. Possible pathway for targeting DNA methylation to specific regions of the genome. A gene destined for long term silencing may first be shut off by histone deacetylation and histone methylation. This then creates an environment favorable for the binding of a chromatin remodeling enzyme (denoted with an *). The chromatin remodeling enzyme binds and repositions or slides nucleosomes in such a way as to directly promote the binding of a DNA methyltransferase or DNMT-associated protein. Alternatively, the remodeling enzyme creates a 'signature' (denoted by black half circle and *), departs, then a DNMT complex binds and methylates DNA. This results in stable long term silencing of the gene. The DNA methylation also recruits methyl-CpG binding proteins and their associated corepressor activities (including histone deacetylase), to further reinforce transcriptional silence and chromatin compaction (not shown). In tumor cells, transcriptional shutdown of key growth regulatory genes may occur before DNA methylation, as is the case in several normal cellular de novo methylation processes. HDAC inhibitors (and possibly chromatin remodeling enzyme and histone methylase inhibitors) can reverse aberrant gene silencing if applied before DNA methylation occurs. Once the region is methylated, an HDAC inhibitor and a DNA methylation inhibitor must be applied to reverse the aberrant gene silencing.

the aberrant histone modifications or chromatin structures, but also the DNA methylation. This is because of the 'permanence' of DNA methylation and its apparent dominance over histone modifications, once established.[45] Thus it will be critical not only to develop novel DNA methyltransferase inhibitors, but also inhibitors of HDACs, HATs, histone methyltransferases, and ATPases. With these inhibitors in hand it may then be possible to design novel treatment regimens to reverse the epigenetic defects at various stages in the tumorigenesis process.

Acknowledgments

K.D.R. is a Cancer Scholar supported by intramural funds from the National Cancer Institute.

References

1. Yuan R. Structure and mechanism of multifunctional restriction endonucleases. Annu Rev Biochem 1981; 50:285-319.
2. Marinus MG. DNA methylation in Escherichia coli. Annu Rev Genet 1987; 21:113-131.
3. Kalousek F, Morris NR. The purification and properties of deoxyribonucleic acid methylase from rat spleen. J Biol Chem 1969; 244(5):1157-1163.
4. Roy PH, Weissbach A. DNA methylase from HeLa cell nuclei. Nucleic Acids Res 1975; 2(10):1669-1684.
5. Gruenbaum Y, Stein R, Cedar H et al. Methylation of CpG sequences in eukaryotic DNA. FEBS Lett 1981; 124(1):67-71.
6. Bestor T, Laudano A, Mattaliano R et al. Cloning and sequencing of a cDNA encoding DNA methyltransferase of mouse cells. J Mol Biol 1988; 203:971-983.
7. Bestor TH. The DNA methyltransferases. Hum Mol Genet 2000; 9(16):2395-2402.
8. Smith SS, Kaplan BE, Sowers LC et al. Mechanism of human methyl-directed DNA methyltransferase and the fidelity of cytosine methylation. Proc Natl Acad Sci USA 1992; 89:4744-4748.
9. Baylin SB, Esteller M, Rountree MR et al. Aberrant patterns of DNA methylation, chromatin formation and gene expression in cancer. Hum Mol Genet 2001; 10(7):687-692.
10. Jones PA, Laird PW. Cancer epigenetics comes of age. Nature Genet 1999; 21:163-166.
11. Robertson KD. DNA methylation, methyltransferases, and cancer. Oncogene 2001; 20:3139-3155.
12. Bird A. DNA methylation patterns and epigenetic memory. Genes Dev 2002; 16:6-21.
13. Yoder JA, Walsh CP, Bestor TH. Cytosine methylation and the ecology of intragenomic parasites. Trends Genet 1997; 13:335-340.
14. Reik W, Dean W, Walter J. Epigenetic reprogramming in mammalian development. Science 2001; 293:1089-1093.
15. Bird A. CpG-rich islands and the function of DNA methylation. Nature 1986; 321:209-213.
16. Okano M, Xie S, Li E. Dnmt2 is not required for de novo and maintenance methylation of viral DNA in embryonic stem cells. Nucleic Acids Res 1998; 26(11):2536-2540.
17. Okano M, Xie S, Li E. Cloning and characterization of a family of novel mammalian DNA (cytosine-5) methyltransferases. Nature Genet 1998; 19:219-220.
18. Aapola U, Shibuya K, Scott HS et al. Isolation and initial characterization of a novel zinc finger gene, DNMT3L, on 21q22.3, related to the cytosine-5-methyltransferase family. Genomics 2000; 65:293-298.
19. Pradhan S, Talbot D, Sha M et al. Baculovirus-mediated expression and characterization of the full-length murine DNA methyltransferase. Nucleic Acids Res 1997; 25(22):4666-4673.
20. Pradhan S, Bacolla A, Wells RD. Recombinant human DNA (cytosine-5) methyltransferase I. Expression, purification, and comparison of de novo and maintenance methylation. J Biol Chem 1999; 274(46):33002-33010.
21. Leonhardt H, Page AW, Weier H et al. A targeting sequence directs DNA methyltransferase to sites of DNA replication in mammalian nuclei. Cell 1992; 71:865-873.
22. Liu Y, Oakeley EJ, Sun L et al. Multiple domains are involved in the targeting of the mouse DNA methyltransferase to the DNA replication foci. Nucleic Acids Res 1998; 26(4):1038-1045.
23. Chuang LS-H, Ian H-I, Koh T-W et al. Human DNA-(cytosine-5) methyltransferase-PCNA complex is a target for p21^Waf1. Science 1997; 277:1996-2000.
24. Robertson KD, Ait-Si-Ali S, Yokochi T et al. DNMT1 forms a complex with Rb, E2F1, and HDAC1 and represses transcription from E2F-responsive promoters. Nature Genet 2000; 25:338-342.
25. Merteinit C, Yoder JA, Taketo T et al. Sex-specific exons control DNA methyltransferase in mammalian germ cells. Development 1998; 125:889-897.
26. Okano M, Bell DW, Haber DA et al. DNA methyltransferases Dnmt3a and Dnmt3b are essential for de novo methylation and mammalian development. Cell 1999; 99:247-257.
27. Robertson KD, Uzvolgyi E, Liang G et al. The human DNA methyltransferases (DNMTs) 1, 3a, and 3b: Coordinate mRNA expression in normal tissues and overexpression in tumors. Nucleic Acids Res 1999; 27(11):2291-2298.
28. Li E, Bestor TH, Jaenisch R. Targeted mutation of the DNA methyltransferase gene results in embryonic lethality. Cell 1992; 69:915-926.
29. Bourc'his D, Xu G-L, Lin C-S et al. Dnmt3L and the establishment of maternal genomic imprints. Science 2001; 294:2536-2539.
30. Krishna TSR, Kong X-P, Gary S et al. Crystal structure of the eukaryotic DNA polymerase processivity factor PCNA. Cell 1994; 79:1233-1243.

31. Kelman Z. PCNA: structure, functions and interactions. Oncogene 1997; 14:629-640.
32. Luo RX, Postigo AA, Dean DC. Rb interacts with histone deacetylase to repress transcription. Cell 1998; 92:463-473.
33. Brehm AB, Miska EA, McCance DJ et al. Retinoblastoma protein recruits histone deacetylase to repress transcription. Nature 1998; 391:597-601.
34. Nielsen SJ, Schneider R, Bauer U-M et al. Rb targets histone H3 methylation and HP1 to promoters. Nature 2001; 412:561-565.
35. Dyson N. The regulation of E2F by pRb-family proteins. Genes Dev 1998; 12:2245-2262.
36. Hanahan D, Weinberg RA. The hallmarks of cancer. Cell 2000; 100:57-70.
37. Pradhan S, Kim G-D. The retinoblastoma gene product interacts with maintenance human DNA (cytosine-5) methyltransferase and modulates its activity. EMBO J 2002; 21(4):779-788.
38. Kennedy BK, Barbie DA, Classon M et al. Nuclear organization of DNA replication in primary mammalian cells. Genes Dev 2000; 14:2855-2868.
39. Wade PA, Gegonne A, Jones PL et al. The Mi-2 complex couples DNA methylation to chromatin remodeling and histone deacetylation. Nature Genet 1999; 23:62-66.
40. Jones PL, Veenstra GJC, Wade PA et al. Methylated DNA and MeCP2 recruit histone deacetylase to repress transcription. Nature Genet 1998; 19:187-191.
41. Ng H-H, Zhang Y, Hendrich B et al. MBD2 is a transcriptional repressor belonging to the MeCP1 histone deacetylase complex. Nature Genet 1999; 23:58-61.
42. Fuks F, Bergers WA, Brehm A et al. DNA methyltransferase Dnmt1 associates with histone deacetylase activity. Nature Genet 2000; 24:88-91.
43. Fuks F, Burgers WA, Godin N et al. Dnmt3a binds deacetylases and is recruited by a sequence-specific repressor to silence transcription. EMBO J 2001; 20(10):2536-2544.
44. Bachman KE, Rountree MR, Baylin SB. Dnmt3a and Dnmt3b are transcriptional repressors that exhibit unique localization properties to heterochromatin. J Biol Chem 2001; 276(34):32282-32287.
45. Cameron EE, Bachman KE, Myohanen S et al. Synergy of demethylation and histone deacetylase inhibition in the reexpression of genes silenced in cancer. Nature Genet 1999; 21:103-107.
46. Rountree MR, Bachman KE, Baylin SB. DNMT1 binds HDAC2 and a new corepressor, DMAP1, to form a complex at replication foci. Nature Genet 2000; 25:269-277.
47. Csankovszki G, Nagy A, Jaenisch R. Synergism of Xist RNA, DNA methylation, and histone hypoacetylation in maintaining X chromosome inactivation. J Cell Biol 2001; 153(4):773-783.
48. Pannell D, Osborne CS, Yao S et al. Retrovirus vector silencing is de novo methylase independent and marked by a repressive histone code. EMBO J 2000; 19(21):5884-5894.
49. Berger SL. An embarrassment of niches: the many covalent modifications of histones in transcriptional regulation. Oncogene 2001; 20:3007-3013.
50. Jenuwein T, Allis CD. Translating the histone code. Science 2001; 293:1074-1080.
51. Peters AHFM, Mermoud JE, Carroll DO et al. Histone H3 lysine 9 methylation is an epigenetic imprint of facultative heterochromatin. Nature Genet 2002; 30:77-80.
52. Mermoud JE, Popova B, Peters AHFM et al. Histone H3 lysine 9 methylation occurs rapidy at the onset of random X chromosome inactivation. Curr Biol 2002; 12:247-251.
53. Tamaru H, Selker EU. A histone H3 methyltransferase controls DNA methylation in Neurospora crassa. Nature 2001; 414:277-283.
54. Jackson JP, Lindroth AM, Cao X et al. Control of CpNpGp DNA methylation by the KRYPTONITE histone H3 methyltransferase. Nature advance online publication. 2002; 17 March.
55. Lachner M, O'Carroll D, Rea S et al. Methylation of histone H3 lysine 9 creates a binding site for HP1 proteins. Nature 2001; 410:116-119.
56. Rea S, Eisenhaber F, O'Caroll D et al. Regulation of chromatin structure by site-specific histone H3 methyltransferases. Nature 2000; 406:593-599.
57. Kouzarides T. Histone methylation in transcriptional control. Curr Opin Genet Dev 2002; 12:198-209.
58. Zhong S, Salomoni P, Pandolfi PP. The transcriptional role of PML and the nuclear body. Nature Cell Biol 2000; 2:E85-E90.
59. Zhong S, Salomoni P, Ronchetti S et al. Promyelocytic leukemia protein (PML) and Daxx participate in a novel nuclear pathway for apoptosis. J Exp Med 2000; 191(4):631-639.
60. Li H, Leo C, Zhu J et al. Sequestration and inhibition of Daxx-mediated transcriptional repression by PML. Mol Cell Biol 2000; 20(5):1784-1796.
61. Müller S, Hoege C, Pyrowolakis G et al. Sumo, ubiquitin's mysterious cousin. Nature Rev Mol Cell Biol 2001; 2:202-210.
62. Di Croce L, Raker AA, Corsaro M et al. Methyltransferase recruitment and DNA hypermethylation of target promoters by an oncogenic transcription factor. Science 2002; 295:1079-1082.

63. Aoki K, Meng G, Suzuki K et al. RP58 associates with condensed chromatin and mediates a sequence-specific transcriptional repression. J Biol Chem 1998; 273(41):26698-26704.

64. Meng G, Inazawa J, Ishida R et al. Structural analysis of the gene encoding RP58, a sequence-specific transrepressor associated with heterochromatin. Gene 2000; 242:59-64.

65. Tatematsu K-i, Yamazaki T, Ishikawa F. MBD2-MBD3 complex binds hemi-methylated DNA and forms a complex containing DNMT1 at the replication foci in late S phase. Genes Cells 2000; 5:677-688.

66. Michaelson JS. The Daxx enigma. Apoptosis 2000; 5:217-220.

67. Ohsawa K, Imai Y, Ito D et al. Molecular cloning and characterization of annexin V-binding proteins with highly hydrophilic peptide structure. J Neurochem 1996; 67:89-97.

68. Doherty AS, Bartolomei MS, Schultz RM. Regulation of stage-specific nuclear translocation of Dnmt1o during preimplantation mouse development. Dev Biol 2002; 242:255-266.

69. Raynal P, Kuijpers G, Rojas E et al. A rise in nuclear calcium translocates annexins IV and V to the nuclear envelope. FEBS Lett 1996; 392:263-268.

70. Vignali M, Hassan AH, Neely KE et al. ATP-dependent chromatin-remodeling complexes. Mol Cell Biol 2000; 20(6):1899-1910.

71. Varga-Weisz P. ATP-dependent chromatin remodeling factors: Nucleosome shufflers with many missions. Oncogene 2001; 20:3076-3085.

72. Havas K, Whitehouse I, Owen-Hughes T. ATP-dependent chromatin remodeling activities. Cell Mol Life Sci 2001; 58:673-682.

73. Eisen JA, Sweder KS, Hanawalt PC. Evolution of SNF2 family of proteins: Subfamilies with distinct sequences and functions. Nucleic Acids Res 1995; 23:2715-2723.

74. Jeddeloh JA, Stokes TL, Richards EJ. Maintenance of genomic methylation requires a SWI2/SNF2-like protein. Nature Genet 1999; 22:94-97.

75. Gibbons RJ, Bachoo S, Picketts DJ et al. Mutations in transcriptional regulator ATRX establish the functional significance of a PHD-like domain. Nature Genet 1997; 17:146-148.

76. Gibbons RJ, McDowell TL, Raman S et al. Mutations in ATRX, encoding a SWI/SNF-like protein, cause diverse changes in the patterns of DNA methylation. Nature Genet 2000; 24:368-371.

77. Jarvis CD, Geiman T, Vila-Storm MP et al. A novel putative helicase produced in early lymphocytes. Gene 1996; 169:203-207.

78. Raabe EH, Abdurrahman L, Behbehani G et al. An SNF2 factor involved in mammalian development and cellular proliferation. Dev Dyn 2001; 221:92-105.

79. Dennis K, Fan T, Geiman T et al. LSH, a member of the SNF2 family, is required for genome-wide methylation. Genes Dev 2001; 15:2940-2944.

80. Clapier CR, Nightingale KP, Becker PB. A critical epitope for substrate recognition by the nucleosomal remodeling ATPase ISWI. Nucleic Acids Res 2002; 30(3):649-655.

81. Tong JK, Hassig CA, Schnitzler GR et al. Chromatin deacetylation by an ATP-dependent nucleosome remodeling complex. Nature 1998; 395:917-921.

82. Wolffe AP. The cancer-chromatin connection. Science and Medicine 1999:28-37.

83. Cairns BR. Emerging roles for chromatin remodeling in cancer biology. Trends Cell Biol 2001; 11(11):S15-S21.

84. Jacobson S, Pillus L. Modifying chromatin and concepts of cancer. Curr Opin Genet Dev 1999; 9:175-184.

85. Ellis NA. DNA helicases in inherited human disorders. Curr Opin Genet Dev 1997; 7:354-363.

86. Qiu C, Sawada K, Zhang X et al. The PWWP domain of mammalian DNA methyltransferase Dnmt3b defines a new family of DNA-binding folds. Nature Struct Biol 2002; 9(3):217-224.

87. Yen R-WC, Vertino PM, Nelkin BD et al. Isolation and characterization of the cDNA encoding human DNA methyltransferase. Nucleic Acids Res 1992; 20(9):2287-2291.

88. Robertson KD, Keyomarsi K, Gonzales FA et al. Differential mRNA expression of the human DNA methyltransferases (DNMTs) 1, 3a, and 3b during the G_0/G_1 to S phase transition in normal and tumor cells. Nucleic Acids Res 2000; 28(10):2108-2113.

89. Yoder JA, Bestor TH. A candidate mammalian DNA methyltransferase related to pmt1 from fission yeast. Hum Mol Genet 1998; 7:279-284.

90. Gowher H, Jeltsch A. Enzymatic properties of recombinant Dnmt3a DNA methyltransferase from mouse: The enzyme modifies DNA in a nonprocessive manner and also methylates nonCpG sites. J Mol Biol 2001; 309:1201-1208.

91. Yokochi T, Robertson KD. Preferential methylation of unmethylated DNA by mammalian de novo DNA methyltransferase Dnmt3a. J Biol Chem 2002; 277:11735-11745.

92. Aoki A, Suetake I, Miyagawa J et al. Enzymatic properties of de novo-type mouse DNA (cytosine-5) methyltransferases. Nucleic Acids Res 2001; 29(17):3506-3512.

93. Bonfils C, Beaulieu N, Chan E et al. Characterization of the human DNA methyltransferase splice variant Dnmt1b. J Biol Chem 2000; 275(15):10754-10760.
94. Hsu D-W, Lin M-J, Lee T-L et al. Two major isoforms of DNA (cytosine-5) methyltransferases in human somatic tissues. Proc Natl Acad Sci USA 1999; 96:9751-9756.
95. Howell CY, Bestor TH, Ding F et al. Genomic imprinting disrupted by a maternal effect mutation in the Dnmt1 gene. Cell 2001; 104:829-838.
96. Zhang X, Verdine GL. Mammalian DNA cytosine-5 methyltransferase interacts with p23 protein. FEBS Lett 1996; 392:179-183.
97. Michaelson JS, Bader D, Kuo F et al. Loss of Daxx, a promiscuously interacting protein, results in extensive apoptosis in early mouse development. Genes Dev 1999; 13:1918-1923.
98. Kang ES, Park CW, Chung JH. Dnmt3b, de novo DNA methyltransferase, interacts with SUMO-1 and Ubc9 through its N-terminal region and is subject to modification by SUMO-1. Biochem Biophys Res Commun 2001; 289:862-868.

CHAPTER 3

DNA Hypo- vs. Hypermethylation in Cancer:
Tumor Specificity, Tumor Progression,
and Therapeutic Implications

Melanie Ehrlich and Guanchao Jiang

Abstract

DNA hypomethylation associated with cancer is probably as frequent as cancer-linked DNA hypermethylation. The hypomethylation of genomic sequences often exceeds hypermethylation so that cancers frequently display lower levels of genomic 5-methylcytosine than do a variety of normal postnatal tissues. Different types of sequences are generally targeted for cancer-related decreases and increases in DNA methylation. Some sequences appear to be more frequently hypomethylated in certain types of cancers than other types. DNA hypomethylation sometimes is evident early in tumorigenesis; however, it also can be associated with tumor progression. Hypomethylation of specific DNA sequences, especially DNA repeats, may serve as a marker for tumorigenesis or tumor progression, as can hypermethylation of unique DNA sequences. Cancer-linked DNA hypomethylation can occur without an association with DNA hypermethylation. Because of this finding and the very frequent targeting of DNA sequences for this hypomethylation in diverse cancers, genomic hypomethylation is likely to contribute to carcinogenesis and not to be just a byproduct of oncogenic transformation. Therefore, caution should be used in development of treatment schemes for cancer involving DNA demethylation because they might result in increased tumor progression.

Introduction

All vertebrate genomes have genetically programmed modification of some of their cytosine residues resulting in a fifth DNA base, 5-methylcytosine (m^5C). The methylation of cytosine residues occurs soon after DNA replication and is mostly, but not only, present in 5'-CG-3' dinucleotides (CpGs).[1] The percentage of cytosine residues in the genome that are methylated is species-specific and tissue-specific.[2,3] The pattern of which CpGs are methylated is also tissue-specific and can be age-specific, although many (but not all) CpG-rich regions, independent of tissue type, show little or no methylation when they overlap promoters or high levels of methylation in unexpressed DNA repeats and CpG-rich intragenic regions.[4-8] Some of the cell type- or developmental stage-specific differences, especially in promoters, probably play a role in helping to keep transcription turned off in certain cell types.[8-14] In addition, DNA methylation is involved in imprinting, X chromosome inactivation, and silencing of tumor suppressor genes in a wide variety of cancers.[4,15,16] Cancers frequently have abnormally high levels of methylation in an average of a few percent of their CpG-rich promoter regions.[17] Often, this cancer-associated hypermethylation downregulates tumor suppressor gene expression and can thus contribute to tumor formation and tumor progression.[4,16]

DNA Methylation and Cancer Therapy, edited by Moshe Szyf. ©2005 Eurekah.com
and Kluwer Academic/Plenum Publishers.

It is now clear that DNA hypermethylation is just as important a source of gene dysfunction leading to cancer formation as are various forms of mutagenesis, namely, point mutagenesis, gene amplification, deletions, and chromosome rearrangements.[17-19] However, decreased methylation of DNA (DNA hypomethylation) in human cancer is also very frequent.[20-23] DNA hypomethylation in cancer often affects more cytosine residues than does hypermethylation so that net losses of genomic m^5C are seen in many human cancers.[22,24-28] In assignments of either hypo- or hypermethylation of DNA from tumors, appropriate control tissue(s) should be used for comparison,[29] and the experiments should not rely only on cell cultures because of changes in DNA methylation upon in vitro cell propagation.[30]

The role of DNA hypomethylation in cancer is less well understood than cancer-associated DNA hypermethylation. Nonetheless, experiments involving DNA methylation inhibitors in vivo and in vitro and *Dnmt1* knockout mice[23,31-34] indicate the importance of DNA hypomethylation to oncogenesis. As described below, the high frequency of cancer-associated DNA hypomethylation, the nature of the affected sequences, and the finding that cancer-linked DNA hypomethylation can occur without an association with cancer-linked DNA hypermethylation are consistent with an independent role for hypomethylation of certain DNA sequences in cancer formation or tumor progression. This review focuses on the specificity of hypomethylation of DNA in cancer, the questions of whether there is a relationship of DNA hypomethylation and hypermethylation in cancer and whether DNA hypomethylation can be correlated with tumor progression, and the implications of cancer-associated DNA hypomethylation for DNA demethylation-directed cancer chemotherapy.

Are There Tumor-Specific DNA Hypomethylation Profiles Like the Tumor-Specific DNA Hypermethylation Profiles?

Hypermethylation in cancer has most often been described for the CpG-rich regions (CpG islands) that overlap many promoters.[17,35,36] While disparate types of cancer may display hypermethylation of the same promoters, nonetheless there are highly significant differences in the methylation frequencies for CpG islands between tumor types.[17] As for DNA hypomethylation associated with cancer, cancer-linked hypermethylation probably involves much overshooting in terms of targets that function in carcinogenesis. For example, *MYOD* at its 5' CpG island is often hypermethylated in some types of cancers, and this is unlikely to be of functional significance.[4] Cancer-associated DNA hypermethylation seems to predominantly occur in the non-repeated DNA sequences of the human genome although hypermethylation of ribosomal RNA genes was found in breast cancer.[37]

In contrast to increased DNA methylation in cancer, cancer-associated hypomethylation of DNA frequently involves repeated DNA elements. The most well-documented interspersed repeat displaying hypomethylation in various cancers are the LINE-1 sequences.[38-41] Most of these human retrotransposon- derived repeats are probably incapable of retrotransposition.[42] They are up to 6 kb in length and comprise about 15% of the human genome. In a study of 73 urothelial carcinomas, about 54% of tumors were hypomethylated in >10% of the LINE-1 repeats compared to normal bladder.[43] Hypomethylation of a much lower copy-number retrotransposon, HERV-K proviruses, significantly paralleled that of LINE-1 repeats.[43] In another study by the same group, 53% of 32 examined prostate adenocarcinomas were found to have LINE-1 hypomethylation.[41] However, no LINE-1 hypomethylation was detected in 34 renal cell carcinomas compared to normal bladder.[43] Therefore, LINE-1 hypomethylation appears to vary in frequency between different types of cancers.

Tandem DNA repeats are also frequently hypomethylated in human cancers, including satellite DNA sequences, which are found in constitutive heterochromatin. Hypomethylation of centromeric satellite α DNA (Satα) and juxtacentromeric (centromere-adjacent) satellite 2 (Sat2) in chromosomes 1 and 16 was demonstrated in various cancers. Sat2 in chromosome 1 was shown to be hypomethylated in 55% of Wilms tumors (29 out of 51 of these pediatric kidney cancers) in one study[6] and 51% of another set of Wilms tumors (18 out of 35) in a second study.[44] In breast adenocarcinomas and ovarian epithelial carcinomas, 44 and 63% of the samples, respectively, were hypomethylated in Sat2 of chromosome 1[45,46] (and M. Ehrlich and L. Dubeau, unpub. data). In all of these tumors, there was a high degree of concordance in hypomethylation of Sat2 of chromosome 1 and that of chromosome 16. Sat DNA hypomethylation in these cancers was defined as less hypomethylation than in any of the examined postnatal somatic tissues, all of which were highly methylated in this sequence as determined by Southern blot analysis with a CpG methylation-sensitive restriction endonuclease. Another group using similar methods found that 69% of hepatocarcinomas (25 out of 36) displayed hypomethylation in Sat2.[47] However, in a study of Sat2 hypomethylation in 30 colorectal cancers and 24 stomach cancers, 25% of the stomach cancers were hypomethylated in both Sat2 (found in the juxtacentromeric heterochromatin of chromosomes 1 and 16) and Sat3 (found in the juxtacentromeric heterochromatin of chromosome 9) but none of the colorectal cancers were hypomethylated in either sequence.[48] Therefore, while a diverse collection of examined cancers have Sat2 hypomethylation at high frequencies, there may be some tumor-specific differences in the incidence of this hypomethylation of constitutive heterochromatic DNA repeats.

Hypomethylation of Satα, the major centromeric DNA sequence, was observed in both chromosome 1 and throughout the centromeres in most examined ovarian epithelial carcinomas[46] (M. Ehrlich and L. Dubeau, unpub. results; M. Ehrlich and Widschwendter, unpub. results). The controls were a variety of normal postnatal somatic tissues. Similarly, in Wilms tumors, this hypomethylation was seen in 90% of samples (51 out of 52 tumors) in one study[6] and 83% (29 out of 35 tumors) in another.[44] In the only reported study of Satα hypomethylation in another type of cancer, breast adenocarcinomas, hypomethylation of this sequence was observed in 19% of the samples (4 out of 21).[45] Further studies will reveal if there are significant differences in the frequency of hypomethylation of Sata sequences between tumor types.

An unrelated pericentromeric DNA repeat (either in the centromere or juxtacentromeric region) that shows cancer-associated hypomethylation is a *Not*I repeat, Y10752 (GenBank Accession Number), isolated by Nagai and coworkers.[49] It is 70% homologous to a previously cloned DNA sequence displaying hypomethylation in sperm and a high degree of methylation in postnatal somatic tissues.[5] As determined by Southern blot analysis with a CpG methylation-sensitive restriction endonuclease, it was hypomethylated in 75% of hepatocellular carcinomas relative to normal liver.[49] It was less frequently hypomethylated in colon and stomach cancers. Y10752 is the same as the sequence NBL2 found by Thoraval et al to be localized to the pericentromeric regions of four chromosomes and frequently hypomethylated in neuroblastomas, but not in glial or lung cancers.[50] Importantly, Y10752/NBL2 and Sat2 share the attributes of being highly methylated in normal postnatal somatic tissues, frequently hypomethylated in certain types of cancers, and always hypomethylated in sperm DNA and in somatic cells from patients with the DNA methyltransferase deficiency ICF syndrome (immunodeficiency, centromeric region instability, and facial anomalies).[6,46,51-56] ICF is associated with only a small percentage of the genome being hypomethylated (7% less m^5C in ICF brain DNA than in normal DNA [52] because of the existence of multiple DNA methyltransferase genes, only one of which has been shown to be mutated in ICF patients.[54,55] While cancers with juxtacentromeric Sat2 hypomethylation frequently have centromeric Satα hypomethylation, most ICF patients have Sat2, but not Satα, hypomethylation.[51,52] A cancer predisposition has not been reported for ICF patients, but there are only low numbers of identified ICF patients (<40), and they frequently die in childhood. The small number and very short average lifespan of ICF patients would preclude detection of a cancer predisposition that was not very high and that did not result in tumors rather quickly.

Some tumor- or proliferation-associated genes have been found to be hypomethylated in human cancers (reviewed in ref. 29). Sometimes, hypomethylation has been shown to be concordant with expression, e.g., the CpG- rich promoter of the *HOX11*, a proto-oncogene and the 5' end of *pS2*, which encodes a pleiotropic factor implicated in the control of cell proliferation.[57,58]

However, these genes were studied only in breast adenocarcinomas and normal breast or endometrium and T-cell lymphoblastic leukemia samples and normal bone marrow. Imprinted genes can also display cancer-associated hypomethylation.[59-63] It is likely that certain of these genes or adjacent sequences are preferentially hypomethylated in a tumor-type specific manner, although this remains to be demonstrated.

One DNA hypomethylation-linked superfamily of genes encodes tumor-specific antigens and includes *MAGE, LAGE,* and *GAGE* families. These genes display testes-specific and cancer-specific expression with no detectable expression in a wide variety of normal postnatal somatic tissues.[64-66] *MAGE-C1, LAGE-1,* and *BAGE* are frequently expressed in melanomas, bladder, breast, and lung carcinomas but not in colorectal carcinomas, renal cell carcinomas, or leukemias.[66-68] Transcription of these genes can be induced in non-expressing cancer cell lines or normal cell lines by treatment with 5-azadeoxycytidine. Furthermore, methylation of Ets motifs in the promoter of *MAGE-A1* interferes with binding of an Ets family transcription factor.[64] The *MAGE-A1* promoter is hypomethylated in *MAGE-A1*-expressing cell lines, but not in non-expressing cell lines and leukocytes.[64,69] The CpG-rich promoters and 5' gene regions of *MAGE-A1* and *LAGE-1* are highly methylated in a variety of normal tissues and mostly unmethylated in sperm.[8] Unmethylated promoter sequences were found for three *MAGE* family genes in lung cancers and, to a lesser extent, in surrounding tissue.[70] Expression was significantly correlated with promoter hypomethylation. Therefore, the *MAGE/LAGE/GAGE* superfamily of genes is expressed and probably concordantly hypomethylated in only certain types of cancers.

Hypomethylation of some gene regions that have no apparent relationship to carcinogenesis has been described, e.g., the gene encoding β-globin in colon and breast adenocarcinomas.[71,72] Also, various amounts of hypomethylation of uncharacterized DNA sequences from human colon, bladder, and prostate cancers compared to analogous apparently normal adjacent tissues were seen in methylation-sensitive arbitrarily primed PCR used *Hpa*II or *Msp*I and *Rsa*I for digestion.[73] However, it is not known if the frequency of hypomethylation of these sequences differs significantly in different types of cancers.

Global hypomethylation of DNA may be a common attribute of diverse cancers.[22] Overall deficiencies in the m^5C content of DNA have been frequently found in many disparate types of cancer, including ovarian epithelial carcinomas vs. cystadenomas or normal postnatal somatic tissues;[24] prostate metastatic tumors vs. normal prostate;[25] leukocytes from B-cell chronic lymphocytic leukemia vs. normal leukocytes;[26] hepatocellular carcinomas vs. matched non-hepatoma liver tissue;[27] cervical cancer and high-grade dysplastic cervical lesions vs. normal cervical tissue or low-grade dysplasia of the cervix;[28] colon adenocarcinomas vs. adjacent normal mucosa,[74] and Wilms tumors vs. various normal postnatal somatic tissues.[75] Some types of cancers, e.g., testicular germ cell seminomas, may display especially large amounts of genomic hypomethylation[22,76A] although this could sometimes be the result of the cell of origin being unusually hypomethylated in its DNA. In these studies, as in many quantitative studies of DNA methylation, the percentage of cells from the tumor sample that are contaminating non-neoplastic cells may vary and affect the results, if the tumor sample has not been microdissected. This is not an important factor for some types of tumors, like Wilms tumors and ovarian epithelial carcinomas, which consist mostly of neoplastic cells, but is important for others, like breast adenocarcinomas.

Is There a Relationship between Cancer-Associated DNA, Hypomethylation and DNA Hypermethylation?

Hypomethylation of some DNA sequences and hypermethylation of other sequences has been found in rat hepatocarcinomas and human breast, colon, and prostate adenocarcinomas.[45,76B,72] Given the prevalence of both of these types of changes in cancers when they have been studied individually, they probably coexist in the vast majority of cancers but simply have not yet been documented to be simultaneously present. Hypermethylation of the *GST P* promoter and hypomethylation of LINE-1 repeats were found in some of the same prostate adenocarcinomas but no significant association was observed between these two types of epigenetic changes in this kind of cancer.[41] Recently, in collaboration with Peter Laird and Emerich Fiala, we have shown that global DNA hypomethylation, satellite DNA hypomethylation, hypomethylation of a promoter, and hypermethylation of CpG island-promoters are present concurrently in many Wilms tumors.[75] Global hypomethylation was significantly associated with Sat2 hypomethylation in the juxtacentromeric heterochromatin of chromosome 1 and with Satα hypomethylation throughout the centromeres. However, there was no significant association between global DNA hypomethylation and CpG island hypermethylation or satellite DNA hypomethylation and CpG island hypermethylation. These results argue against DNA hypomethylation being just a bystander during carcinogenesis or just a provoker of DNA hypermethylation. Therefore, given the prevalence of DNA hypomethylation in cancer, it is very likely that it plays an independent role in carcinogenesis.

The role of DNA hypomethylation in cancer may involve increasing chromosome rearrangements[6,47,52,77] and also altering gene expression. Besides standard *cis* effects when hypomethylation includes transcription regulatory regions, there may be *trans* effects involving cancer-associated hypomethylation of satellite DNA in heterochromatin. There are precedents for centromeric heterochromatin interacting with early lymphogenesis genes and the β-globin locus in a manner that downregulates expression of those genes either directly or by acting as a reservoir for transcription control proteins that preferentially bind to them as well as to certain gene promoters.[78-80] In parallel, it has been suggested that the hypomethylation of juxtacentromeric heterochromatin of ICF syndrome lymphoid cells interferes with normal downregulation of certain genes by disrupting heterochromatin-euchromatin interactions in *trans*.[81] Alternatively, there may be *cis* effects of hypomethylation of repeated DNA sequences on neighboring gene regions, which could result in alterations in levels of transcription regulatory factors or signal transduction molecules and, thus, many downstream effects. Therefore, hypomethylation of certain repeated DNA sequences in tumors might be another source of gene dysregulation in cancer, perhaps in conjunction with cancer-linked alterations in expression from elsewhere in the genome.

Is DNA Hypomethylation, Like DNA Hypermethylation, Sometimes Associated with Tumor Progression?

Hypermethylation of a subset of CpG islands is progressive in some types of cancer although sometimes this regional hypermethylation occurs very early in tumorigenesis, and other times it can serve as a significant indicator of survival.[16,82,83] Also, DNA hypomethylation is sometimes associated with tumor progression as seen in studies of repeated DNA sequences. In a study of 31 hepatocellular carcinomas by Itano and coworkers,[84] the degree of hypomethylation of either of two repetitive sequences was significantly correlated with postoperative recurrence of hepatocellular carcinoma and was a better predictor than conventional factors. These repeats were the above-mentioned 1.4-kb Y10752/NBL2 sequence that is found in the pericentromeric regions of chromosomes 13, 14, 21, and 9[50] and a 13-kb repeat present in tandem about 200 times on 8q21. Hypomethylation was determined by quantitating the corresponding radioactive spots relative to reference spots by restriction landmark genomic scanning (RLGS) involving two-dimensional electrophoresis of DNA doubly digested with the CpG methylation-sensitive *Not*I and the CpG methylation-insensitive *Pvu*II. The 1.4-kb Y10752/NBL2 repeat (CNIC) and the 13-kb repeat (HTRS) were fully methylated at the examined sites in normal liver. Hypomethylation of Y10752/NBL2 and HTRS was also significantly correlated with the presence of hepatitis B or C antibodies in the serum. Itano and coworkers suggest that viral infection could influence the tumors' malignant potential and tumor recurrence indirectly by predisposing to DNA hypomethylation. In an earlier study, this group showed that the number of spots that appear in RLGS profiles of hepatocellular carcinomas but not in the profiles of normal liver was also a significant and independent prognostic indicator of postoperative occurrence of the disease.[85] Their appearance specifically in the hepatocarcinoma RLGS profile is presumably because they are hypomethylated only in the tumors.

In a prostate cancer study, LINE-1 hypomethylation had a highly significant relationship with lymph node involvement for prostate adenocarcinomas.[41] Recently, we have shown that hypomethylation of both Satα centromeric and Sat2 juxtacentromeric repeats is significantly associated with tumor grade and decreased survival in primary ovarian carcinomas (M. Ehrlich and M. Widschwendter, unpub. results). In collaboration with Louis Dubeau, we also demonstrated that there is a significant association of malignant potential and hypomethylation of Sat2 DNA in the juxtacentromeric heterochromatin of chromosomes 1 and 16 in a comparison of benign ovarian cystadenomas, low malignant potential tumors, and carcinomas.[46] Moreover, there was also a significant association of Sat2 hypomethylation with global hypomethylation of the genome in these neoplasms, as determined by Southern blot analysis for satellite hypomethylation and high-performance liquid chromatography of DNA digested to deoxynucleosides for global hypomethylation. These studies suggest that one of the carcinogenesis-promoting advantages of global hypomethylation is that it is often linked to hypomethylation of satellite DNA sequences. Global hypomethylation might also be related to hypomethylation of interspersed DNA repeats, including retroelements.[43]

Not only is the degree of global genomic hypomethylation in ovarian epithelial neoplasms associated with the degree of malignancy,[24] but also it is significantly associated with the grade of cervical neoplasia,[28] with multifocal vs. unifocal hepatocellular carcinomas,[86] and the disease stage, tumor size, and histological grade for breast tumors.[87] However, the latter three studies were done by incorporation of methyl groups in vitro, and it is desirable to see confirmation with a direct analysis of m^5C levels. A study of global levels of DNA methylation at *Hpa*II sites in breast cancer, as determined by the extent of smearing of DNA fragments in *Hpa*II digests upon electrophoresis, revealed no significant association of hypomethylation with tumor grade although almost all of the tumors were hypomethylated compared to normal breast tissue.[88] The quantitation of DNA methylation by this method might have been complicated by different amounts of degradation of the DNA during isolation.

As for CpG island hypermethylation in cancers, cancer-associated hypomethylation of certain DNA sequences in some types of cancers occurs early in tumorigenesis and, in others, only later. Previously, we found that 21 benign tumors from the breast, ovary, uterus, thyroid, or brain had an average genomic m^5C content that was the same as that from all 15 of the various normal human tissues (0.89 mol%; percentage of the bases as m^5C) examined.[22] The analogous values for the 20 metastases and 62 primary tumors were 0.78 and 0.83 mol%, respectively.

In contrast, there is evidence for the early appearance of DNA hypomethylation during some types of tumorigenesis. Goelz et al[71] found hypomethylation of 3-4 of 10 examined genes by Southern blot analysis with CpG methylation-sensitive restriction endonucleases in adenomatous colon polyps from seven patients. Five of those patients had colon cancers also displaying hypomethylation of the same genes. In collaboration with Andy Feinberg and Charles Gehrke, we compared colon tumors and adjacent apparently normal tissue. We found that there was a reproducible decrease in global DNA methylation levels of about 8% in eight colonic polyps from colon cancer patients.[74] Ribieras and coworkers observed hypomethylation of the A-γ globin gene in two samples of lobular carcinoma in situ, an early stage of breast cancer.[72] However, we found that only one of 19 samples from abnormal but nonmalignant breast tissue showed moderate-to-strong hypomethylation of Sat2 compared with almost half of 25 examined breast adenocarcinomas displaying moderate-to-strong hypomethylation.[45] The benign samples were tissues displaying mild or moderate fibrocystic changes, fibroadenoma, gynecomastia, or benign phylloides tumor. How early DNA hypomethylation can be detected during tumorigenesis probably depends on the DNA sequence being examined, the type of tumor, and the individual tumor sample, as is often the case for CpG island hypermethylation.

Might There Be Deleterious Consequences of Introducing DNA Hypomethylation in the Genome As a Cancer Therapy?

The DNA methylation inhibitors 5-azacytidine, 5-aza-2'-deoxycytidine (decitabine), and 5,6-dihydro-5- azacytidine have been used as cancer chemotherapeutic agents in clinical trials on various neoplasms, including refractory acute leukemia;[89] myelodysplastic syndrome;[90] advanced non-small cell lung cancer;[91] malignant mesothelioma;[92] accelerated or blast phase of chronic myeloid leukemia;[93] advanced ovarian or cervical carcinoma;[94,95] malignant melanomas; and colorectal, head and neck, and renal carcinomas.[96] For solid tumors, usually little or no clinical efficacy and often no disease stabilization was seen, but many toxic effects were observed.[89,91,94-96] Combination therapy on malignant mesothelioma, which showed a low response to 5,6- dihydro-5-azacytidine alon,[92] did not improve the response rate (17%) and increased the toxicity.[97] There has been considerable attention recently to testing the efficacy of treatment of high-risk myelodysplastic syndrome (MDS) with 5-azacytidine or 5-aza-2'-deoxycytidine.[98-100] Only supportive care is standard treatment for high-risk MDS. Upon 5-aza-2'-deoxycytidine treatment of 61 high-risk MDS patients displaying cytogenetic abnormalities, 31% of the patients had major cytogenetic responses.[101] These patients had a significantly longer mean survival than the cytogenetic non-responders although almost all the patients died by 34 months after initiation of treatment. Most definitive, although still with the drawbacks inherent in clinical studies on cancers with poor prognoses,[99] was a randomized Phase III study of 191 MDS patients.[90] In the group of patients receiving supportive care, only 5% showed improvement during the course of the study. In contrast, 37% of the 5-azacytidine-treated patients were classified as improved; 16%, as partial responders; and 7% as complete responders. However, there was no significant difference in overall survival between the 5-azacytidine-treated patients and the control patients.

The best outcomes of treatment with 5-azacytidine or 5-aza-2'-deoxycytidine were seen for patients with refractory or relapsed acute leukemia.[93] The response rates were 33-89%, and complete remission rates were 27-45%, alone or in combined therapy regimes. In one study on patients with refractory acute leukemia, 8 out of 11 patients achieved complete remission after treatment with 5-aza-2'-deoxycytidine in combination with Amsacrine.[102] By digestion of DNA to deoxyribonucleosides and high performance liquid chromatography, global DNA methylation levels were determined on bone marrow samples from patients who still had more than 70% leukemic cells. These samples were taken just before treatment and after 3-6 days of treatment. Seven days after treatment, there were decreases in global genomic methylation in bone marrow samples, but there was no relation between the extent of drug-associated hypomethylation and the clinical response. 5-Azacytidine and 5-aza-2'-deoxycytidine are DNA replication inhibitors as well as inhibitors of DNA methyltransferase activity once they are incorporated into DNA. These drugs cause various other metabolic disturbances that may be unrelated to their induction of DNA hypomethylation, but may contribute toward their effectiveness as antileukemia drugs.[103,104]

There is insufficient long-term survival data on cancer patients treated with DNA methylation inhibitors (either the above three drugs or newly developed drugs, including those based upon antisense therapy) to evaluate whether there are long-term risks associated with these treatments. Given the evidence for increased chromosome rearrangements associated with DNA hypomethylation[6,47,51,52,105-107] and for significant associations of DNA hypomethylation with tumor progression, this is a possibility that should not be overlooked in developing new protocols for therapeutic DNA demethylation in cancer patients. This is especially pertinent to patients who do not have a poor prognosis. It may be that studies of DNA methylation changes in cancer will have their most practical applications in cancer diagnostics and prognostics and in aiding in the design of transcription therapy protocols aimed at transcription factors, chromatin proteins, or histone modifications.[108,109]

Acknowledgements

This work was supported in part by NIH Grant CA81506.

References

1. Ramsahoye BH, Biniszkiewicz D, Lyko F et al. Non-CpG methylation is prevalent in embryonic stem cells and may be mediated by DNA methyltransferase 3a. Proc Natl Acad Sci USA 2000; 97(10):5237-42.
2. Gama-Sosa MA, Midgett RM, Slagel VA et al. Tissue-specific differences in DNA methylation in various mammals. Biochim Biophys Acta 1983740(2):212-19.
3. Ehrlich M, Gama-Sosa M, Huang L-H et al. Amount and distribution of 5-methylcytosine in human DNA from different types of tissues or cells. Nucleic Acids Res 1982; 10:2709-21.
4. Issa JP. Hypermethylator Phenotypes in Aging and Cancer. In: Ehrlich M, ed. DNA alterations in cancer: genetic and epigenetic alterations. Natick: Eaton Publishing; 2000; 311-22.
5. Zhang X-Y, Loflin PT, Gehrke CW et al. Hypermethylation of human DNA sequences in embryonal carcinoma cells and somatic tissues but not sperm. Nucleic Acids Res 1987; 15:9429-49.
6. Qu G, Grundy PE, Narayan A et al. Frequent hypomethylation in Wilms tumors of pericentromeric DNA in chromosomes 1 and 16. Cancer Genet Cytogenet 1999; 109:34-39.
7. Jones PA. The DNA methylation paradox. Trends Genet 1999; 15(1):34-7.
8. De Smet C, Lurquin C, Lethe B et al. DNA methylation is the primary silencing mechanism for a set of germ line- and tumor-specific genes with a CpG-rich promoter. Mol Cell Biol 1999; 19(11):7327-35.
9. Lee PP, Fitzpatrick DR, Beard C et al. A critical role for Dnmt1 and DNA methylation in T cell development, function, and survival. Immunity 2001; 15(5):763-74.
10. Kroft TL, Jethanandani P, McLean DJ et al. Methylation of CpG dinucleotides alters binding and silences testis- specific transcription directed by the mouse lactate dehydrogenase C promoter. Biol Reprod 2001; 65(5):1522-27.
11. White GP, Watt PM, Holt BJ et al. Differential patterns of methylation of the IFN-gamma promoter at CpG and non-CpG sites underlie differences in IFN-gamma gene expression between human neonatal and adult CD45RO- T cells. J Immunol 2002; 168(6):2820-27.
12. Singal R, Wang SZ, Sargent T et al. Methylation of promoter proximal-transcribed sequences of an embryonic globin gene inhibits transcription in primary erythroid cells and promotes formation of a cell type-specific methyl cytosine binding complex. J Biol Chem 2002; 277(3):1897- 905.
13. Takizawa T, Nakashima K, Namihira M et al. DNA methylation is a critical cell-intrinsic determinant of astrocyte differentiation in the fetal brain. Dev Cell 2001; 1(6):749-58.
14. Lee DU, Agarwal S, Rao A. Th2 lineage commitment and efficient IL-4 production involves extended demethylation of the IL-4 gene. Immunity 2002; 16(5):649-60.
15. Attwood JT, Yung RL, Richardson BC. DNA methylation and the regulation of gene transcription. Cell Mol Life Sci 2002; 59(2):241-57.
16. Baylin SB, Herman JG. Epigenetics and Loss of Gene Function in Cancer. In: Ehrlich M, ed. DNA Alterations in Cancer: Genetic and Epigenetic Alterations. Natick: Eaton Publishing; 2000:293-309.
17. Costello JF, Fruhwald MC, Smiraglia DJ, Rush LJ, Robertson GP, Gao X, et al. Aberrant CpG-island methylation has non-random and tumour-type-specific patterns. Nat Genet 2000;24(2):132-38.
18. Yan PS, Efferth T, Chen HL et al. Use of CpG island microarrays to identify colorectal tumors with a high degree of concurrent methylation. Methods 2002; 27(2):162-9.
19. Esteller M, Herman JG. Cancer as an epigenetic disease: DNA methylation and chromatin alterations in human tumours. J Pathol 2002; 196(1):1-7.
20. Feinberg AP, Vogelstein B. Hypomethylation distinguishes genes of some human cancers from their normal counterparts. Nature 1983; 301(5895):89-92.
21. Feinberg AP, Vogelstein B. Hypomethylation of ras oncogenes in primary human cancers. Biochem. Biophys. Res Commun 1983; 111(1):47-54.
22. Gama-Sosa MA, Slagel VA, Trewyn RW et al. The 5- methylcytosine content of DNA from human tumors. Nucleic Acids Res 1983; 11:6883-94.
23. Ehrlich M. DNA hypomethylation and cancer. In: Ehrlich M, ed. DNA Alterations in Cancer: Genetic and Epigenetic Changes. Natick: BioTechniques Books, Eaton Publishing; 2000:273-91.
24. Cheng P, Schmutte C, Cofer KF et al. Alterations in DNA methylation are early, but not initial, events in ovarian tumorigenesis. Br J Cancer 1997; 75:396-402.
25. Bedford MT, van Helden PD. Hypomethylation of DNA in pathological conditions of the human prostate. Cancer Res 1987; 47(20):5274-76.
26. Wahlfors J, Hiltunen H, Heinonen K et al. Genomic hypomethylation in human chronic lymphocytic leukemia. Blood 1992; 80(8):2074-80.

27. Lin CH, Hsieh SY, Sheen IS et al. Genome-wide hypomethylation in hepatocellular carcinogenesis. Cancer Res 2001; 61(10):4238-43.
28. Kim Y-I, Giuliano A, Hatch KD et al. Global DNA hypomethylation increases progressively in cervical dysplasia and carcinoma. Cancer 1994; 74:893-99.
29. Ehrlich M. DNA methylation in cancer: too much, but also too little. Oncogene 2002:in press.
30. Smiraglia DJ, Rush LJ, Fruhwald MC et al. Excessive CpG island hypermethylation in cancer cell lines versus primary human malignancies. Hum Mol Genet 2001; 10(13):1413-19.
31. Carr BI, Reilly G, Smith SS et al. The tumorigenicity of 5-azacytidine in the male Fisher rat. Carcinogenesis 1984;5:1583-90.
32. Thomas GA, Williams ED. Production of thyroid tumours in mice by demethylating agents. Carcinogenesis 199213:1039-42.
33. Denda A, Rao PM, Rajalakshmi S et al. 5-azacytidine potentiates initiation induced by carcinogens in rat liver. Carcinogenesis 1985; 6(1):145-46.
34. Trinh BN, Long TI, Nickel AE et al. DNA methyltransferase deficiency modifies cancer susceptibility in mice lacking DNA mismatch repair. Mol Cell Biol 2002; 22(9):2906-17.
35. Esteller M, Corn PG, Baylin SB et al. A gene hypermethylation profile of human cancer. Cancer Res 2001; 61(8):3225-29.
36. Dai Z, Lakshmanan RR, Zhu WG et al. Global methylation profiling of lung cancer identifies novel methylated genes. Neoplasia 2001; 3(4):314-23.
37. Yan PS, Rodriguez FJ, Laux DE et al. Hypermethylation of ribosomal DNA in human breast carcinoma. Br J Cancer 2000; 82(3):514-17.
38. Dante R, Dante-Paire J, Rigal D et al. Methylation patterns of long interspersed repeated DNA and alphoid repetitive DNA from human cell lines and tumors. Anticancer Res 1992; 12(2):559-63.
39. Jurgens B, Schmitz-Drager BJ, Schulz WA. Hypomethylation of L1 LINE sequences prevailing in human urothelial carcinoma. Cancer Res 1996; 56(24):5698-703.
40. Takai D, Yagi Y, Habib N et al. Hypomethylation of LINE1 retrotransposon in human hepatocellular carcinomas, but not in surrounding liver cirrhosis. Jpn J Clin Oncol 2000; 30(7):306-09.
41. Santourlidis S, Florl A, Ackermann R et al. High frequency of alterations in DNA methylation in adenocarcinoma of the prostate. Prostate 1999; 39(3):166-74.
42. Sassaman DM, Dombroski BA, Moran JV et al. Many human L1 elements are capable of retrotransposition. Nat Genet 1997; 16(1):37-43.
43. Florl AR, Lower R, Schmitz-Drager BJ et al. DNA methylation and expression of LINE-1 and HERV-K provirus sequences in urothelial and renal cell carcinomas. Br J Cancer 1999;80(9):1312-21.
44. Ehrlich M, Hopkins N, Jiang G et al. Satellite hypomethylation in karyotyped Wilms tumors. Cancer Genet Cytogenet 2003; 141:97-105.
45. Narayan A, Ji W, Zhang X-Y et al. Hypomethylation of pericentromeric DNA in breast adenocarcinomas. Int J Cancer 1998; 77:833-38.
46. Qu G, Dubeau L, Narayan A et al. Satellite DNA hypomethylation vs. overall genomic hypomethylation in ovarian epithelial tumors of different malignant potential. Mut Res 1999; 423:91-101.
47. Wong N, Lam WC, Lai PB et al. Hypomethylation of chromosome 1 heterochromatin DNA correlates with q-arm copy gain in human hepatocellular carcinoma. Am J Pathol 2001; 159(2):465-71.
48. Kanai Y, Ushijima S, Kondo Y et al. DNA methyltransferase expression and DNA methylation of CPG islands and peri-centromeric satellite regions in human colorectal and stomach cancers. Int J Cancer 2001; 91(2):205-12.
49. Nagai H, Kim YS, Yasuda T et al. A novel sperm-specific hypomethylation sequence is a demethylation hotspot in human hepatocellular carcinomas. Gene 1999; 237(1):15-20.
50. Thoraval D, Asakawa J, Wimmer K et al. Demethylation of repetitive DNA sequences in neuroblastoma. Genes Chromosomes Cancer 1996; 17(4):234-44.
51. Jeanpierre M, Turleau C, Aurias A et al. An embryonic-like methylation pattern of classical satellite DNA is observed in ICF syndrome. Hum Mol Genet 1993;2:731-35.
52. Tuck-Muller CM, Narayan A, Tsien F et al. DNA hypomethylation and unusual chromosome instability in cell lines from ICF syndrome patients. Cytogenet. Cell Genet 2000; 89:121-28.
53. Kondo T, Comenge Y, Bobek MP et al. Whole-genome methylation scan in ICF syndrome: hypomethylation of non-satellite DNA repeats D4Z4 and NBL2. Hum Mol Gen 2000; 9:597-604.
54. Hansen RS, Wijmenga C, Luo P et al. The DNMT3B DNA methyltransferase gene is mutated in the ICF immunodeficiency syndrome. Proc Natl Acad Sci USA 1999; 96(25):14412-17.
55. Wijmenga C, Hansen RS, Gimelli G et al. Genetic variation in ICF syndrome: evidence for genetic heterogeneity. Hum Mutat 2000; 16(6):509-17.
56. Gowher H, Jeltsch A. Molecular enzymology of the catalytic domains of the Dnmt3a and Dnmt3b DNA methyltransferases. J Biol Chem 2002; 277(23):20409-14.

57. Martin V, Ribieras S, Song-Wang XG et al. Involvement of DNA methylation in the control of the expression of an estrogen-induced breast-cancer-associated protein (pS2) in human breast cancers. J Cell Biochem 1997; 65(1):95-106.
58. Watt PM, Kumar R, Kees UR. Promoter demethylation accompanies reactivation of the HOX11 proto- oncogene in leukemia. Genes Chromosomes Cancer 2000; 29(4):371-77.
59. Sullivan MJ, Taniguchi T, Jhee A et al. Relaxation of IGF2 imprinting in Wilms tumours associated with specific changes in IGF2 methylation. Oncogene 1999; 18(52):7527-34.
60. Takai D, Gonzales FA, Tsai YC et al. Large scale mapping of methylcytosines in CTCF-binding sites in the human H19 promoter and aberrant hypomethylation in human bladder cancer. Hum Mol Genet 2001; 10(23):2619-26.
61. Tycko B. Genomic Imprinting and Human Neoplasia. In: Ehrlich M, ed. DNA and Alterations in Cancer: Genetic and Epigenetic Alterations. Natick: Eaton Publishing; 2000:333-49.
62. Schwienbacher C, Gramantieri L, Scelfo R et al. Gain of imprinting at chromosome 11p15: A pathogenetic mechanism identified in human hepatocarcinomas. Proc Natl Acad Sci USA 2000; 97(10):5445-49.
63. Malik K, Salpekar A, Hancock A et al. Identification of differential methylation of the WT1 antisense regulatory region and relaxation of imprinting in Wilms' tumor. Cancer Res 2000; 60(9):2356-60.
64. De Smet C, De Backer O, Faraoni I et al. The activation of human gene MAGE-1 in tumor cells is correlated with genome-wide demethylation. Proc Natl Acad Sci USA 1996; 93(14):7149-53.
65. De Backer O, Arden KC, Boretti M et al. Characterization of the GAGE genes that are expressed in various human cancers and in normal testis. Cancer Res. 1999; 59(13):3157-65.
66. Lethe B, Lucas S, Michaux L et al. LAGE-1, a new gene with tumor specificity. Int J Cancer 1998; 76(6):903-08.
67. Lucas S, De Smet C, Arden KC et al. Identification of a new MAGE gene with tumor-specific expression by representational difference analysis. Cancer Res 1998; 58(4):743-52.
68. Boel P, Wildmann C, Sensi M et al. BAGE: a new gene encoding an antigen recognized on human melanomas by cytolytic T lymphocytes. Immunity 1995; 2(2):167-75.
69. Serrano A, Garcia A, Abril E et al. Methylated CpG points identified within MAGE-1 promoter are involved in gene repression. Int J Cancer 1996; 68(4):464-70.
70. Jang SJ, Soria JC, Wang L et al. Activation of melanoma antigen tumor antigens occurs early in lung carcinogenesis. Cancer Res 2001; 61(21):7959-63.
71. Goelz SE, Vogelstein B, Hamilton SR et al. Hypomethylation of DNA from benign and malignant human colon neoplasms. Science 1985; 228(4696):187-90.
72. Ribieras S, Song-Wang XG, Martin V et al. Human breast and colon cancers exhibit alterations of DNA methylation patterns at several DNA segments on chromosomes 11p and 17p. J Cell Biochem 1994; 56(1):86-96.
73. Liang G, Salem CE, Yu MC et al. DNA methylation differences associated with tumor tissues identified by genome scanning analysis. Genomics 1998; 53(3):260-68.
74. Feinberg AP, Gehrke CW, Kuo KC et al. Reduced genomic 5-methylcytosine content in human colonic neoplasia. Cancer Res 1988; 48:1159-61.
75. Ehrlich M, Jiang G, Fiala ES et al. Hypomethylation and hypermethylation in Wilms tumors. Oncogene 2002; 21:6694-6702.
76A. Smiraglia DJ, Szymanska J, Kraggerud SM et al. Distinct epigenetic phenotypes in seminomatous and nonseminomatous testicular germ cell tumors. Oncogene 2002; 21:3909-16.
76B. Ushijima T, Morimura K, Hosoya Y et al. Establishment of methylation-sensitive-representational difference analysis and isolation of hypo- and hypermethylated genomic fragments in mouse liver tumors. Proc Natl Acad Sci USA 1997;94(6):2284-89.
77. Mitelman F, Johansson B, Mertens F. Database of Chromosome Aberrations in Cancer. http://cgap.nci.nih.gov/Chromosomes/Mitelman 2001.
78. Sabbattini P, Lundgren M, Georgiou A et al. Binding of Ikaros to the lambda5 promoter silences transcription through a mechanism that does not require heterochromatin formation. EMBO J 2001; 20(11):2812-22.
79. Cobb BS, Morales-Alcelay S, Kleiger G et al. Targeting of Ikaros to pericentromeric heterochromatin by direct DNA binding. Genes Dev 2000; 14(17):2146-60.
80. Francastel C, Magis W, Groudine M. Nuclear relocation of a transactivator subunit precedes target gene activation. Proc Natl Acad Sci USA 2001; 98(21):12120-25.
81. Ehrlich M, Buchanan K, Tsien F et al. DNA methyltransferase 3B mutations linked to the ICF syndrome cause dysregulation of lymphocyte migration, activation, and survival genes. Hum Mol Genet 2001; 10:2917-31.
82. Salem C, Liang G, Tsai YC et al. Progressive increases in de novo methylation of CpG islands in bladder cancer. Cancer Res 2000; 60(9):2473-76.

83. Fruhwald MC, O'Dorisio MS, Dai Z et al. Aberrant promoter methylation of previously unidentified target genes is a common abnormality in medulloblastomas—implications for tumor biology and potential clinical utility. Oncogene 2001; 20(36):5033-42.

84. Itano O, Ueda M, Kikuchi K et al. Correlation of postoperative recurrence in hepatocellular carcinoma with demethylation of repetitive sequences. Oncogene 2002; 21(5):789-97.

85. Itano O, Ueda M, Kikuchi K et al. A new predictive factor for hepatocellular carcinoma based on two- dimensional electrophoresis of genomic DNA. Oncogene 2000; 19(13):1676-83.

86. Shen L, Fang J, Qiu D et al. Correlation between DNA methylation and pathological changes in human hepatocellular carcinoma. Hepatogastroenterology 1998; 45(23):1753-59.

87. Soares J, Pinto AE, Cunha CV et al. Global DNA hypomethylation in breast carcinoma: correlation with prognostic factors and tumor progression. Cancer 1999; 85(1):112-18.

88. Bernardino J, Roux C, Almeida A et al. DNA hypomethylation in breast cancer: an independent parameter of tumor progression? Cancer Genet Cytogenet 1997;97(2):83-89.

89. Case DC, Jr. 5-azacytidine in refractory acute leukemia. Oncology 1982; 39(4):218-21.

90. Silverman LR, Demakos EP, Peterson BL et al. Randomized controlled trial of azacitidine in patients with the myelodysplastic syndrome: a study of the cancer and leukemia group B. J Clin Oncol 2002; 20(10):2429-40.

91. Holoye PY, Dhingra HM, Umsawasdi T et al. Phase II study of 5,6-dihydro-5-azacytidine in extensive, untreated non- small cell lung cancer. Cancer Treat Rep 1987; 71(9):859-60.

92. Vogelzang NJ, Herndon JE 2nd, Cirrincione C et al. Dihydro- 5-azacytidine in malignant mesothelioma. A phase II trial demonstrating activity accompanied by cardiac toxicity. Cancer and Leukemia Group B Cancer 1997; 79(11):2237-42.

93. Santini V, Kantarjian HM, Issa JP. Changes in DNA methylation in neoplasia: pathophysiology and therapeutic implications. Ann Intern Med 2001; 134(7):573-86.

94. Sessa C, ten Bokkel Huinink W, Stoter G et al. Phase II study of 5-aza-2'- deoxycytidine in advanced ovarian carcinoma. The EORTC Early Clinical Trials Group. Eur J Cancer 1990; 26(2):137-38.

95. Vermorken JB, Tumolo S, Roozendaal KJ et al. 5-aza-2'-deoxycytidine in advanced or recurrent cancer of the uterine cervix. Eur J Cancer 1991; 27(2):216-17.

96. Abele R, Clavel M, Dodion P et al. The EORTC Early Clinical Trials Cooperative Group experience with 5-aza- 2'-deoxycytidine (NSC 127716) in patients with colo- rectal, head and neck, renal carcinomas and malignant melanomas. Eur J Cancer Clin Oncol 1987; 23(12):1921-24.

97. Samuels BL, Herndon JE 2nd, Harmon DC et al. Dihydro-5- azacytidine and cisplatin in the treatment of malignant mesothelioma: a phase II study by the Cancer and Leukemia Group B. Cancer 1998; 82(8):1578-84.

98. Silverman LR. Targeting hypomethylation of DNA to achieve cellular differentiation in myelodysplastic syndromes (MDS). Oncologist 2001; 6(Suppl 5):8-14.

99. Kantarjian HM. Treatment of myelodysplastic syndrome: questions raised by the azacitidine experience. J Clin Oncol 2002; 20(10):2415-16.

100. Kornblith AB, Herndon JE, 2nd, Silverman LR, et al. Impact of azacytidine on the quality of life of patients with myelodysplastic syndrome treated in a randomized phase III trial: a Cancer and Leukemia Group B study. J Clin Oncol 2002; 20(10):2441- 52.

101. Lubbert M, Wijermans P, Kunzmann R et al. Cytogenetic responses in high-risk myelodysplastic syndrome following low-dose treatment with the DNA methylation inhibitor 5-aza-2'- deoxycytidine. Br J Haematol 2001; 114(2):349-57.

102. Richel DJ, Colly LP, Kluin-Nelemans JC et al. The antileukaemic activity of 5-Aza-2 deoxycytidine (Aza-dC) in patients with relapsed and resistant leukaemia. Br J Cancer 1991; 64(1):144-48.

103. Davidson S, Crowther P, Radley J et al. Cytotoxicity of 5-aza-2'-deoxycytidine in a mammalian cell system. Eur J Cancer 1992; 28:362-68.

104. Juttermann R, Li E, Jaenisch R. Toxicity of 5-aza-2'-deoxycytidine to mammalian cells is mediated primarily by covalent trapping of DNA methyltransferase rather than DNA demethylation. Proc Natl Acad Sci USA 1994; 91:11797-801.

105. Hernandez R, Frady A, Zhang X-Y et al. Preferential induction of chromosome 1 multibranched figures and whole-arm deletions in a human pro-B cell line treated with 5-azacytidine or 5-azadeoxycytidine. Cytogenet Cell Genet 1997; 76:196-201.

106. Ji W, Hernandez R, Zhang X-Y et al. DNA demethylation and pericentromeric rearrangements of chromosome 1. Mutat Res 1997; 379:33-41.

107. Chen RZ, Pettersson U, Beard C et al. DNA hypomethylation leads to elevated mutation rates. Nature 1998; 395(6697):89-93.

108. Pandolfi PP. Transcription therapy for cancer. Oncogene 2001; 20(24):3116-27.

109. Chung D. Histone modification: the 'next wave' in cancer therapeutics. Trends Mol Med 2002; 8(4):S10-11.

CHAPTER 4

DNA Methylation in Urological Cancers

Wolfgang A. Schulz and Hans-Helge Seifert

Abstract

Urological cancers are a diverse group with different alterations of DNA methylation. In all urological cancers, DNA hypermethylation of specific genes has been described. In contrast, methylation of repetitive sequences is often diminished, resulting in decreased overall methylation levels ("global hypomethylation"). Altered imprinting is also found. Testicular tumors are derived from more or less immature germ cells whose methylation patterns they often reflect. Subtypes can be distinguished by the extents of global hypomethylation and hypermethylation. Renal cell carcinomas typically display hypermethylation restricted to specific genes important for tumor development and progression. By comparison, methylation patterns are more severely disturbed in prostate and bladder cancers in which hypermethylation of multiple genes coexists with genome-wide hypomethylation. Causes of altered methylation may also differ. Hypermethylation could be incidental in renal cancers, but is more likely caused by primary defects in the methylation machinery in bladder and prostate cancers, which are still undefined. However, potential influences by diet and by chemical carcinogens need to be better understood. DNA hypermethylation acts as an important mechanism in the silencing of tumor suppressor genes. Global hypomethylation often correlates with chromosomal instability. The mechanism underlying this association is not understood. Hypermethylation of multiple genes has been detected in urine, ejaculate, blood and tissue biopsies. DNA methylation assays can improve detection, monitoring, staging and classification of urological cancers and in the near future could be employed to select patient-adapted therapies. In contrast, efficacy, application range and risk of inducing tumor progression of drugs targeting DNA methylation are yet to be determined in urological cancers. The diversity of these cancers require a carefully adapted approach to optimal exploitation of their DNA methylation alterations.

An Overview of Urological Cancers

Urological cancers constitute a diverse group of tumors different in origin, biology, clinical course and treatment options (Fig. 1).

Testicular cancers are usually diseases of younger men. Most share a common origin from germ cells that have become aberrant at different stages of development. Unlike many other carcinomas, they respond well to chemotherapy.

The more common urological malignancies befall older people. This is most evident for prostate carcinoma which is now the most prevalent lethal cancer of older men in industrialized countries. This adenocarcinoma is derived from the secretory epithelium of the gland. Organ-confined cases can be cured by surgical removal of the prostate (prostatectomy) or by radiotherapy. Since testosterone is essential for the proliferation of normal and many transformed prostate cells, many locally advanced and metastatic tumors respond to androgen deprivation, although this treatment is not curative. Because of the difficulty to clinically distinguish organ-confined from nonorgan confined disease, curative treatment fails in up to 40% of

DNA Methylation and Cancer Therapy, edited by Moshe Szyf. ©2005 Eurekah.com and Kluwer Academic/Plenum Publishers.

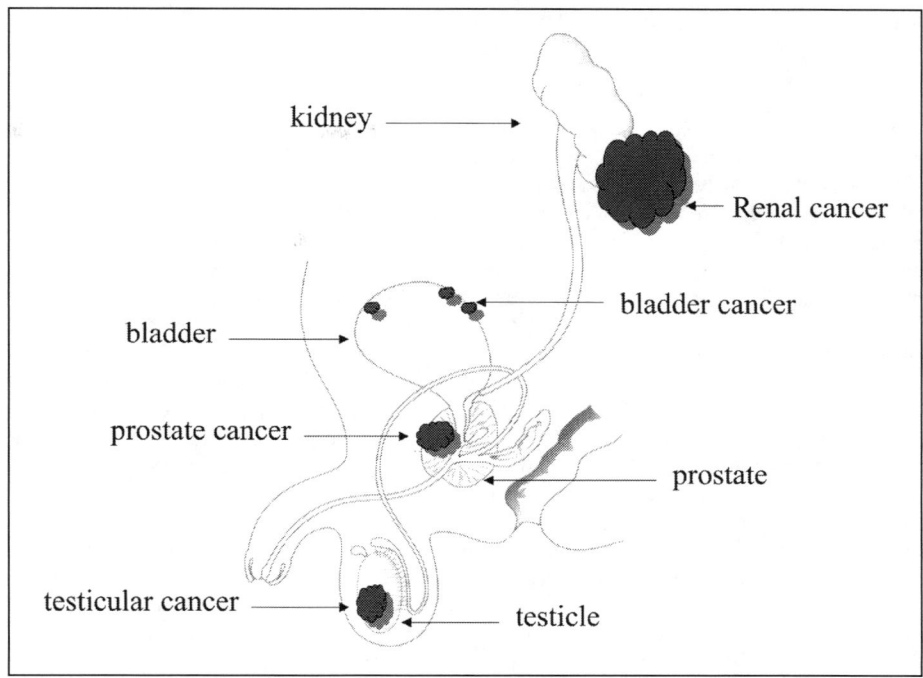

kidney

Renal cancer

bladder cancer

bladder

prostate cancer

prostate

testicular cancer

testicle

Figure 1. The male genitourinary tract and its most important associated cancers.

the patients. Therefore goals of prostate cancer research are not only to detect prostate carcinoma early, but also to identify those patients who will definitively benefit from curative treatment options.

Renal cell carcinomas (RCC) are histologically more diverse. Clear-cell carcinoma, the most common variety, is derived from the proximal tubules, whereas other histological types originate from other segments of Henle's loop. Treatment of organ-confined renal cancer by nephrectomy achieves a five-year survival rate of up to 85%, but the prospects for metastatic disease are dire.

Although most frequently located in the urinary bladder (therefore: "bladder cancer"), transitional cell carcinoma (TCC) also develop in other segments of the urothelium lining the renal pelvis, ureter, urinary bladder, and upper part of the urethra. Urothelial carcinoma retains morphological and biochemical urothelial differentiation to some degree and represents the main histological type in industrialized countries; the second most frequent type (5% of all cases) is squamous cell carcinoma. Renal and bladder cancers affect both genders, but are more frequent in males. Because urothelial carcinomas are heterogeneous in their clinical behaviour, they may comprise at least two distinct, albeit related diseases. Papillary superficial carcinomas are encountered in about 80% of diagnosed cases and, although they recur in up to 50-70% of the patients, rarely progress to higher stages. They can be successfully treated by local transurethral surgery, often combined with instillation of cytostatic drugs or BCG, a mycobacterial strain, to prevent recurrences and progression. Solid carcinomas derive mainly from highly dysplastic carcinoma in situ and invade the muscular tissue layers underlying the urothelium. Upon presentation, 30 to 60% have already metastasized to the local or regional lymph nodes. Treatment comprises radical cystectomy with local lymph node dissection and urinary diversion. Adjuvant chemotherapy improves the survival of some patients with lymph node metastases. In advanced disease, chemotherapy is the treatment of choice with or without palliative cystectomy. Thus, as for prostate carcinoma, the challenge is not only to detect the presence of TCC or monitor for recurrence, but also to predict its clinical behavior as a basis for rational treatment choice.

Over the last decade, molecular biological analysis has identified many genetic and epigenetic changes during the development and progression of these cancers. For many changes, it is already understood how they bear on the biology and the clinical characteristics of the cancer. Insights into the mechanisms of carcinogenesis are beginning to aid prevention and early detection. New techniques for detection, monitoring, and differential diagnosis are gradually entering the clinic. For instance, an improved classification of kidney cancers is based on the pattern of chromosomal alterations.[1] New molecular markers for detection of bladder cancer cells in the urine to supplement cytological analysis are now tested in the clinic.[2,3] Following the recognition that mutations in the p53 tumor suppressor gene are decisive for the clinical course of bladder cancer, an international study investigates the use of adjuvant chemotherapy according to the p53 status.[4] Analysis of molecular alterations is helping to identify suitable targets to improve efficacy and specifity of drug therapy or establish genetic therapies.

Alterations in DNA methylation have been observed in all urological malignancies.[5] In all urological malignancies, DNA hypermethylation of CpG-rich promoter regions has been recognized as a mechanism of inactivation of individual genes. Nevertheless, in many urological malignancies—although with important exceptions—decreases in overall DNA methylation were found. This leads to global hypomethylation of repetitive sequences in the genome which contain most of the methylcytosine in normal somatic cells. Altered genomic imprinting is another change relating to DNA methylation. It may be particularly important in testicular cancers. A detailed description of these findings will be presented in the following section. We will then go on to discuss current attempts at understanding the causes and consequences of these alterations. A systematic description of DNA methylation changes in human cancers may become feasible and we will discuss initial studies to recognize patterns of methylation changes for tumor classification and prognosis. Diagnostics exploiting DNA methylation changes in individual genes seems even closer and may soon enter the clinic. The final section will argue that these diagnostic improvements and new agents acting on DNA methylation and associated changes in chromatin structure hold the promise of improved, rational therapy.

A Description of DNA Methylation Changes in Urological Malignancies

Testicular Cancers

The most important group of testicular cancers are germ cell cancers (GCC) which can be classified into seminomas and nonseminomas. Seminomas retain the morphology of spermatogonia but are blocked in their terminal differentiation. Nonseminomas maintain different extents of pluripotency and present features of various embryonic and somatic tissues, often mixed. They are accordingly classified into embryonal carcinoma, choriocarcinoma, teratoma etc. Testicular GCC usually contain aneuploid genomes and—in spite of their histological differences—share one common chromosomal alteration, i.e., gain of 12p through isochromosome formation or amplification.[6-8] This leads regularly to overexpression of Cyclin D2 which contributes to aberrant cell cycle regulation and of *DAD-R* which may suppress apoptosis. Testicular GCC are clinically important as the most frequent tumors in young adult males. They are also extremely interesting from a basic science point of view, because of their pluripotency. Although chemotherapy is very efficacious in these tumors, its molecular basis is not fully understood. Part of the explanation may be that most testicular tumors retain functional p53.

Seminomas and nonseminomas present clearly distinct DNA methylation patterns.[9] Seminomas are hypomethylated throughout their genome with rare gene-specific hypermethylation, whereas nonseminomas usually display a lower degree of genome-wide hypomethylation, but wide-spread hypermethylation. One of the most consistently hypermethylated genes is *CDKN2A* leading to decreased expression of the cell cycle inhibitor p16[INK4A].[10] Much work has focused on imprinted genes, mostly the *IGF2/H19* pair.[11-14] Very frequently, apparent loss of imprinting has been observed. A plausible explanation for this finding is that GCCs are derived from stages of spermatocyte development in which old imprints have already been erased and no new or only paternal imprints are established. Thus, methylation and imprinting patterns of some GCC may not be truly aberrant, but rather reflect the methylation pattern of the cell of origin. Indeed, more detailed investigations have revealed imprinting patterns in different GCC subtypes closely corresponding to various stages of germ cell development.[8,14] However, LOI at *IGF2* is not only a marker of tumor development, but also means overexpression of a potent growth and survival factor. Defects in imprinting and DNA methylation may therefore fundamentally contribute to the development of testicular cancers.

Prostate Carcinoma

Prostate adenocarcinoma is regarded as a paradigmatic hormone-dependent cancer. Advanced disease is palliatively treated by pharmacological or surgical androgen deprivation. This prolongs survival, but the cancer gradually evolves towards androgen independence, becoming hypersensitive to residual low androgen levels or activating alternate pathways to permit cell proliferation.[15] In prostate carcinoma, specific chromosomal aberrations recur, e.g., loss of 6q, 7q, 8p, 10q, 13q, 16q and gain of 8q, but few tumor suppressor genes or dominant oncogenes have been adequately defined.[16]

Alterations in DNA methylation seem particularly important in prostate cancer (Table 1). The most consistent case is hypermethylation of the *GSTP1* gene. Hypermethylation of the gene promoter initially discovered by Lee et al[17] has been confirmed by several groups worldwide with all available techniques.[18-21] In summary, *GSTP1* hypermethylation is found in more than 80% of all prostate cancers, is established early in cancer development, probably even in preneoplastic high-grade PIN (prostate intraepithelial neoplasia) and—unusually—often affects both alleles. Other genes with confirmed hypermethylation include the *CD44*[22-24] and the *CDH1* (E-cadherin) genes[25-27] encoding cell surface proteins crucial for cell adhesion that are down-regulated during invasion and metastasis. Not unexpectedly, methylation of many genes related to hormone and growth factor response has been found to be altered in prostate cancer, notably androgen and estrogen receptor genes,[28-32] the *RARB2* gene encoding a retinoic acid receptor,[33] *TIG1* also involved in retinoic acid response,[34] as well as *ARA70*,[35] *BMP-6*,[36] *NEP*,[37] *RASSF1A*[38] and inhibin α.[39] An illustrative case is *EDNRB*, encoding the endothelin receptor B, in which methylation is altered, but not in regulatory sequences.[40,41] The growth behavior of prostate carcinomas suggests that initiating events target cell survival rather than cell cycle regulation. Relevant changes may comprise increased methylation of a surprise candidate tumor suppressor gene, *CAV-1*, located at 7q31.1 and encoding caveolin,[42] and *TNFRSF6*, encoding the FAS death receptor. Hypermethylation of *TNFRSF6* is more likely a consequence rather than cause of its down-regulation.[43] DNA methylation does probably not contribute to the frequent down-regulation of *PTEN*.[44] The most consistent alteration in cell cycle regulation seems down-regulation of p27[KIP1] during tumor progression, but this is not caused by promoter hypermethylation or mutation.[45] Controversial results have been reported on the frequency of genetic and epigenetic changes in *RB1* and *CDKN2A* that are inactivated in many other human tumors. However, all data agree that promoter hypermethylation of these genes is rare in prostate carcinoma.[46,47]

Global hypomethylation in prostate cancer had been discovered already in the 1980s when many cases were advanced upon presentation.[48] Recent studies have clarified that genome-wide hypomethylation is almost universal in metastatic, androgen-refractory cancers, but lacking in many early-stage tumors.[19,49] Surprisingly, hypomethylation of LINE-1 sequences which reflects global methylation, correlated with metastasis, but even better with alterations on chromosome 8 which quite certainly predisposes to systemic disease. Thus, global DNA hypomethylation may characterize a subgroup of particularly aggressive prostate cancers.

Table 1. DNA methylation alterations in urological cancers at a glance

Tumor	Genes Affected by Hypermethylation		Global Hypomethylation	Specifics
	As Well In...	Specifically		
Testicular Germ Cell Cancers	CDKN2A		Strong in semi-nomas; moderate in non-seminomas	Frequently loss of imprinting at IGF2/H19 a.o.; Gene-specific hypermethylation mostly in non-seminomas
Prostate Carcinoma	RARB2, RASSF1A, CDH1	GSTP1, ER, AR, CD44, TIG1, ARA70, BMP6, NEP, Inhibin α, EDNRB, CAV-1, TNFRSF6	Pronounced only in an aggressive subfraction	GSTP1 hypermethylation in >80% of cases; Global hypomethylation possibly correlated to prognosis
Renal Cell Carcinoma	RASSF1A, CDH1, TIMP3, (GSTP1)	VHL	Only in cell lines	
Bladder carcinoma	CDKN2A, RARB2, CDH1, DAPK, RASSF1A, APC	DBCCR1, PAX6	Common	Imprinting at IGF2/H19 often disturbed

The table summarizes the data from the section *A Description of DNA Methylation Changes in Urological Malignancies*. Results from "genome-wide screens" are discussed in the section *A Global View of DNA Methylation Alterations in Urological Cancers*.

Renal Carcinomas

The crucial genetic change in clear-cell renal carcinoma (CC-RCC) is loss of VHL function. *VHL* located on chromosome 3p25 is a bona fide tumor suppressor gene.[50] Both copies are defunct in sporadic as well as hereditary cases, in which one mutated allele is inherited. The second allele is inactivated in tumors by deletion, point mutation or occassionally promoter hypermethylation.[51] The same mechanisms combine in sporadic cases, but generally one allele is lost by deletion of chromosome 3p.[52] Many breaks take place at the 3p14 fragile site also destroying the *FHIT* gene.[53] Loss of 3p is consistent enough to allow differential diagnosis of clear-cell vs. other RCCs.[1] Very likely loss of 3p targets other genes beyond *VHL*, *FHIT* representing an obvious candidate. Another likely target is *RASSF1A* at 3p21.3.[54-56] Hypermethylation and down-regulation of *RASSF1A* expression are frequent in CC-RCC and several other cancers. Further recurrent chromosomal aberrations characterize CC-RCC progression, but few genes involved are definitely characterized. Loss of E-cadherin may be involved, since LOH on 16q and hypermethylation of *CDH1* are found.[57] Hypermethylation of the metalloproteinase inhibitor TIMP3 may contribute to invasion.[58] In some cases, *GSTP1* hypermethylation is observed,[59] but expression of the related GSTA isozymes is more consistently down-regulated.[60] Several other instances of altered methylation have been reported in renal cancer cell lines, but have not been corroborated in tumor tissue. Thus, RCC cell lines, but not tumor tissues display LINE-1 hypomethylation.[61] It appears that relatively few DNA methylation alterations take place in RCC. Therefore, those changes that do occur may be functionally significant. For instance, the *CA9* gene encoding a cell surface and nuclear carboanhydratase is strongly activated and its promoter hypomethylated in RCCs.[62] This activation can be traced to VHL inactivation in CC-RCC which causes stabilization of the HIF1α transcription factor that induces *CA9* expression.[63] Therefore, hypomethylation of this gene probably results from its constitutive activation.

Bladder Cancer

Two important regulatory systems are inactivated in almost all advanced bladder cancers.[64,65] Regulation of the cell cycle is disturbed by loss of RB1 or p16[INK4A], or by cyclin D1 overexpression. Control of genomic integrity by p53 is compromised by *TP53* mutations, *CDKN2A* (p14[ARF1]) inactivation, or occasionally overexpression of HDM2. Accordingly, loss of 13q, 17p, 9p and amplification on 12q are found. Additional recurrent chromosomal alterations include loss of chromosome 9q, 11p and 8p and gains of 5p, 6p, 8q, 20q. Tumor suppressors and oncogenes at these locations are actively sought. Low-grade papillary tumors contain much fewer genetic alterations, sometimes only losses on chromosome 9.[66]

A wide range of genes have been reported as hypermethylated in bladder cancer. Hypermethylation can affect both promoters and exon 2 of *CDKN2A*.[67-69] Of note, each of these changes has different functions. For instance, the frequent hypermethylation of exon 2 is not related to decreased transcription, whereas exon 1α hypermethylation actually associated with p16[INK4A] down-regulation is much rarer.[70,71] Likewise, methylation of the *PAX6* intragenic CpG island is a good tumor-marker, but unrelated to gene silencing.[72] Another important lesson from *CDKN2A* is that DNA hypermethylation, point mutations and homozygous deletion can equally inactivate one gene.[73] Other genes hypermethylated as well in bladder cancer comprise *RARB2*, *CDH1*, *RASSF1A*, and *APC*.[71,74-76] Hypermethylation of *DBCCR1*, a candidate tumor suppressor gene on 9q, may be specific for TCC, but the exact relationship between hypermethylation and expression requires further investigation.[68,78] The proapoptotic DAPK may be a particularly frequent target in bladder cancer.[71,79] Decreased DNA methylation in the *MDR1* promoter during chemotherapy of bladder cancer may contribute to drug resistance.[80] Hypermethylation of *DBCCR1* and *CDH1* is also observed in nonmalignant aging bladder tissues.[74,81]

Global hypomethylation affecting various repetitive sequences including LINEs, endogenous retroviruses and satellites appears at early stages of TCC.[61,82] Global hypomethylation may contribute to disturbed expression of imprinted genes on chromosome 11p15.5, *IGF2*, *H19* and *CDKN1C*.[83-85] This region, however, is also subject to frequent LOH. More detailed investigations of this issue are needed.[86]

A Global View of DNA Methylation Alterations in Urological Cancers

The results presented in the previous section suggest that although hypermethylation of some genes (*VHL*, *DBCCR1*) occurs only in specific urological cancers, many genes are subject to DNA hypermethylation in several tumor types (Table 1). Such genes may interfere with universal properties of tumor cells, such as uncontrolled proliferation (*CDKN2A*) or decreased adhesion (*CDH1*). Nevertheless, some genes of general importance in human cancers seem to become hypermethylated never (*PTEN*) or rarely (*RB1*). The molecular basis for this difference has begun to be addressed by studies on *CDKN2A*[67,87] and *GSTP1*.[88-91] Moreover, several genes, like *CDKN2A*, are inactivated by hypermethylation, point mutation, or homozygous deletion in different human cancers, but the contribution of each mechanism differs widely.[73] The molecular basis for these differences is also unknown.

Beyond the investigation of individual genes, several new methods developed recently aim at surveying methylation changes in cancers (see other contributions in this volume). More comprehensive surveys of DNA methylation in human cancers are expected to provide differential diagnostic and prognostic information for the clinic. In addition, answering basic questions such as how many genes are hypermethylated overall and how many are specific for certain tumors will provide insight into the mechanisms underlying altered methylation (cf. below). Rapid methods, such as MS-PCR and MS-SNuPE, have already been employed to investigate several genes in the same urological cancer.[68,71,77,79] A larger number of genes can be studied by array-based techniques, directed at genes deemed important in a cancer or at CpG-islands.[93,94] Restriction landmark genome scanning (RLGS) also mostly identifies changes in CpG islands, without a priori selection of genes.[10,95] Arbitrarily-primed PCR is also largely unbiased.[76,96] In a completely different approach DNA methylation inhibitors were combined with mRNA expression profiling.[97]

So far, the data from these more comprehensive methylation analysis techniques largely confirm that urological cancers differ considerably in the extent of methylation changes, as suggested by analyses of individual genes. Analysis of DNA methylation in germ cell tumors by RLGS has confirmed the fundamental difference between seminomas and nonseminomas, i.e., global hypomethylation in seminomas vs. hypermethylation of multiple CpG-islands with less extensive hypomethylation in nonseminomas.[10] A comparison of different urological malignancies by a bisulfite-modification based chip technique showed the smallest difference between tumor and normal tissue in the kidney,[94] as expected. Conversely, wide-spread and progressive deterioration of methylation patterns was confirmed by different methods in prostate and bladder carcinoma. DNA hypermethylation in bladder cancer may affect up to 7000 CpG-islands[96] in spite of almost all classes of CpG-containing repetitive sequences being hypomethylated.[60,98] Since on average methylation alterations appear to increase in advanced tumors, attempts have been made to correlate altered methylation indices with prognosis in bladder and prostate cancers.[71,79,92] This approach deserves to be pursued, but it should be considered that tumor progression in the same tumor type can be driven by different mechanisms altering tumor suppressor and oncogene function. Tumor progression in different subtypes of colon cancer appears to be driven preferentially by chromosomal instability, point mutations or by DNA methylation alterations.[99] A DNA methylation index might be most useful in the last subgroup, designated CIMP+ and exhibiting widespread hypermethylation together with a particular constellation of genetic changes.[100,101] Similar subgroups have not yet been defined in urological cancers, although they may well exist. For instance, the widely used bladder carcinoma cell line T24 resembles CIMP+ colon carcinomas in displaying extensive hypermethylation, including the p16[INK4A] promoter, and a *RAS* mutation. The same pattern may be present in some bladder carcinoma tissues.

DNA methylation surveys may also yield novel insights. It was already apparent from the investigation of individual genes that many genes related to steroid hormone action display altered methylation in prostate cancer (Table 1). More unexpected, a large group of genes hypermethylated in T24 belonged to the interferon response pathway.[97]

Causes of Altered DNA Methylation in Urological Cancers

The causes of altered methylation may be distinct in different cancers (Fig. 2). In some cancers, apparently aberrant methylation may in fact largely reflect the methylation pattern of the affected stem cell. Thus, different germ cell tumors display methylation patterns at imprinted genes corresponding to distinct stages of germ cell development.[8,10,14] Conversely, failure to set up proper methylation patterns of mature cells may underlie blocked differentiation. Clear-cell renal carcinomas may exemplify a group of cancers displaying a limited number of methylation changes. Some of these could be caused by incidental errors for which there is strong selection during tumor development because they lead to inactivation of crucial tumor suppressor genes such as *VHL*. Others such as *CA9* hypomethylation[62] may be secondary to alterations in transcriptional activators. In contrast, advanced prostate and bladder cancers are typically characterized by severely disturbed DNA methylation patterns which most likely are caused by defects in the regulation of DNA methylation. Of note, each individual DNA methylation alteration in such cancers may influence the tumor phenotype, but does not have to be essential.

It is obviously of great interest which defects underlie such grossly aberrant methylation patterns. Presently, four different hypotheses are considered.

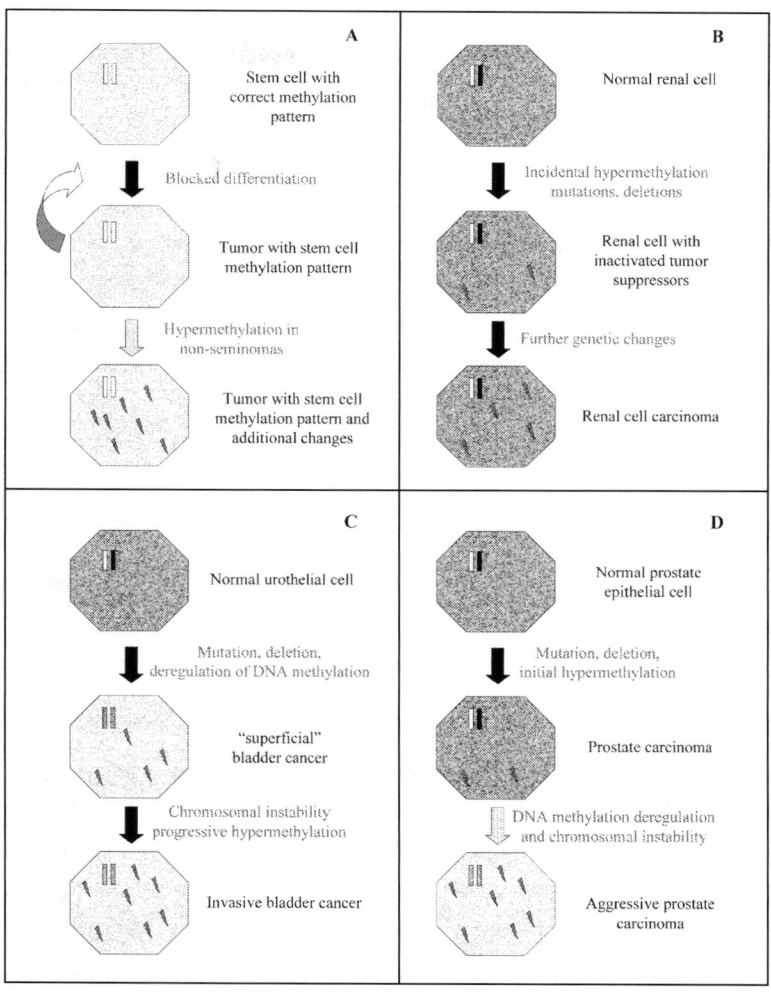

Figure 2. DNA methylation alterations in urological cancers. A simplified summary of the course, causes and consequences of altered DNA methylation in four major urological malignancies. A) Testicular germ cell cancers originate from stem cells with decreased methylation levels and stage-specific imprinting. In cancers, this methylation pattern is maintained and gene expression from hypomethylated (imprinted) genes may contribute to tumor growth. Further methylation changes occur in nonseminomas including hypermethylation contributing to the aberrant phenotype and progression. B) Renal cell carcinomas (clear-cell subtype) usually develop by genetic or epigenetic inactivation of crucial tumor suppressor genes. Methylation patterns are otherwise little disturbed; in particular, global hypomethylation does not occur. C) In bladder cancers deregulation of DNA methylation takes place during the initial stages and contributes to tumor suppressor inactivation by hypermethylation and possibly to genomic instability by hypomethylation of repetitive sequences. Methylation at imprinted genes can be altered and further instances of hypermethylation may be associated with progression. D) Initial development of prostate carcinoma may be characterized by frequent, but selective DNA hypermethylation; deregulation of DNA methylation including global hypomethylation is restricted to a aggressive subfraction. In all panels, the background in the octagon cells signifies global methylation levels (dark = normal; light = lowered); the two small rectangles symbolize imprinted genes (black & white: diparental, white: uniparental, striped: disturbed either way). Lightnings indicate aberrant hypermethylation. Stippled arrows designate changes occurring only in a subtype of the cancer.

1. Aberrant methylation patterns could be due to dysfunction of enzymes patterning DNA methylation. Thus far, only DNA methyltransferases have been investigated. DNMT1 seems to be deregulated in bladder cancer cells[102] and often expressed at levels inadequate for proliferating cells.[102a] This may partly explain the prevalence of global hypomethylation. Since only moderate increases in the putative de novo methyltransferases DNMT3A and DNMT3B were found, CpG-island hypermethylation remains unexplained. Very few data are available for prostate cancer cell lines and almost none on tissues.[103]

2. Aberrant DNA methylation patterns might result from defects in factors directing DNA methylation (see other contributions). In vitro, neither known DNA methyltransferases nor demethylases display enough sequence specifity to account for cell-type specific DNA methylation patterns. DNA methyltransferases interact with other chromatin proteins, notably histone modifying enzymes and methyl-CpG-binding proteins (MBD). Very little is known about these proteins in urological malignancies. One interesting protein interacting with chromatin modulators as well as DNMT1 is RB1 which is deregulated or even lacking in almost all bladder cancers[64,65] and in some prostate cancers.[16] Since RB1 may coordinate DNMT1 action with DNA replication and chromatin state inheritance,[104,105] RB1 disfunction might disturb this intricate coordination, favoring global hypomethylation and progressive DNA hypermethylation.

3. Defects in methyl group metabolism could promote aberrant DNA methylation patterns by limiting S-adenosylmethionine supply, more precisely, by decreasing the ratio of S-adenosylmethionine to S-adenosylhomocysteine. This ratio depends on dietary supply with methyl-providing compounds and vitamins, specifically folic acid. The importance of methyl group metabolism is well-known from animals experiments, but only recently different lines of research have converged to provide evidence for its importance in man. Folic acid deficiency, which is surprisingly prevalent in industrialized countries, is reflected in a decreased intracellular SAM:SAH ratio which in turn is linked to decreased methylcytosine in leukocytes.[106,107] In tumor cells, this metabolic misbalance may be exacerbated by rapid proliferation and insufficient blood supply.[108] Inadequate nutrition is known to synergize with low activity alleles of enzymes of methyl-group metabolism to cause cardiovascular disease.[109] The best studied and most prevalent polymorphism is 677A/V in methylene-tetrahydrofolate reductase (*MTHFR*). A weak association of the less active allele with prostate cancer was found in a pilot study, although none in bladder cancer.[110,111] Clearly, the interaction between diet and genetic predisposition needs to be investigated in larger studies. The "methyl group metabolism" hypothesis suggests straightforward means for prevention of bladder and prostate cancers.[112] Although far from proven, it may already have elicited changes in the treatment of prostate cancer patients who now often receive vitamin supplements.

4. There is some evidence that certain carcinogens may affect DNA methylation. These comprise acrolein,[113] monomethyl- and dimethylsulfate,[114] acting directly on DNMTs, or arsenic[115] which appears to affect DNA methylation indirectly. Chemical carcinogenesis is well documented and particularly relevant in bladder cancer.[116]

Consequences of Altered DNA Methylation in Urological Cancers

There is general agreement that DNA hypermethylation at CpG-rich promoters fundamentally contributes to the changes in chromatin that cause gene silencing. Accordingly, in cell lines from urological tumors, hypermethylated genes can often be reactivated by DNA methylation inhibitors. Reactivation of hypermethylated cell cycle inhibitor genes, e.g., *CDKN2A*, by the DNA methylation inhibitor 5'-deoxy-azacytidine (5-dazaC) has been shown to be associated with cell cycle prolongation.[117] However, since many genes are hypermethylated in advanced bladder and prostate carcinomas, determining which gene is responsible for cessation of cell growth is not trivial. In a recent study, about 1% of all genes investigated by microarray hybridization substantially increased in activity when T24 bladder carcinoma cells were treated with the inhibitor, twice as many as in a fibroblast cell line.[97] Thus, several hundred genes may be silenced by DNA hypermethylation in T24 and could influence cell growth.

The consequences of global hypomethylation are less well understood.[118] Although it was initially thought that hypomethylation might reactivate growth-promoting oncogenes, only few instances of hypomethylation of proto-oncogenes have been reported, e.g., for the *MYC* gene in bladder cancer.[119] However, DNA methylation changes in such cases rather seem to represent a consequence of altered gene activity.[62] Whether DNA hypomethylation disturbs imprinting is also not fully clear.[85,86] Hypomethylation occurs together with loss of imprinting in testicular cancers,[10] but the relation is likely coincidental rather than causal.

Hypomethylation of repetitive sequences could be more relevant, since DNA methylation may participate in their repression.[120] Decreases in the methylation of endogenous retroviral and retrotransposon sequences in tumor cells can be substantial. For instance, the most active endogenous retro-proviruses in the human genome, HERV-K, are almost completely demethylated in testicular cancers[121-123] and advanced bladder cancers.[61] Nevertheless, because of the tissue-specificity of their LTR they are only expressed in testicular cancer cells.[61] Although no mature retroviruses are produced, retroviral proteins could be used as serum markers for seminoma.[123,124] Since nonLTR retrotransposons do not contain tissue-specific promoters, their hypomethylation permits expression also in cancers from somatic tissues. In model experiments, retrotransposition depended on activity of the LINE-1 encoded reverse transcriptase and endonuclease, and occurred at an up to 1:250 rate per cell generation.[125] The actual frequency in cancer cells is evidently much lower.[126] Another potential adverse effect of retroelement transcription is transcriptional interference (discussed in refs. 126 and 127). Homologous recombinations between repetitive elements in the germ-line are a common cause of genetic disease, but no good estimates exist for their frequency in cancers.[128] Recombination between repetitive sequences depends on accessibility controlled by chromatin structure and DNA methylation. The most prevalent sequence families in the human genome, ALU SINEs and L1 LINEs, are highly methylated in normal somatic cells.[120] In addition, individual elements differ by mutations and truncations. These alterations reduce homology and represent a second obstacle to illegitimate recombination. This obstacle does not exist for CpG-rich satellites[129] which are hypomethylated in many human cancers including bladder cancer. Hypomethylation of satellites in the human hereditary ICF syndrome, caused by DNMT3B mutations, is associated with a propensity towards radial chromosome formation, predominantly in cell types with recombinase activity such as preB cells.[118] In urological cancers, satellite demethylation could therefore also be associated with chromosomal instability. Indeed, there is a good correlation between the extent of DNA hypomethylation and chromosomal instability in bladder and prostate cancers,[49,70] but a causal relationship is not established.

DNA Methylation in Diagnosis and Therapy of Urological Cancers

Diagnostic procedures based on altered DNA methylation are expected to soon enter the clinic. They will be used to detect urological cancers, to provide differential diagnosis and staging, and to monitor clinical course. Their most immediate application will likely be in the choice of individualized treatment.

Diagnostics based on DNA methylation alterations has several advantages. DNA is chemically more stable than RNA and many proteins. DNA from tumor cells can be isolated from blood, biopsies, and surgical tissue specimens, but in urology also noninvasively from urine and ejaculate. Assays have become more robust and can be performed on material used for histopathological diagnosis, with or without microdissection. Moreover, detection of hypermethylation can be very sensitive due to low—or ideally lacking—methylation at CpG-islands in normal tissues.

An ideal marker for cancer detection should be highly specific, sensitive and be prevalent in the risk population. Few markers fulfil all three criteria. In urology, serum markers are currently used for detection and monitoring of testicular cancers (alpha-fetoprotein, human gonadotropin, and placental alkaline phosphatase) and of prostate carcinoma (prostate-specific antigen, PSA). The testicular cancer markers are specific and sensitive, but cover only a subset of tumors. HERV-K proteins[121-124] may provide an alternative. PSA serum levels increase in most prostate cancer patients, but this increase is not very specific. Levels of PSA > 10 ng/ml indicate a high probability of prostate cancer and levels <2.5 ng/ml (depending on age) of its absence, but a large 'grey zone' necessitates additional tests and biopsies. Moreover, PSA levels are only loosely related to tumor stage and grade. For detection and monitoring of bladder cancer, traditional cytopathology is becoming supplemented by tests detecting tumor-specific proteins or cytogenetic aberrations in tumor cells shed into the urine.[2,3] Since sensitivities and specificities of these assays are still limited, unpleasant and costly cystoscopy must often be performed. No routine serum or urine assays are available to detect and classify common renal cancers. Since biopsies are not advisable, subtype and degree of malignancy of renal cancers are often only revealed during surgery.

The most promising DNA methylation marker in patients with urological malignancies is *GSTP1* hypermethylation in prostate cancer, which has even earned its first specialized review.[130] In many respects, *GSTP1* hypermethylation complements PSA detection which identifies most cases, if a low cut-off is used, but does not discriminate well between malignant and benign tissue changes. *GSTP1* hypermethylation is found in 70-95% prostate cancers, but not in benign hyperplastic tissue.[17-21,131] The *GSTP1* promoter is only methylated in liver among normal tissues, although occasionally in other tumors, a.o. renal carcinoma.[59] Several assays and applications to diagnostic samples from various sources have been described.[20,21,132-134] It is now necessary to test their validity and practicability in clinical practice. Detection of *GSTP1* hypermethylation per se does not distinguish different stages of prostate carcinoma, but can identify prostate carcinoma cells in lymph nodes, blood, bone marrow and metastases of unknown origin.

Other instances of hypermethylation in urological cancers occur in a smaller proportion of the tumors and seem not as useful for initial detection. Detection of altered methylation may rather yield information on tumor stage. For instance, global DNA hypomethylation characterizes a particularly aggressive subfraction of prostate carcinoma.[49] Likewise, once identified in a tumor, methylation markers may be used to monitor its recurrence or progression. For instance, *ARF1* hypermethylation in primary bladder cancers can be detected in serum of patients with recurrences.[69] Other methylation patterns may predict response to chemotherapy, e.g., that of the *MDR1* gene in bladder cancer.[80]

Detection of multiple methylation markers should yield a more complete molecular profile permitting cancer classification, staging, subtyping and prediction of clinical course and response to various therapies. A pioneer study has indicated that this may be feasible for testicular cancers.[10] These results may help to answer one of the most pressing clinical questions in this tumor, i.e., which patients with stage I disease carry occult metastases. Bladder and prostate cancers can be classified according to the number of methylation changes detectable by MS-PCR assays.[71,77,79,92] It needs to be determined whether new subclasses can be defined by methylation analysis and whether such classification improves on current histological diagnosis. DNA methylation chip techniques may simplify such analyses. Even a methylation chip not specifically developed for urological malignancies allowed to distinguish between different urological cancers and their corresponding normal tissues.[94] Distinctions between clinically relevant subgroups will certainly require more specifically designed chips and algorithms.

Therapy directly targeting altered methylation in urological malignancies seems more remote than diagnosis. Drugs altering DNA methylation or histone acetylation have been applied in tissue culture and animal models of urological malignancies and reported to be efficacious. Bladder carcinoma cell lines treated with 5-aza-dC reexpress the cell cycle inhibitor p16[INK4A] and slow down proliferation.[117] In prostate carcinoma cell lines, histone deacetylase inhibitors may even induce apoptosis.[135-139] One preliminary phase II study was performed with 5-aza-dC.[140] Unexpectedly, procainamide was shown to inhibit DNA methylation and diminish

GSTP1 hypermethylation in a prostate cancer xenograft.[141] Thus, DNA methylation inhibitors, probably together with drugs targeting histone modifications, seem useful to treat urological cancers. Several questions are open, however.

1. The range of urological cancers reacting to drugs targeting DNA methylation is only vaguely defined. For instance, many renal carcinoma cell lines tolerate >10-fold higher concentrations of 5-aza-dC than those efficacious in the T24 cell line.

2. While drugs under development aim to inhibit DNA methylation,[142] global hypomethylation is already present in many urological cancers. It is not known, how this affects their action.[118] Conceivably, tumors with pronounced hypomethylation could be particularly sensitive, since somatic cells appear to require a minimal level of methylation. Overall methylation in urological cancer cells never falls short of 30% of normal levels. Knocking-out DNMT1 function in somatic cells causes cell death, which is partially dependent on p53.[143] On the other hand, if global hypomethylation indeed promoted genomic instability rather than was merely associated with it, DNA methylation inhibitors might carry a risk of inducing progression.

3. In clinical practice, DNA methylation inhibitors would likely supplement rather than replace existing treatments. Therefore, their interaction with established chemotherapy regimens needs to be addressed, e.g., in bladder cancer.

In spite of these reservations, opportunities for drugs targeting DNA methylation in urological malignancies are plentiful. In testicular cancers, established chemotherapies are very efficient. Improvements could be made with regard to toxicity and in the treatment of very advanced nonseminomas. In bladder cancer, current chemotherapy protocols are only moderately efficacious and could be improved on. Since DNA methylation changes are so prevalent in prostate cancers, this cancer constitutes a promising target for DNA methylation inhibitors. Recent improvements in chemotherapy suggest prospects for a fundamental change in prostate cancer treatment strategy to which DNA methylation inhibitors might well contribute. Finally, we have previously[5] suggested that the most straightforward application of DNA methylation inhibitors in urology might be in those renal carcinomas tumors that display hypermethylation of VHL.

References

1. Kovacs G, Askhtar M, Beckwith BJ et al. The Heidelberg classification of renal cell tumours. J Pathol 1997; 183:131-133.
2. Han KR, Pantuck AJ, Belldegrun AS et al. Tumor markers for the early detection of bladder cancer. Front Biosci 2002; 7:e19-26.
3. Tiguert R, Fradet Y. New diagnostic and prognostic tools in bladder cancer. Curr Opin Urol 2002; 12:239-243.
4. Cote RJ, Esrig D, Groshen S et al. p53 and treatment of bladder cancer. Nature 1997; 385:123-125
5. Schulz WA. DNA methylation in urological malignancies. Int J Oncol 1998; 13:151-167.
6. Chaganti RSK, Houldsworth J. Genetics and biology of adult human male germ cell tumors. Cancer Res 2000; 60:1475-1482.
7. Looijenga LHJ, Osterhuis JW. Pathobiology of testicular germ cell tumors: Views and news. Anal Quant Cytol Histol 2002; 24:263-279.
8. Heidenreich A, Srivastava S, Moul JW et al. Molecular genetic parameters in pathogenesis and prognosis of testicular germ cell tumors. Eur Urol 2000; 37:121-135.
9. Peltomaki P. DNA methylation changes in human testicular cancer. Biochim Biophys Acta 1991; 1096:187-196.
10. Smiraglia DJ, Szymanska J, Kraggerud SM et al. Distinct epigenetic phenotypes in seminomatous and nonseminomatous testicular germ cell tumors. Oncogene 2002; 21:3909-3916.
11. van Gurp RJ, Oosterhuis JW, Kalscheuer V et al. Biallelic expression of the H19 and IGF2 genes in human testicular germ cell tumors. J Natl Cancer Inst 1994; 86:1070-1075.
12. Verkerk AJ, Ariel I, Dekker MC et al. Unique expression patterns of H19 in human testicular cancers of different etiology. Oncogene 1997; 14:95-107.
13. Ross JA, Schmidt PT, Perenteis JP et al. Genomic imprinting of H19 and insulin-like growth factor-2 in pediatric germ cell tumors. Cancer 1999; 85:1389-1394.

14. Schneider DT, Schuster AE, Fritsch MK et al. Multipoint imprinting analysis indicates a common precursor cell for gonadal and nongonadal pediatric germ cell tumors. Cancer Res 2001; 61:7268-7276.
15. Feldman BJ, Feldman D. The development of androgen-independent prostate cancer. Nature Rev Cancer 2001; 1:34-45.
16. Dong JT. Chromosomal deletions and tumor suppressor genes in prostate cancer. Cancer Metast Rev 2001; 20:173-193.
17. Lee WH, Morton RA, Epstein JI et al. Cytidine methylation of regulatory sequences near the pi-class glutathione S-transferase gene accompanies human prostatic carcinogenesis. Proc Natl Acad Sci USA 1994; 91:11733-11737.
18. Millar DS, Ow KK, Paul CL et al. Detailed methylation analysis of the glutathione S-transferase pi (GSTP1) gene in prostate cancer. Oncogene 1999; 18:1313-1324.
19. Santourlidis S, Florl A, Ackermann R et al. High frequency of alterations in DNA methylation in adenocarcinoma of the prostate. Prostate 1999; 39:166-174.
20. Goessel C, Muller M, Heicappell R et al. DNA-based detection of prostate cancer in blood, urine, and ejaculates. Ann N Y Acad Sci 2001; 945:51-58.
21. Jeronimo C, Usadel H, Henrique R et al. Quantitation of GSTP1 methylation in nonneoplastic prostatic tissue and organ-confined prostate adenocarcinoma. J Natl Cancer Inst 2001; 93:1747-1752.
22. Lou W, Krill D, Dhir R et al. Methylation of the CD44 metastasis suppressor gene in human prostate cancer. Cancer Res 1999; 59:2329-2331.
23. Verkaik NS, Trapman J, Romijn JC et al. Down-regulation of CD44 expression in human prostatic carcinoma cell lines ist correlated with DNA hypermethylation. Int J Cancer 1999; 80:439-443.
24. Vis AN, Oomen M, Schroder FH et al. Feasibility of assessment of promoter methylation of the CD44 gene in serum of prostate cancer patients. Mol Urol 2001; 5:199-203.
25. Graff JR, Herman JG, Lapidus RG et al. E-cadherin expression is silenced by DNA hypermethylation in human breast and prostate carcinomas. Cancer Res 1995; 44:5195-5199.
26. Kallakury BV, Sheehan CE, Winn-Deen E et al. Decreased expression of catenins (alpha and beta), p120CTN, and E-cadherin cell adhesion proteins and E-cadherin gene promoter methylation in prostatic adenocarcinomas. Cancer 2001; 92:2786-2795.
27. Li L-C, Zhao H, Nakajima K et al. Methylation of the E-cadherin gene promoter correlates with progession of prostate cancer. J Urol 2001; 166:705-709.
28. Jarrard DF, Kinoshita H, Shi Y et al. Methylation of the androgen receptor promoter CpG island is associated with loss of androgen receptor expression in prostate cancer cells. Cancer Res 1998; 58:5310-5314.
29. Kinoshita H, Shi Y, Sandefur C et al. Methylation of the androgen receptor minimal promoter silences transcription in human prostate cancer. Cancer Res 2000; 60:3623-3630.
30. Li LC, Chui R, Nakajima K et al. Frequent methylation of estrogen receptor in prostate cancer: Correlation with tumor progression. Cancer Res 2000; 1:702-706.
31. Sasaki M, Tanaka Y, Perinchery G et al. Methylation and inactivation of estrogen, progesterone and androgen receptors in prostate cancer. J Natl Cancer Inst 2002; 94:384-390.
32. Nojima D, Li LC, Dahiya A. CpG hypermethylation of the promoter region inactivates the estrogen receptor-beta gene in patientes with prostate carcinoma. Cancer 2001; 92:2076-2083.
33. Nakayama T, Watanabe M, Yamanaka M et al. The role of epigenetic modifications in retinoic acid receptor beta2 gene expression in human prostate cancer. Lab Invest 2001; 81:1049-1057.
34. Jing C, El-Ghany MA, Beesley C et al. Tazarotene-induced Gene 1 (TIG1) expression in prostate carcinomas and its relationship to tumorigenicity. J Natl Cancer Inst 2002; 94:482-490.
35. Tekur S, Lau KM, Long J et al. Expression of RFG/ELE1alpha/ARA70 in normal und malignant prostatic epithelial cell cultures and lines: Regulation by methylation and sex steroids. Mol Carcinog 2001; 30:1-13.
36. Tamada H, Kitazawa R, Gohji K et al. Epigenetic regulation of human bone morphogenetic protein 6 gene expression in prostate cancer. J Bone Min Res 2001; 16:487-496.
37. Usmani BA, Shen R, Janeczko M et al. Methylation of the neutral endopeptidase gene promoter in human prostate cancers. Clin Cancer Res 2000; 6:1664-1670.
38. A Kuzmin I, Gillespie JW, Protopopov A et al. The RASSF1A tumor suppressor gene is inactivated in prostate tumors and suppresses growth of prostate carcinoma cells. Cancer Res 2002; 62:3498-3502.
39. Schmitt JF, Millar DS, Pedersen JS. Hypermethylation of the inhibin alpha-subunit gene in prostate carcinoma. Mol Endocrinol 2002; 16:213-220
40. Nelson JB, Lee W-H, Nguyen SH et al. Methylation of the 5'CpG island of the endothelin B receptor gene is common in human prostate cancer. Cancer Res 1997; 57:35-37.

41. Pao MM, Tsutsumi M, Liang G et al. The endothelin receptor B (EDNRB) promoter displays heterogeneous, site specific methylation patterns in normal and tumor cells. Hum Mol Genet 2001; 10:903-910.
42. Cui J, Rohr LR, Swanson G et al. Hypermethylation of the caveolin-1 gene promoter in prostate cancer. Prostate 2001; 46:249-256.
43. Santourlidis S, Warskulat U, Florl AR et al. Hypermethylation of the APT1 (FAS, CD95/Apo-1) gene promoter at rel/NFκB sites in prostatic carcinoma. Mol Carcinogen 2001; 32:36-43.
44. Whang YE, Wu X, Suzuki H et al. Inactivation of the tumor suppresor PTEN/MMAC1 in advanced human prostate cancer through loss of expression. Proc Natl Acad Sci U S A 1998; 95:5246-5350.
45. Kibel AS, Christopher M, Faith DA et al. Methylation and mutational analysis of p27(kipl) in prostate carcinoma. Prostate 2001; 48:248-253.
46. Nguyen TT, Nguyen CT, Gonzales FA et.al. Analysis of cyclin-dependent kinase inhibitor expression and methylation patterns in human prostate cancers. Prostate 2000; 43:233-242.
47. Konishi N, Nakamura M, Kishi M et al. Heterogeneous methylation and deletion patterns of the INK4a/ARF locus within prostate carcinomas. Am J Pathol 2002; 160:1207-1214.
48. Bedford MT, van Helden PD. Hypomethylation of DNA in pathological conditions of the human prostate. Cancer Res 1987; 47:5274-5276.
49. Schulz WA, Elo JP, Florl AR et al. Genome-wide DNA hypomethylation is associated with alterations on chromosome 8 in prostate carcinoma. Genes Chromosomes Cancer 2002; 35:58-65.
50. Clifford SC, Maher ER. Von Hippel-Lindau disease: Clinical and molecular perspectives. Adv Cancer Res 2001; 82:85-105.
51. Prowse AH, Webster AR, Richards FM et al. Somatic inactivation of the VHL gene in Von Hippel-Lindau disease tumors. Am J Hum Genet 1997; 60:765-771.
52. Kondo K, Yao M, Yoshida M et al. Comprehensive mutational analysis of the VHL gene in sporadic renal cell carcinoma: Relationship to clinicopathological parameters. Genes Chromosomes Cancer 2002; 34:58-68.
53. Huebner K, Garrison PN, Barnes LD et al. The role of the FHIT/FRA3B locus in cancer. Annu Rev Genet 1998; 32:7-31.
54. Dreijerink K, Braga E, Kuzmin I et al. The candidate tumor suppressor gene, RASSF1A, from human chromosome 3p21.3 is involved in kidney tumorgenesis. Proc Natl Acad Sci USA 2001; 98:7504-7509.
55. Morrissey C, Martinez A, Zatyka M et al. Epigenetic inactivation of the RASSF1A 3p21.3 tumor suppressor gene in both clear cell and papillary renal cell carcinoma. Cancer Res 2001; 61:7277-7281.
56. Yoon JH, Dammann R, Pfeifer GP. Hypermethylation of the CpG island of the RASSF1A gene in ovarian and renal cell carcinomas. Int J Cancer 2001; 94:212-217.
57. Nojima D, Nakajima K, Li LC et al. CpG methylation of promoter region inactivates E-cadherin gene in renal cell carcinoma. Mol Carcinog 2001; 32:19-27.
58. Bachmann KE, Hermann JG, Corn PG et al. Methylation-associated silencing of the tissue inhibitor of metalloproteinase-3 gene suggest a suppressor role in kidney, brain, and other human cancers. Cancer Res 1999; 59:798-802.
59. Esteller M, Corn PG, Urena JM et al. Inactivation of glutathione S-transferase P1 gene by promoter hypermethylation in human neoplasia. Cancer Res 1998; 58:4515-4518.
60. Eickelmann P, Ebert T, Warskulat U et al. Expression of NAD(P)H:quinone oxidoreductase and glutathione S-transferases Alpha and Pi in human renal cell carcinoma and in kidney cancer-derived cell lines. Carcinogenesis 1994; 15:219-225.
61. Florl AR, Loewer R, Schmitz-Dräger BJ et al. DNA methylation and expression of L1 LINE and HERV-K provirus sequences in urothelial and renal cell carcinoma. Br J Cancer 1999; 80:1312-1321.
62. Cho M, Uemura H, Kim SC. Hypomethylation of the MN/CA9 promoter and upregulated MN/CA9 expression in human renal cell carcinoma. Br J Cancer 2001; 85:563-567
63. Wykoff CC, Beasley NJ, Watson PH et al. Hypoxia-inducible expression of tumor-associated carbonic anhydrases. Cancer Res 2000; 60:7075-7083.
64. Knowles MA. What we could do now: Molecular pathology of bladder cancer. Mol Pathol 2001; 54:215-221.
65. Rabbani F, Cordon-Cardo C. Mutation of cell cycle regulators and their impact on superficial bladder cancer. Urol Clin North Am 2000; 27:83-102.
66. Hartmann A, Schlake G, Zaak D et al. Occurrence of chromosome 9 and p53 alterations in multifocal dysplasia and carcinoma in situ of human urinary bladder. Cancer Res 2002; 62:809-818.
67. Gonzalgo ML, Hayashida T, Bender CM et al. The role of DNA methylation in expression of the p19/p16 locus in human bladder cancer cell lines. Cancer Res 1998; 58:1245-1252.
68. Salem C, Liang G, Tsai YC et al. Progressive increases in de novo methylation of CpG island in bladder cancer. Cancer Res 2000; 60:2473-2476.

69. Dominguez G, Carballido J, Silva J et al. p14ARF promoter hypermethylation in plasma DNA as an indicator of disease recurrence in bladder cancer patients. Clin Cancer Res 2002; 8:980-985.

70. Florl AR, Franke KH, Niederacher D et al. DNA methylation and the mechanisms of CDKN2A inactivation in transitional cell carcinoma of the urinary bladder. Lab Invest 2000; 80:1513-1522.

71. Maruyama R, Toyooka S, Toyooka KO et al. Abberrant promoter methylation profile of bladder cancer and its relationship to clinicopathological features. Cancer Res 2001; 61:8659-8663.

72. Salem CE, Markl ID, Bender CM et al. PAX6 methylation and ectopic expression in human tumor cells. Int J Cancer 2000; 87:179-185.

73. Ruas M, Peters G. The p16INK4a/CDKN2A tumor suppressor and its relatives. Biochim Biophys Acta 1998; 1378:F115-177.

74. Bornman DM, Mathew S, Alsruhe J et al. Methylation of the E-cadherin gene in bladder neoplasia and in normal urothelial epithelium from elderly individuals. Am J Pathol 2001; 159:831-835.

75. Lee MG, Kim HY, Byun DS et al. Frequent epigenetic inactivation of RASSF1A in human bladder carcinoma. Cancer Res 2001; 61:6688-6692.

76. Markl ID, Cheng J, Liang G et al. Global and gene-specific epigenetic patterns in human bladder cancer genomes are relatively stable in vivo and in vitro over time. Cancer Res 2001; 61:5875-5884.

77. Chan MWY, Chan LW, Tang NLS et al. Hypermethylation of multiple genes in tumor tissues und voided urine in urinary bladder cancer patients. Clin Cancer Res 2002; 8:464-470.

78. Habuchi T, Luscombe M, Elder PA et al. Structure and methylation-based silencing of a gene (DBCCR1) with a candidate bladder cancer tumor suppressor region at 9q32-q33. Genomics 1998; 48:277-288.

79. Tada Y, Wada M, Taguchi K et al. The association of death-associated protein kinase hypermethylation with early recurrence in superficial bladder cancers. Cancer Res 2002; 62:4618-4627.

80. Tada Y, Wada M, Kuroiwa K et al. MDR1 gene overexpression and altered degree of methylation at the promoter region in bladder cancer during chemotherapeutic treatment.Clin Cancer Res 2000; 6:4618-4627.

81. Habuchi T, Takahashi T, Kakinuma H et al. Hypermethylation at 9q32-33 tumour suppressor region is age-related in normal urothelium and an early and frequent alteration in bladder cancer. Oncogene 2001; 20:531-537.

82. Jürgens B, Schmitz-Dräger BJ, Schulz WA. Hypomethylation of L1 LINE sequences prevailing in human urothelial carcinoma. Cancer Res 1996; 56:5698-5703.

83. Cooper MJ, Fischer M, Komitowski D et al. Developmentally imprinted genes as markers for bladder tumor progression. J Urol 1996; 155:2120-2127.

84. Oya M, Schulz WA. Decreased expression of p57(KIP2)mRNA in human bladder cancer. Br J Cancer 2000; 83:626-631.

85. Takai D, Gonzales FA, Tsai YC et al. Large scale mapping of methylcytosines in CTCF-binding sites in the human H19 promoter and aberrant hypomethylation in human bladder cancer. Hum Mol Genet 2001; 10:2619-2626.

86. Scelfo RA, Schwienbacher C, Veronese A et al. Loss of methylation at chromosome 11p15.5 is common in human adult tumors. Oncogene 2002; 21:2564-2572.

87. Velicescu M, Weisenberger DJ, Gonzales FA et al. Cell division is required for de novo methylation of CpG islands in bladder cancer cells. Cancer Res 2002; 62:2378-2384

88. Millar DS, Paul CL, Molloy PL et al. A distinct sequence (ATAAA)n separates methylated and unmethylated domains at the 5′-end of the GSTP1 CpG island. J Biol Chem 2000; 275:24893-24899.

89. Lin X, Tascilar M, Lee WH et al. GSTP1 CpG island hypermethylation is responsible for the absence of GSTP1 expression in human prostate cancer cells. Am J Pathol 2001; 159:1815-1826.

90. Singal R, van Wert J, Bashambu M. Cytosine methylation represses glutathione S-transferase P1 (GSTP1) gene expression in human prostate cancer cells. Cancer Res 2001; 61:4820-4826.

91. Song JZ, Stirzaker C, Harrison J et al. Hypermethylation trigger of the glutathione-S-transferase gene (GSTP1) in prostate cancer cells. Oncogene 2002; 21:1048-1061.

92. Maruyama R, Toyooka S, Toyooka KO et al. Aberrant promoter methylation profile of prostate cancers and its relationship to clinicopathological features. Clin Cancer Res 2002; 8:514-519.

93. Gitan RS, Shi H, Chen CM et al. Methylation-specific oligonucleotide microarray: A new potential for high-throughput methylation analysis. Genome Res 2002; 12:158-164.

94. Adorján P, Distler J, Lipscher E et al. Tumour class prediction and discovery by microarray-based DNA methylation analysis. Nucleic Acids Res 2002; 30:e21

95. Kawai J, Hirose K, Fushiki S et al. Comparison of DNA methylation patterns among mouse cell lines by restriction landmark genomic scanning. Mol Cell Biol 1994; 14:7421-7427.

96. Liang G, Salem CE, Yu MC et al. DNA methylation differences associated with tumor tissues identified by genome scanning analysis. Genomics 1998; 53:260-268.

97. Liang G, Gonzales FA, Jones PA et al. Analysis of gene induction in human fibroblasts and bladder cancer cells exposed to the methylation inhibitor 5-aza-2′-deoxycytidine. Cancer Res 2002; 62:961-966.

98. Kimura F, Florl AR, Seifert HH et al. Destabilisation of chromosome 9 in transitional cell carcinoma of the urinary bladder. Br J Cancer 2001; 85:1887-1893.

99. Shibata D, Aaltonen LA. Genetic predisposition and somatic diversification in tumor development and progression. Adv Cancer Res 2001; 80:83-114.

100. Toyota M, Ahuja N, Suzuki H et al. Aberrant methylation in gastric cancer associated with the CpG island methylator phenotype. Cancer Res 1999; 59:5438-5442.

101. Toyota M, Ohe-Toyota M, Ahuja N et al. Distinct genetic profiles in colorectal tumors with or without the CpG island methylator phenotype. Proc Natl Acad Sci USA 2000; 97:710-715.

102. Robertson KD, Keyomarsi K, Gonzales FA et al. Differential mRNA expression of the human DNA methyltransferases (DNMTs) 1, 3a and 3b during the G(0)/G(1) to S phase transition in normal and tumor cells. Nucleic Acids Res 2000; 28:2108-2113.

102a. Kimura F, Siegert HH, Florl AR et al. Decrease of DNA methyltransferase I expression relative to cell proliferation in transitional cell carcinoma. Int J Cancer 2003; 104:568-578.

103. Patra SK, Patra A, Zhao H et al. DNA methyltransferase and demethylase in human prostate cancer. Mol Carcinogen 2002; 33:163-171.

104. Robertson KD, Ait-Si-Ali S, Yokochi T et al. DNMT1 forms a complex with Rb, E2F1 and HDAC1 and represses transcription from E2F-responsive promoters. Nat Genet 2000; 25:338-342.

105. Pradhan S, Kim GD. The retinoblastoma gene product interacts with maintenance human DNA (cytosine-5) methyltransferase and modulates its activity. EMBO J 2002; 21:779-788.

106. Yi P, Melnyk S, Pogribna M et al. Increase in plasma homocysteine associated with parallel increases in plasma S-adenosylhomocysteine and lymphocyte DNA hypomethylation. J Biol Chem 2000; 275:29318-29323.

107. Friso S, Choi SW, Girelli D et al. A common mutation in the 5,10-methylene-tetrahydrofolate reductase gene affects genomic DNA methylation through an interaction with folate status. Proc Natl Acad Sci USA 2002; 99:5606-5611.

108. Piyathilake CJ, Johanning GL, Macaluso M et al. Localized folate and vitamin B-12 deficiency in squamous cell lung cancer is associated with global DNA hypomethylation. Nutr Cancer 2000; 37:99-107.

109. Van den Dyver IB. Genetic effects of methylation diets. Annu Rev Nutr 2002; 22:255-282.

110. Kimura F, Franke KH, Steinhoff C et al. Methyl group metabolism gene polymorphisms and susceptibility toward prostatic carcinoma. Prostate 2000; 45:225-231.

111. Kimura F, Florl AR, Steinhoff C et al. Polymorphic methyl group metabolism genes in patients with transitional cell carcinoma of the urinary bladder. Mut Res Genomics 2001; 458:49-54.

112. Fair WR, Fleshner NE, Heston W. Cancer of the prostate: A nutritional disease? Urology 1997; 50:840-848.

113. Cox R, Goorha S, Irving CC. Inhibition of DNA methylase activity by acrolein. Carcinogenesis 1988; 9:463-465.

114. Chuang LS-H, Tan EH-H, Oh H-K et al. Selective depletion of human DNA-methyltransferase DNMT1 proteins by sulfonate-derived methylating agents. Cancer Res 2002; 62:1592-1597.

115. Mass MJ, Wang L. Arsenic alters cytosine methylation patterns of the promoter of the tumor suppressor gene p53 in human lung cells: A model for a mechanism of carcinogenesis. Mutat Res 1997; 386:263-277.

116. Rehn L. Blasengeschwülste bei Fuchsinarbeitern. Arch Klin Chir 1895[*sic!*]; 50:588-590.

117. Bender CM, Pao MM, Jones PA. Inhibition of DNA methylation by 5-aza-2′-deoxycytidine suppresses the growth of human tumor cell lines. Cancer Res 1998; 58:95-101.

118. Ehrlich M. DNA methylation in cancer: Too much, but also too little. Oncogene 2002; 21:5400-5413..

119. Del Senno L, Maestri I, Piva R et al. Differential hypomethylation of the c-myc protooncogene in bladder cancers at different stages and grades. J Urol 1989; 142:146-149.

120. Yoder JA, Walsh CP, Bestor TH. Cytosine methylation and the ecology of intragenomic parasites. Trends Genet 1997; 13:335-340.

121. Lower R, Lower J, Kurth R. The viruses in all of us: Characteristics and biological significance of human endogenous retrovirus sequences. Proc Natl Acad Sci USA 1996; 93:5177-5184.

122. Herbst H, Sauter M, Mueller-Lantzsch N. Expression of human endogenous retrovirus K elements in germ cell and trophoblastic tumors. Am J Pathol 1996; 149:1727-1735.

123. Vinogradova T, Leppik L, Kalinina E et al. Selective Differential Display of RNAs containing interspersed repeats: Analysis of changes in the transcription of HERV-K LTRs in germ cell tumors. Mol Genet Genomics 2002; 266:796-805.
124. Herbst H, Kuhler-Obbarius C, Lauke H, et al. Human endogenous retrovirus (HERV)-K transcripts in gonadoblastomas and gonadoblastoma-derived germ cell tumours. Virchows Arch 1999; 434:11-15.
125. Moran JV, Holmes SE, Naas TP et al. High frequency retrotransposition in cultured mammalian cells. Cell 1996; 87:917-927.
126. Ostertag EM, Kazazian Jr HH. Biology of mammalian L1 retrotransposons. Annu Rev Genet 2001; 35:501-538.
127. Jones PA. The DNA methylation paradox. Trends Genet 1999; 15:34-37.
128. Shaffer LG, Lupski JR. Molecular mechanisms for constitutional chromosomal rearrangements in humans. Annu Rev Genet 2000; 34:297-329.
129. Qu GZ, Grundy PE, Narayan A et al. Frequent hypomethylation in Wilms tumors of pericentromeric DNA in chromosomes 1 and 16. Cancer Genet Cytogenet 1999; 109:34-39.
130. Jimenez RE, Fischer AH, Petros JA et al. Glutathione S-transferase pi gene methylation: The search for a molecular marker of prostatic adenocarcinoma. Adv Anat Pathol 2001; 7:382-389.
131. Brooks JD, Weinstein M, Lin X et al. CG island methylation changes near the GSTP1 gene in prostatic intraepithelial neoplasia. Cancer Epidemiol Biomarkers Prev 1998; 7:531-536.
132. Suh CI, Shanafelt T, May DJ. Comparison of telomerase activity and GSTP1 promoter methylation in ejaculate as potential screening tests for prostate cancer. Mol Cell Probes 2000; 14:211-217.
133. Cairns P, Esteller M, Hermann JG et al. Molecular detection of prostate cancer in urine by GSTP1 hypermethylation. Clin Cancer Res 2001; 7:2727-2730.
134. Goessl C, Muller M, Heicappell R et al. DNA-based detection of prostate cancer in urine after prostatic massage. Urology 2001; 58:335-338.
135. Carducci MA, Nelson JB, Chan-Tack KM et al. Phenylbutyrate induces apoptosis in human prostate cancer and is more potent than phenylacetate. Clin Cancer Res 1996; 2:379-387.
136. Ellerhorst J, Nguyen T, Cooper DN et al. Induction of differentiation and apoptosis in the prostate cancer cell line LNCaP by sodium butyrate and galectin-1. Int J Oncol 1999; 14:225-232.
137. Melchior SW, Brown LG, Figg WD et al. Effects of phenylbutyrate on proliferation and apoptosis in human prostate cancer cells in vitro and in vivo. Int J Oncol 1999;14:501-508.
138. Izbicka E, MacDonald JR, Davidson K et al. 5,6-Dihydro-5'-azacytidine (DHAC) restores androgen responsiveness in androgen-insensitive prostate cancer cells. Anticancer Res 1999; 19:1285-1291.
139. Maier S, Reich E, Martin R et al. Tributyrin induces differentiation, growth arrest and apoptosis in androgen-sensitive and androgen-resistant human prostate cancer cell lines. Int J Cancer 2000; 88:245-251.
140. Thibault A, Figg WD, Bergan RC et al. A phase II study of 5-aza-2'deoxycytidine (decitabine) in hormone independent metastatic (D2) prostate cancer. Tumori 1998; 84:87-89.
141. Lin X, Asgari K, Putzi MJ et al. Reversal of GSTP1 CpG island hypermethylation and reactivation of π-class glutathione S-transferase (GSTP1) expression in human prostate cancer cells by treatment with procainamide. Cancer Res 2001; 61:8611-8616.
142. Strathdee G, Brown R. Epigenetic cancer therapies: DNA methyltransferase inhibitors. Expert Opin Investig Drugs 2002; 11:747-754.
143. Jackson-Grusby L, Beard C, Possemato R et al. Loss of genomic methylation causes p53-dependent apoptosis and epigenetic deregulation. Nat Genet 2001; 27:31-39.

CHAPTER 5

DNA Methylation in Colorectal Cancer

Jeremy R. Jass, Vicki L.J. Whitehall, Joanne Young and Barbara A. Leggett

Abstract

In this chapter, it is pointed out that colorectal cancer is a heterogeneous disease. The case is made for a 'serrated pathway' of neoplasia that would evolve relatively rapidly through the early acquisition of DNA instability. DNA hypermethylation is likely to be of critical importance in driving this pathway. Inhibition of apoptosis is conceived as the first step. Thereafter, methylation of one of several DNA repair genes would result in a state of tolerated hypermutability. It remains to be shown whether this model applies to a small subset of colorectal cancers or in fact explains the great majority given the overall low risk of progression for an individual adenoma initiated by mutation of *APC*.

Introduction

Colorectal cancer is not only an important disease in terms of its frequency and contribution to human suffering but also because it provides an instructive model for neoplasia in general. There are two reasons why colorectal cancer has served as a successful model. First, the precancerous stages present as mucosal lesions (polyps) that can be identified and removed with relative ease. Second, the discovery of the genetic mechanisms underlying rare, familial forms of large bowel cancer has generated insights that are relevant to the common form of the disease. Specifically, inheritance of a recessive cancer predisposition gene explains familial cancer whereas inactivation of precisely the same gene at the somatic level initiates the evolution of sporadic forms of the same type of cancer.

The most well documented forms of familial colorectal cancer are familial adenomatous polyposis (FAP)[1] and hereditary nonpolyposis colorectal cancer (HNPCC).[2] Both are inherited as autosomal dominant conditions in which affected subjects develop colorectal cancer at a young age. Cancers also are likely to be multiple and affected individuals are prone to develop cancers in particular sites outside the large intestine. The main clinical difference between the two conditions is with respect to the precancerous polyp or adenoma. In the case of FAP, many hundreds if not thousands of adenomas develop in the mucosal lining of the large intestine by the second decade (Fig. 1). Left untreated, one or more of these will develop into a cancer, generally by the fourth or fifth decade. In the case of HNPCC, adenomas are only a little more common than cancers and the time required for an adenoma to convert into a cancer is relatively short. From these observations it has been deduced that the genetic lesion in the *APC* gene (responsible for the condition FAP) serves to initiate adenomas whereas the genetic lesion in the genes responsible for HNPCC acts to accelerate tumor progression. The genes underlying HNPCC are DNA mismatch repair genes of which *hMLH1* and *hMSH2* are the most frequently implicated.

DNA Methylation and Cancer Therapy, edited by Moshe Szyf. ©2005 Eurekah.com and Kluwer Academic/Plenum Publishers.

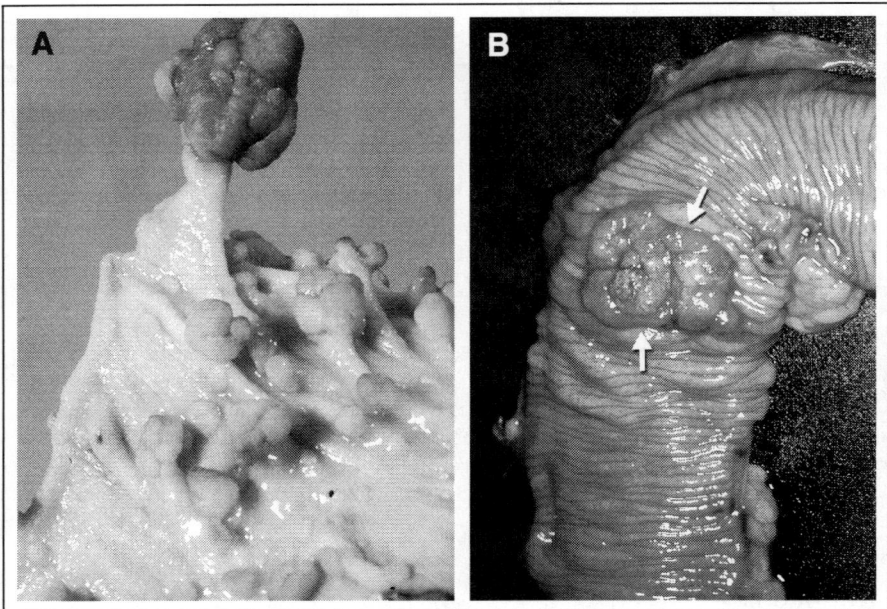

Figure 1. Close up view of multiple adenomas in colonic mucosa from subject with familial adenomatous polyposis (FAP) (A) versus a single cancer with no adenomas in a total colectomy specimen from a subject with hereditary nonpolyposis colorectal cancer (HNPCC) (B).

Relevance of Familial Models of Colorectal Cancer to Sporadic Neoplasia

In certain respects the conditions FAP and HNPCC provide models that serve only as 'caricatures' of sporadic colorectal neoplasia. In FAP, the ratio of adenoma to carcinoma is approximately 1000 to 1,[1] in HNPCC it is 1 to1,[3] whereas in sporadic colorectal cancer it is around 30 to1.[4] To put it another way, an individual adenoma in a subject with FAP is highly unlikely to develop into a cancer. The opposite is true in HNPCC whereas sporadic adenomas fall somewhere in between these extremes. These simple clinical observations indicate that morphologically similar adenomas may be biologically very different. It should be stressed that the prevailing dogma views all adenomas as being created equal following the advent *APC* inactivation. The occurrence of subsequent genetic alterations is then deemed to determine the timing of progression. It is now clear, however, that: (1) adenomas are not the only precancerous lesions that occur in the colorectum,[5] and (2) sporadic adenomas may be initiated by genetic alterations apart from inactivation of *APC*, for example by K-*ras* mutation.[6] In addition, while it is clear that inactivation of *APC* is sufficient to initiate an adenoma and allow it to grow to a certain size,[7] other coexisting or modifying factors may determine the likelihood of subsequent malignant conversion.

DNA Microsatellite Instability

The lesson of HNPCC indicates the nature of an additional key factor that modifies the magnitude of risk of progression. Loss of a DNA mismatch repair gene such as *hMLH1* or *hMSH2* has two outcomes: (1) inhibition of apoptosis,[8] and (2) generation of mutations at the DNA level. The term 'genetic instability' may be applied to a population of cells that continues to proliferate in the face of accumulating DNA damage, failure of damage and/or mismatch repair and the subsequent generation of widespread alterations in the genome. The type of genetic instability that occurs in HNPCC is extreme and can be extrapolated to only about 10% of sporadic colorectal cancers. Because DNA microsatellite markers are used to detect frameshift mutations that arise as a consequence of loss of DNA mismatch repair proficiency, such cancers are described as microsatellite instability-high (MSI-H).[9,10] When a panel of microsatellite markers is used to demonstrate DNA instability in such tumors, it is usual to detect instability in over 50% of markers with the mean frequency being around 80%.[11]

It may be inferred that less extreme forms of genetic instability must be acquired in the case of most if not all of the remaining sporadic colorectal cancers. One such type of less extreme instability is also detected with the aid of microsatellite markers and is known as MSI-low (MSI-L). By using a very large panel of markers it has been shown that MSI-L is distributed as a nonrandom quantitative trait. That is, some cancers show more instability than would be expected by chance whereas others show less instability (or no instability) with this type of biomarker.[12] Some MSI-L cancers may show instability in over 30% of markers, making them technically MSI-H according to the current NCI criteria.[9] However, these cancers differ fundamentally from bona fide MSI-H cancers in a number of respects: (1) Instability is rarely detected in mononucleotide markers which are highly sensitive for MSI-H status, (2) There is no deficiency with respect to any known DNA mismatch repair gene, (3) Mutation of *APC*, K-*ras* and *TP53* is frequent, and (4) Loss of heterozygosity at chromosomes 5q, 17p and 18q is frequent.[13,14] The inappropriate inclusion of such cases amongst the genuine MSI-H cancers serves as a continuing source of confusion.

It should be pointed out that some have argued that an increased mutation rate is unnecessary to explain carcinogenesis in general, that genetic instability is likely to be disadvantageous with respect to neoplastic evolution (for example by increasing the probability of apoptosis), and that genetic instability occurs late in the process of neoplastic evolution, if it occurs at all.[15] However, these arguments were based on the premise that virtually all colorectal cancers arise in traditional adenomas that are in turn initiated by an exclusive mechanism, namely inactivation of *APC*. As noted above, this premise is incorrect. Most now accept that Darwinian natural selection is insufficient on its own to explain the stepwise accumulation of key genetic alterations that underlies malignancy.

Mechanisms Underlying Microsatellite Instability in Colorectal Cancer

Factors that may contribute to the acquisition of genetic instability include: (1) presence of DNA damaging agents, (2) lack of mechanisms for repairing damaged DNA, (3) lack of mechanisms for repairing mismatches between template and replicated DNA strands, and (4) lack of checkpoint mechanisms that would normally trigger cell cycle arrest and/or apoptosis in the face of abnormal DNA content or abnormal cell function (for example as a consequence of a mutation). It is clear that genetic instability is a two-edged sword. Whilst it will increase the mutation rate and therefore increase the likelihood of oncogenic mutation it will also predispose to apoptosis (the price of immortality for one cell is the death of many). This is why genetic instability must involve not merely an increased mutation rate but also the inhibition of apoptosis.

The term 'hyperplasia' means an increase in cell numbers. Hyperplasia can arise either through an increase in the rate of cell production or through a reduction in the rate of cell death (or both). The hyperplasia characterizing colorectal adenomas arises through *APC* inactivation leading in turn to increased cellular proliferation. A second type of polyp may occur within the colorectum that is known as a hyperplastic polyp. There are different mechanisms for initiating hyperplastic polyps, but inhibition of apoptosis is an important unifying mechanism. K-*ras* mutation is a frequent initiating genetic alteration in hyperplastic polyps.[6] Oncogenic K-*ras* leads to the inhibition of apoptosis through, for example, the down-regulation of the apoptosis receptor Fas (CD95).[16] It may be surmised, for the reasons given above, that cell populations in polyps that arise through inhibition of apoptosis may be better able to tolerate a state of genetic instability than cell populations in polyps that arise through a state of dysregulated cellular proliferation. Two mechanisms have been identified that not only underlie genetic instability but also occur selectively within hyperplastic polyps. These are: (1) inactivation of the DNA mismatch repair gene *hMLH1*[17] and (2) inactivation of the DNA repair gene O-6-methylguanine DNA methyltransferase (*MGMT*).[18] In sporadic colorectal neoplasia, the inactivation of both genes occurs not by mutation or loss but by the epigenetic mechanism of gene silencing through promoter hypermethylation.[19,20]

The Methylator Phenotype in Colorectal Neoplasia

Just as microsatellite markers may be used to establish the presence of the 'mutator' phenotype in colorectal cancer (and other malignancies), so may anonymous DNA sequences that are prone to methylation of their cytosine nucleotides be used to establish the existence of a 'methylator' phenotype. DNA methylation occurs preferentially within dense clusters of CpG sites known as CpG islands. Therefore reference is made to a CpG island methylator phenotype (CIMP).[21] Markers used to establish CIMP have been termed Methylated IN Tumors or MINTs.[22] If a panel of MINT markers is employed (for example MINTs 1, 2, 12, and 31) around 20% of colorectal cancers show extensive methylation (implicating 3 or 4 loci).[23]

Cancers showing methylation of MINTs will also show methylation and therefore silencing of tumor suppressor genes. Examples of genes that are silenced by this mechanism in colorectal cancer are the DNA repair genes *hMLH1*[19] and *MGMT*,[18] the cell cycle regulator *p16INK4a*,[21] the apoptosis regulator *p14ARF*,[24] the angiogenesis factor *THBS1*,[21] the anti-adhesion gene *HPP1*,[25] and the growth regulating gene *COX-2*.[26] This list is not exhaustive, but it does highlight the existence of a cluster of altered genes within the subset of colorectal cancer showing DNA methylation that is radically different from the traditional genes highlighted in the Vogelstein model.[27]

About 50% of cancers with the 'methylator' phenotype also show the 'mutator' phenotype. In this group of MSI-H cancers there is methylation of the DNA mismatch repair gene *hMLH1* in the vast majority of cases. Studies in which the MSI-H phenotype is carefully established by appropriate testing and in which familial examples (HNPCC) are excluded, indeed show that mutation of the traditional genes of the Vogelstein model, specifically *APC*, K-*ras* and *TP53*, is rarely observed.[14] The remaining cancers with the methylator phenotype but lacking both the mutator phenotype and methylation of *hMLH1* are less well characterized, though patterns are now beginning to emerge. These tumors show frequent methylation of the DNA repair gene *MGMT*,[28] mutation of K-*ras*[22] and low-level microsatellite instability (MSI-L).[29] Inactivation of *MGMT* is the likely explanation for the increased frequency of K-*ras* mutation and MSI-L. The DNA repair protein MGMT repairs methylguanine (mG) adducts which are highly mutagenic. When DNA is replicated during the S-phase of the cell cycle, mG is mistaken for A and this results in mG being mismatched with T. In a second round of cell division (assuming the mismatch is not repaired), A in inserted opposite T. This will create a G:C to A:T transition mutation. In cancers in which *MGMT* is inactivated by promoter methylation, mG:T mismatches will be frequent and it so happens that most K-*ras* mutations are G to A transitions.[20] Similarly, most *TP53* mutations are C to T transitions.[30] The explanation for MSI-L probably lies in the fact that mG:T mismatches are unstable and difficult to repair. The attempts at excision and resynthesis will yield the same mismatch resulting in so-called 'futile cycles of repair'.[31] This cycle could be broken in two ways: (1) apoptosis as signaled by the mismatch repair heterodimer hMSH2-hMSH6[32] or (2) excision of the mismatch with a permanent deletional frameshift mutation replacing the unstable mismatch. The latter would be detected as a small deletion and would be particularly common within microsatellite repeats which are intrinsically prone to aberrant resynthesis (hence the detection of MSI-L) (Fig. 2).

DNA methylation is not restricted to the 20% of colorectal cancers showing extensive methylation of MINT markers. Cancers may show methylation of a small proportion of MINT markers. Additionally, individual tumor suppressor genes may be silenced by methylation in

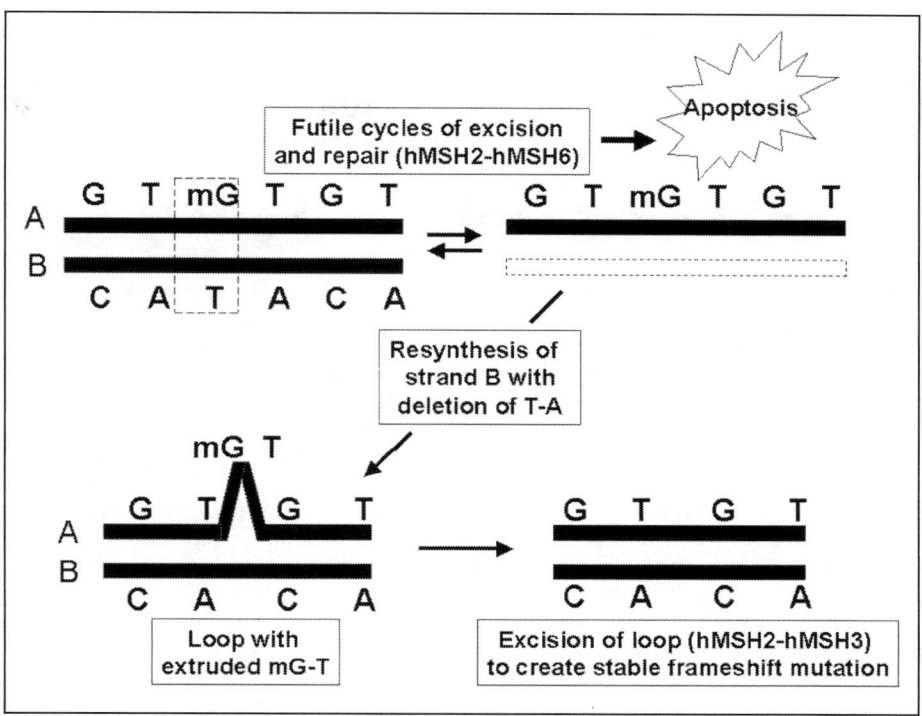

Figure 2. The generation of methylG:T mismatches may lead to 'futile cycles' of DNA repair. When excision of the mismatched sequence is followed by deletion instead of futile resynthesis, the unstable mismatch site is replaced by a stable frameshift mutation. This is detected as MSI-L.

cancers that show no evidence of MINT methylation. This might occur in tumors in which methylation of specific genes is subject to strong selection pressures. An example would be in the condition HNPCC in which one copy of the gene *hMLH1* is already silenced (through inheritance of a germline mutation) and a major growth advantage would result from the methylation of the wild-type copy.[11] Additionally, methylation of the DNA repair gene *MGMT* occurs independently of the status of MINT methylation.[28] Certain genes such as ER are methylated as an age-related phenomenon in normal tissues as well as neoplasms.[33] These have been termed type A genes to distinguish them from type C genes that are methylated in a cancer-specific manner. This distinction is not absolute insofar as methylation of type C genes occurs in normal mucosa but perhaps on a sporadic crypt-by-crypt basis.[34]

The fact that colorectal cancer is a genetically heterogeneous disease has emerged relatively recently. The separation of cancers showing microsatellite instability versus chromosomal instability was initially confounded by the failure to distinguish MSI-H from MSI-L status (see above). Once this distinction was made, the pendulum swung in the opposite direction and bowel cancer became two nonoverlapping diseases with well-defined clinical, pathological and molecular differences. It is now clear that this is an oversimplification. For example, as noted above, the 'nonMSI-H' group includes a subset that is as heavily methylated as the MSI-H group. It has been shown that heavily methylated cancers share a number of morphological features regardless of MSI status. These features include the presence of round, clear nuclei with prominent large nucleoli and the production of extracellular mucin.[28] Other morphological features are not shared, for example extensively methylated nonMSI-H cancers are more likely to show diffuse infiltration and are less likely to show tumor heterogeneity than MSI-H cancers with extensive methylation.[28] Colorectal cancer can therefore be viewed as a disease spectrum, but a spectrum in which particular types of test may reveal discontinuities. Tests that reveal discontinuities inform us about key, rate-limiting steps in the evolution of colorectal neoplasia. Acquisition of the states of apoptosis inhibition and genetic instability are likely to be two such fundamental steps. The validity of this concept can be tested by cDNA chip array analysis in which the test in question is used as an interrogation criterion for hierarchical clustering.

Are Subjects Genetically Predisposed to Methylation of Colorectal Mucosa?

DNA methylation was originally viewed as an age-related stochastic process and this may well be correct with respect to genes that are methylated within normal colorectal mucosa.[33] Nevertheless, there are grounds for believing that subjects developing colorectal cancers with the methylator phenotype are genetically predisposed to do so. Extensive DNA methylation is usually a feature in MSI-H cancers. MSI-H cancers are age-related but are prone to be associated with bowel cancer multiplicity and, unlike colorectal cancers in general, are more common in females.[14] Some earlier onset MSI-H cancers develop in a background of hyperplastic polyposis.[17] Hyperplastic polyps may in turn show DNA methylation.[35] When DNA methylation is found in one hyperplastic polyp in a subject with hyperplastic polyposis, it is likely to be found in the other polyps from the same individual.[35] Such concordance is not seen in the case of subjects with multiple adenomas.[36] There are instances of hyperplastic polyposis occurring in a familial setting,[37-39] although this is not commonly described. Taken together, these observations are consistent with the hypothesis that the predisposition to DNA methylation within tumors can be inherited as a relatively weakly penetrant trait that is expressed in older aged subjects. Environmental and/or additional genetic factors may magnify this trait to account for the development of multiple lesions, earlier age of onset and/or positive family history. The underlying mechanism could relate to defective DNA repair within the promoter region that serves to initiate de novo methylation as a form of reaction to nonrepaired DNA damage. Alternatively, the defect could apply to genes responsible for the control of maintenance and/or de novo methylation. It would therefore be reasonable to conduct, for example, association studies for polymorphisms in genes responsible for DNA damage repair and/or the control of DNA methylation in subjects who develop colorectal cancer with the methylator phenotype.

Morphologic Counterparts of the Methylator Phenotype

As noted above, colorectal cancers showing mutator and/or methylator phenotypes are distinguished morphologically (histologically) from cancers lacking these phenotypes. The associated morphologic features relate to: (1) the nucleus which is round, clear and contains a single large nucleolus, (2) differentiation which is often either poor and/or characterized by production of secretory mucin, (3) presence of intra-epithelial T-lymphocytes, (4) a circumscribed or expanding growth pattern, and (5) marked morphological heterogeneity.[28] The question arises as to whether these features are explained by the mutator or by the methylator phenotype? Some of the features are also seen in HNPCC cancers in which the mutator but not the methylator phenotype is present. Features frequent in both HNPCC and sporadic MSI-H cancers include: (1) presence of intra-epithelial lymphocytes, (2) poor differentiation, and (3) expanding growth pattern. Whilst HNPCC cancers may be mucinous, poorly differentiated and show tumor heterogeneity, these features are significantly more frequent in sporadic MSI-H cancers.[11] On the other hand, methylator cancers without the mutator phenotype showed less tumor heterogeneity and were more likely to show diffuse infiltration than methylator cancers with the mutator phenotype.[28] It may be hypothesized that the main features explained by hypermethylation rather than the mutator phenotype relate to nuclear parameters that are visible at the light microscopic level and production of secretory mucin. It is reasonable to suggest that nuclear features may be influenced by the presence of extensive DNA methylation either directly or indirectly, but this observation necessitates further research. Another issue that needs to be explored is whether the morphological appearances of methylator cancers are presaged by the appearances of the putative epithelial precursor lesions. As noted above, hyperplastic polyps (that are initiated by inhibition of apoptosis) may the precursors of lesions driven by methylation of the DNA repair genes *MGMT* and *hMLH1*. Whereas adenomas show a reduction in expression of secretory mucin, hyperplastic polyps are phenotypically similar to MSI-H cancers in showing increased expression of both intestinal mucin MUC2 and gastric mucin MUC5AC.[40,41] In addition, hyperplastic polyps have round, clear nuclei with prominent nucleoli whereas adenomas have elongated, hyperchromatic nuclei.

Serrated Pathway of Colorectal Neoplasia

Apart from the observations noted in the preceding section, the evidence linking hyperplastic polyps rather than adenomas to sporadic MSI-H cancers is strong: (1) dysplastic subclones within hyperplastic polyps show the same spectrum of genetic alteration as sporadic MSI-H cancers, (2) conventional adenomas and sporadic MSI-H cancers show very different genetic alterations, (3) residual lesions adjacent to sporadic MSI-H cancers fall within the spectrum of 'serrated polyps' (see below) rather than conventional adenomas, (4) hyperplastic polyps occur with increased numbers in patients with sporadic MSI-H cancer.[17,42-44] For the preceding reasons, DNA methylation has been linked to a specific morphologic phenotype characterized by glandular serration (in which intestinal crypts assume a saw-tooth configuration). The serrated spectrum includes microscopic aberrant crypt foci (ACF), hyperplastic polyps, admixed polyps, serrated adenomas and at least some villous adenomas.[5,14,42,45]

It has been suggested that serration is the result of inhibition of apoptosis and particularly the type of apoptosis that is triggered by the shedding of surface epithelial cells (anoikis).[14] The question arises as to whether anoikis could be inhibited by a genetic mechanism involving silencing of a gene by methylation. Should this be the case it would mean that the early stages of colorectal neoplasia would (at least in some instances) be driven by the sequential hypermethylation and silencing first of an anti-apoptotic gene (causing serrated hyperplasia) and then of a DNA repair gene. This two-step model would then serve as the preferred mechanism for initiation colorectal neoplasia in subjects prone to DNA methylation as a result of genetic and/or environmental predisposing factors. To answer the question, a genome-wide search for differentially methylated DNA was undertaken in hyperplastic polyps versus normal mucosa using arbitrarily-primed methylation-sensitive PCR. This approach led to the cloning of a gene named *HPP1*.[25] *HPP1* shows homology with the anti-adhesion gene *SPARC* and is expressed by the surface epithelium of normal large bowel mucosa. Loss of expression and hypermethylation of *HPP1* is observed in colorectal cancers as well as hyperplastic polyps.[25]

Can DNA Methylation Be Reversed Spontaneously?

DNA methylation is regarded as a permanent, heritable state.[46] It is conceivable that a particular methylated site may on occasion fail to be transmitted in the methylated form by one cell to a daughter cell. It is also possible that methylation (for example of *MGMT*) in the early stages of neoplastic evolution is advantageous in that it increases the probability of mutation of key cancer genes such as K-*ras* and *TP53*.[20,30] Thereafter, however, the progressive accumulation of DNA damage may be disadvantageous and a cell that does not inherit methylated *MGMT* may have a selective advantage over the majority that do. This scenario is in fact likely to occur because methylation is more common in adenomas than cancers.[36] Furthermore, it is possible to observe within malignant polyps loss of expression of MGMT in the benign component and reexpression of the DNA repair protein in the malignant component (Fig. 3). This means that, unlike the situation as applies to mutated genes, epigenetic alterations may be transitory ('hit and run'). Therefore, the absence of an epigenetic change in a cancer does not exclude the possibility that such a change did in fact occur within a precancerous stage of evolution. Since investigators can rarely obtain human lesions in transition from

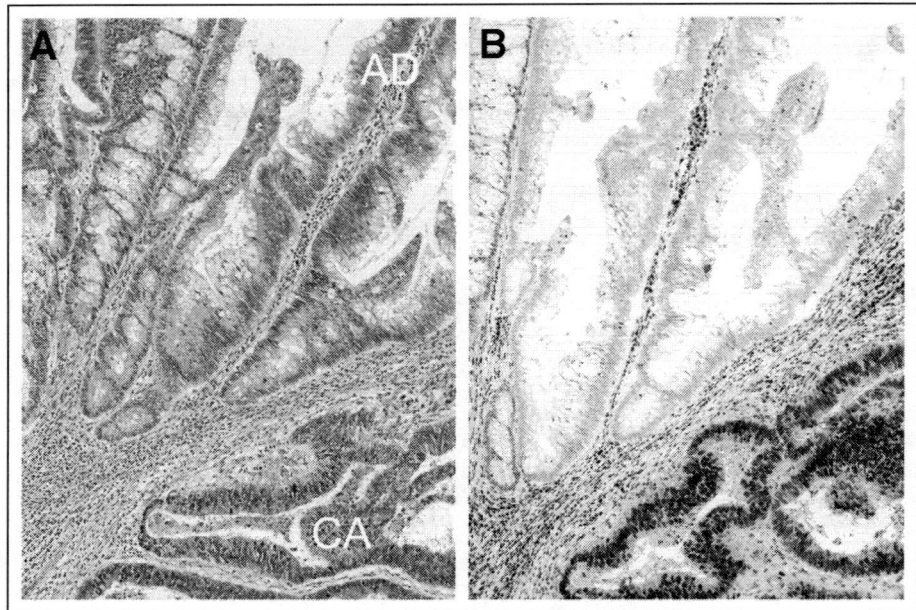

Figure 3. Focus of adenocarcinoma arising in a mixed villous adenoma/serrated adenoma (A: Haematoxylin and eosin). There is loss of expression of MGMT within the nuclei of the adenoma but reexpression of MGMT within the nuclei of the cancer (B: Immunostaining for MGMT, clone MT3.1, NeoMarkers).

benign to malignant, this phenomenon would be an example of nature making life exceedingly difficult for the researcher.

Recently, mutant *TP53* has been linked mechanistically with DNA demethylation.[47] This observation provides an elegant model to explain what may be happening at the critical transition from polyp to colorectal carcinoma. Loss of the DNA repair enzyme MGMT predisposes to *TP53* mutation.[30] This in turn, however, could result in demethylation of *MGMT* and reexpression of the protein (Fig. 3). This model would also provide one explanation for the generally noted negative correlation between *TP53* mutation and DNA methylation.[48] Such a mechanism also cautions against the uncritical use of demethylating agents in cancer therapy since at early stages of cancer evolution this approach could be counterproductive.

References

1. Bussey HJR. Familial Polyposis Coli. Baltimore: Johns Hopkins Press; 1975.
2. Lynch HT, Smyrk T, Lynch JF. Overview of natural history, pathology, molecular genetics and mangement of HNPCC (Lynch syndrome). Int J Cancer 1996; 69:38-43.
3. Järvinen HJ, Aarnio M, Mustonen H et al. Controlled 15-year trial on screening for colorectal cancer in families with hereditary nonpolyposis colorectal cancer. Gastroenterology 2000; 118:829-834.
4. Pollock AM, Quirke P. Adenoma screening and colorectal cancer. The need for screening and polypectomy is unproved. Br J Med 1991; 303:3-4.
5. Jass JR. Serrated route to colorectal cancer: back street or super highway? J Pathol 2001; 193:283-285.
6. Takayama T, Ohi M, Hayashi T et al. Analysis of K-ras, APC, and beta-catenin in aberrant crypt foci in sporadic adenoma, cancer, and familial adenomatous polyposis. Gastroenterology 2001; 121:599-611.
7. Lamlum H, Papadopoulou A, Ilyas M et al. APC mutations are sufficient for the growth of early colorectal adenomas. Proc Natl Acad Sci USA 2000; 97:2225-2228.

8. Fishel R. The selection for mismatch repair defects in hereditary nonpolyposis colorectal cancer: revising the mutator hypothesis. Cancer Res 2001; 61:7369-7374.

9. Boland CR, Thibodeau SN, Hamilton SR et al. A National Cancer Institute Workshop on microsatellite instability for cancer detection and familial predisposition: Development of international criteria for the determination of microsatellite instability in colorectal cancer. Cancer Res 1998; 58:5248-5257.

10. Dietmaier W, Wallinger S, Bocker T et al. Diagnostic microsatellite instability: Definition and correlation with mismatch repair protein expression. Cancer Res 1997; 57:4749-4756.

11. Young J, Simms LA, Biden KG et al. Features of colorectal cancers with high-level microsatellite instability occurring in familial and sporadic settings: Parallel pathways of tumorigenesis. Am J Pathol 2001; 159:2107-2116.

12. Halford S, Sasieni P, Rowan A et al. Low-level microsatellite instability occurs in most colorectal cancers and is a nonrandomly distributed quantitative trait. Cancer Res 2002; 62:53-57.

13. Jass JR, Biden KG, Cummings M et al. Characterisation of a subtype of colorectal cancer combining features of the suppressor and mild mutator pathways. J Clin Pathol 1999; 52:455-460.

14. Jass JR, Young J, Leggett BA. Evolution of colorectal cancer: Change of pace and change of direction. J Gastroenterol Hepatol 2002; 17:17-26.

15. Tomlinson I, Bodmer W. Selection, the mutation rate and cancer: Ensuring that the tail does not wag the dog. Nature Med 1999; 5:11-12.

16. Fenton RG, Hixon JA, Wright PW et al. Inhibition of Fas (CD95) expression and Fas-mediated apoptosis by oncogenic. Ras Cancer Res 1998; 58:3391-3400.

17. Jass JR, Iino H, Ruszkiewicz A et al. Neoplastic progression occurs through mutator pathways in hyperplastic polyposis of the colorectum. Gut 2000; 47:43-49.

18. Esteller M, Hamilton SR, Burger PC et al. Inactivation of the DNA repair gene O^6-methylguanine-DNA methyltransferase by promoter hypermethylation is a common event in primary human neoplasia. Cancer Res 1999; 59:793-797.

19. Kane MF, Loda M, Gaida GM et al. Methylation of the hMLH1 promoter correlates with lack of expression of hMLH1 in sporadic colon tumors and mismatch repair-defective human tumor cell lines. Cancer Res 1997; 57:808-811.

20. Esteller M, Toyota M, Sanchez-Cespedes M et al. Inactivation of the DNA repair gene 06-Methylguanine-DNA Methyltransferase by promoter hypermethylation is associated with G to A mutations in K-ras in colorectal tumorigenesis. Cancer Res 2000; 60:2368-2371.

21. Toyota M, Ahuja N, Ohe-Toyota M et al. CpG island methylator phenotype in colorectal cancer. Proc Natl Acad Sci USA 1999; 96:8681-8686.

22. Toyota M, Ohe-Toyota M, Ahuja N et al. Distinct genetic profiles in colorectal tumors with or without the CpG island methylator phenotype. Proc Natl Acad Sci USA 2000; 97:710-715.

23. Hawkins N, Norrie M, Cheong K et al. CpG island methylation in sporadic colorectal cancer and its relationship to microsatellite instability. Gastroenterology 2002; 122:1376-1387.

24. Robertson KD, Jones PA. The human ARF cell cycle regulatory gene promoter is a CpG island which can be silenced by DNA methylation and down-regulated by wild-type p53. Mol Cell Biol 1998; 18:6457-6473.

25. Young JP, Biden KG, Simms LA et al. HPP1: A transmembrane protein commonly methylated in colorectal polyps and cancers. Proc Natl Acad Sci USA 2001; 98:265-270.

26. Toyota M, Shen L, Ohe-Toyota M et al. Aberrant methylation of the Cyclooxygenase 2 CpG island in colorectal tumors. Cancer Res 2000; 60:4044-4048.

27. Vogelstein B, Fearon ER, Hamilton SR et al. Genetic alterations during colorectal-tumor development. N Engl J Med 1988; 319:525-532.

28. Whitehall VLJ, Wynter CVA, Walsh MD et al. Morphological and molecular heterogeneity within nonmicrosatellite instability-high colorectal cancer. Cancer Res 2002; 62:6011-6014.

29. Whitehall VLJ, Walsh MD, Young J et al. Methylation of 0-6-Methylguanine DNA Methyltransferase characterises a subset of colorectal cancer with low level DNA microsatellite instability. Cancer Res 2001; 61:827-830.

30. Esteller M, Risques RA, Toyota M et al. Promoter hypermethylation of the DNA repair gene O^6-methylguanine-DNA methyltransferase is associated with the presence of G:C to A:T transition mutations in p53 in human colorectal tumorigenesis. Cancer Res 2001; 61:4689-4692.

31. Fink D, Aebi S, Howell SB. The role of DNA mismatch repair in drug resistance. Clin Cancer Res 1998; 4:1-6.

32. Berardini M, Mazurek A, Fishel R. The effect of O^6-methylguanine DNA adducts on the adenosine nucleotide switch functions of hMSH2-hMSH6 and hMSH2-hMSH3. J Biol Chem 2000; 275:27851-27857.

33. Issa J-PJ, Ottaviano YL, Celano P et al. Methylation of the oestrogen receptor CpG island links ageing and neoplasia in human colon. Nat Genet 1994; 7:536-540.
34. Yatabe Y, Tavare S, Shibata D. Investigating stem cells in human colon by using methylation patterns. Proc Natl Acad Sci USA 2001; 98:10839-10844.
35. Chan AO-O, Issa J-PJ, Morris JS et al. Concordant CpG island methylation in hyperplastic polyposis. Am J Pathol 2002; 160:529-536.
36. Rashid A, Shen L, Morris JS et al. CpG island methylation in colorectal adenomas. Am J Pathol 2001; 159:1129-1135.
37. Jeevaratnam P, Cottier DS, Browett PJ et al. Familial giant hyperplastic polyposis predisposing to colorectal cancer: A new hereditary bowel cancer syndrome. J Pathol 1996; 179:20-25.
38. Jass JR, Cottier DS, Pokos V et al. Mixed epithelial polyps in association with hereditary nonpolyposis colorectal cancer providing an alternative pathway of cancer histogenesis. Pathology 1997; 29:28-33.
39. Rashid A, Houlihan S, Booker S et al. Phenotypic and molecular characteristics of hyperplastic polyposis. Gastroenterology 2000; 119:323-332.
40. Biemer-Hüttmann A-E, Walsh MD, McGuckin MA et al. Immunohistochemical staining patterns of MUC1, MUC2, MUC4, and MUC5AC mucins in hyperplastic polyps, serrated adenomas, and traditional adenomas of the colorectum. J Histochem Cytochem 1999; 47:1039-1047.
41. Biemer-Hüttmann A-E, Walsh MD, McGuckin MA et al. Mucin core protein expression in colorectal cancers with high levels of microsatellite instability indicates a novel pathway of morphogenesis. Clin Cancer Res 2000; 6:1909-1916.
42. Jass JR, Young J, Leggett BA. Hyperplastic polyps and DNA microsatellite unstable cancers of the colorectum. Histopathology 2000; 37:295-301.
43. Hawkins NJ, Ward RL. Sporadic colorectal cancers with microsatellite instability and their possible origin in hyperplastic polyps and serrated adenomas. J. Natl Cancer Inst 2001; 93:1307-1313.
44. Mäkinen MJ, George SMC, Jernvall P et al. Colorectal carcinoma associated with serrated adenoma - prevalence, histological features, and prognosis. J Pathol 2001; 193:286-294.
45. Park S-J, Rashid A, Lee J-H et al. Frequent CpG island methylation in serrated adenomas of the colorectum. Am J Pathol 2003; 162:815-822.
46. Pfeifer GP, Steigerwald SD, Hansen RS et al. Polymerase chain reaction-aided genomic sequencing of an X chromosome-linked CpG island: methylation patterns suggest clonal inheritance, CpG site autonomy, and an explanation of activity state stability. Proc Natl Acad Sci USA 1990; 87:8252-8256.
47. Nasr AF, Nutini M, Palombo B et al. Mutations of TP53 induce loss of DNA methylation and amplification of the TROP1 gene. Oncogene 2003; 22:1668-1677.
48. Shen L, Yutaka K, Hamilton SR et al. p14 methylation in human colon cancer is associated with microsatellite instability and wild-type p53. Gastroenterology 2003; 124:626-633.

CpG Island Hypermethylation of Tumor Suppressor Genes in Human Cancer:
Concepts, Methodologies and Uses

Michel Herranz and Manel Esteller

Abstract

Aberrations in the DNA methylation patterns are nowadays recognized as a hallmark of human cancer. One of the most characteristic changes is the hypermethylation of CpG islands of tumor suppressor genes associated with their transcriptional silencing. The target genes are distributed in all cellular pathways (apoptosis, DNA repair, cell cycle, cell adherence, etc.). They are "classical" tumor suppressor genes with associated familial cancers (BRCA1, hMLH1, p16^{INK4a}, VHL, etc.) and putative new tumor suppressor genes which loss may contribute to the transformed phenotype (MGMT, p14^{ARF}, GSTP1, RARB2, etc.). A tumor-type specific profile of CpG island hypermethylation exist in human cancer that allows the use of these aberrantly hypermethylated loci as biomarkers of the malignant disease. The development of new technologies for the careful study of the DNA methylation patterns, and their genetic partners in accomplishing gene silencing, may also provide us with new drugs for the epigenetic treatment of human tumors.

Concepts

Gene Inactivation by Promoter Hypermethylation in Human Cancer: Overview

The inheritance of information based on gene expression levels is known as epigenetics, as opposed to genetics, which refers to information transmitted on the basis of gene sequence. The main epigenetic modification in mammals, and in particular in humans, is the methylation in the cytosine nucleotide residue (Fig. 1). We can consider that about 3-4% of all cytosines are methylated in normal human DNA. Gross alterations such as the aneuploidy state, deletions (loss of heterozigosity) or gains (gene amplification) of genomic material, and small changes (point mutations, small insertions or deletions) in multiple genes are present through the genome of a neoplastic cell. But the malignant cell has also aquired a different epigenotype. In a healthy cell, the DNA methylation patterns are conserved through cell divisions allowing the expression of the particular set of cellular genes necessary for that cell type and blocking the expression of exogenous inserted sequences.

The presence of the CpG dinucleotide in the human genome is suppressed by a statistical criterion.[1] The evolutionary proposed reason for this lack of CpGs in our genome is to avoid spontaneous deamination in the germline. However, approximately half of the human gene promoter regions contain CpG-rich regions, known as "CpG islands".[1] Although the majority of CpG islands are associated with "house-keeping" genes, some of them are located in

DNA Methylation and Cancer Therapy, edited by Moshe Szyf. ©2005 Eurekah.com and Kluwer Academic/Plenum Publishers.

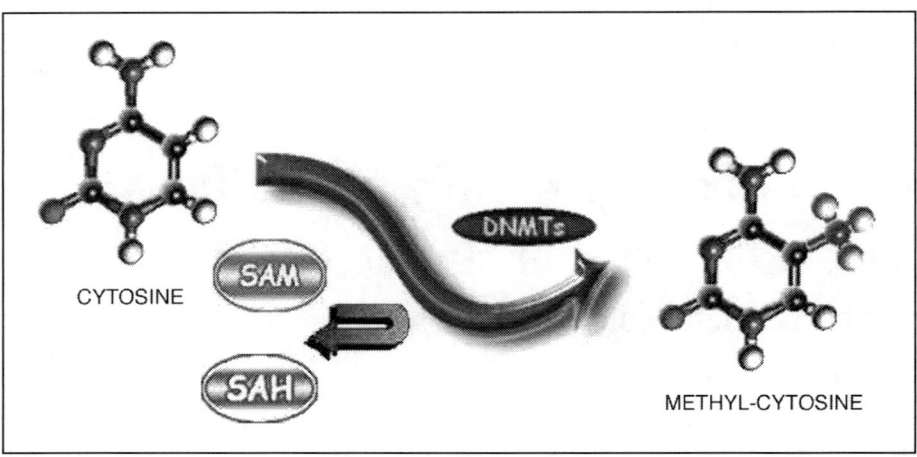

Figure 1. Cytosine methylation.

"tissue-specific genes". It should also be noted that although the most significant proportion of CpG islands is located in the 5′-unstranslated regions and the first exon of the genes, certain CpG islands could occasionally be found within the body of the gene, or even in the 3′-region. CpG islands in these atypical locations are more prone to methylation.[2] Hypermethylation of CpG islands located in the promoter regions of tumor suppressor genes is now definitely established as an important mechanism for gene inactivation (Fig. 2).

CpG island hypermethylation has been described in every tumor type. Many cellular pathways are inactivated by this type of epigenetic lesion: DNA repair (hMLH1, MGMT), cell cycle (p16^{INK4a}, p15^{INK4b}, p14^{ARF}), cell adherence (CDH1, CDH13), apoptosis (DAPK), detoxification (GSTP1), hormonal response (RARB2, ER). etc.[3] However, we still do not know the mechanisms of aberrant methylation and why certain genes are selected over others. Hypermethylation is not an isolated layer of epigenetic control, but is linked to the other pieces of the puzzle such as methyl-binding proteins, DNA methyltransferases, histone deacetylases and histone methyltransferases, but our understanding of the degree of specificity of these epigenetic layers in the silencing of specific tumor suppressor genes remains incomplete. Careful functional and genetic studies are necessary to determine which hypermethylation events are truthfully relevant for human tumorigenesis. The development of CpG island hypermethylation profiles for every form of human tumors has yielded valuable pilot clinical data in monitoring and treating cancer patients based on our knowledge of DNA methylation.

A Brief History of CpG Island Hypermethylation

The first discovery of methylation in a CpG island of a tumor suppressor gene in a human cancer was that of the Retinoblastoma (Rb) gene in 1989,[4] only a few years after the first oncogene mutation (H-ras) was discovered in a human primary tumor. In 1994 the idea that CpG island promoter hypermethylation could be a mechanism to inactivate genes in cancer was reborn as a result of the discovery that the Von Hippel-Lindau (VHL) gene also undergoes methylation-associated inactivation.[5] However, the true origin of the current period of research in cancer epigenetic silencing was perhaps the discovery that CpG island hypermethylation was a common mechanism of inactivation of the tumor suppressor gene p16^{INK4a} in human cancer.[6,7,8] The introduction of powerful and accessible techniques, such as sodium bisulfite modification[9] and Methylation-Specific PCR,[10] provided keys to start the game. From that time, the list of candidate genes with putative aberrant methylation of their CpG islands has grown exponentially.[3]

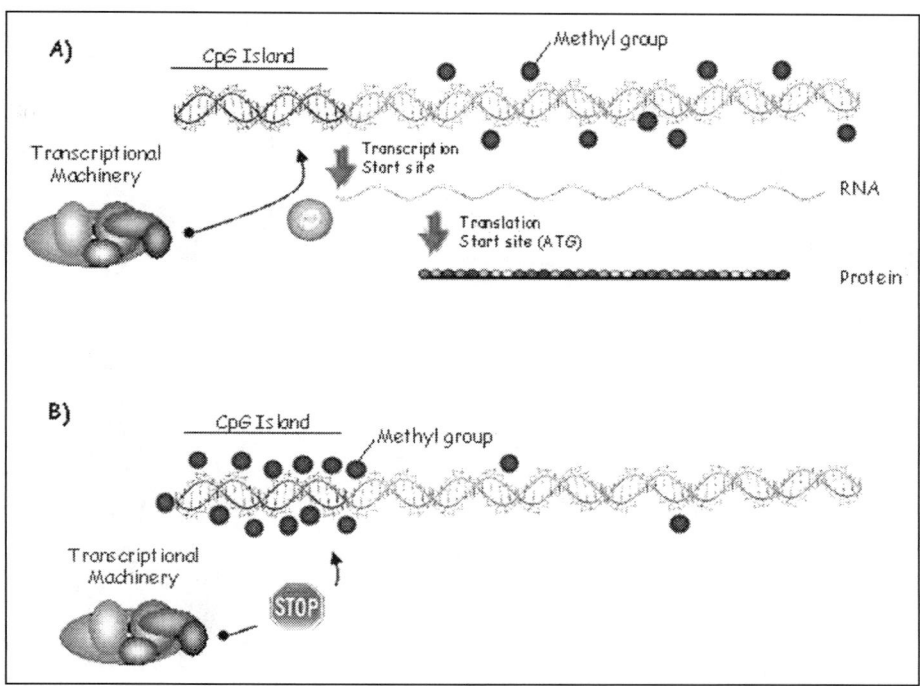

Figure 2. Methylation status in tumor suppressor genes. A) Normal cell; B) Tumoral cell.

Gene Hypermethylation Profile of Human Cancer

We know that cancer is a disease of multiple pathways and genetic lesions and all of them are necessary to develop a fully established tumor The existence of genetic alterations affecting genes involved in cellular proliferation and death, such as *p53* and *K-ras*, is one of the most common features of tumor cells. Recently, the inactivation of tumor suppressor genes by pro-moter hypermethylation has been added to this scenario. The presence of CpG island pro-moter hypermethylation affects genes involved in cell cycle (p16^{INK4a}, p15^{INK4b}, Rb, p14^{ARF}), DNA repair (BRCA1, hMLH1, MGMT), cell-adherence (CDH1, CDH13), apoptosis (DAPK, TMS1), carcinogen-metabolism (GSTP1), hormonal response (RARB2, ER), etc. Figure 3 shows the most relevant hypermethylated genes in human cancer reported so far and their chromosomal localization. In most of cases, methylation involves loss of expression, absence of a coding mutation and restoration of transcription by the use of demethylating agents.

A profile of CpG island hypermethylation exists in accordance with the tumor type.[3] For example BRCA1 hypermethylation is characteristic of breast and ovarian tumors,[11] but it does not occur in other tumor types.[3] hMLH1 methylation-mediated silencing occurs in colorectal, gastric and endometrial neoplasms, but in almost none of the other solid tumors.[3] This cau-tiously respected pattern of epigenetic inactivation is not only a property of the sporadic tu-mors, but also neoplasms appearing in inherited cancer syndromes display CpG island hypermethylation specific to the tumor type.[11] We can call this the "Methylotype", for the pattern of analogy with the genetic term "Genotype". There are also tumor types that have more methylation of the known CpG islands than others: for example the most permethylated tumor types are originated from the gastrointestinal tract (esophagus, stomach, colon), while significantly less hypermethylation has been reported in other types such as ovarian tumors.[3] There is a clear gradient of the distribution of tumors with different degrees of CpG island methylation: from tumors with few hypermethylated CpG islands to neoplasms with a very

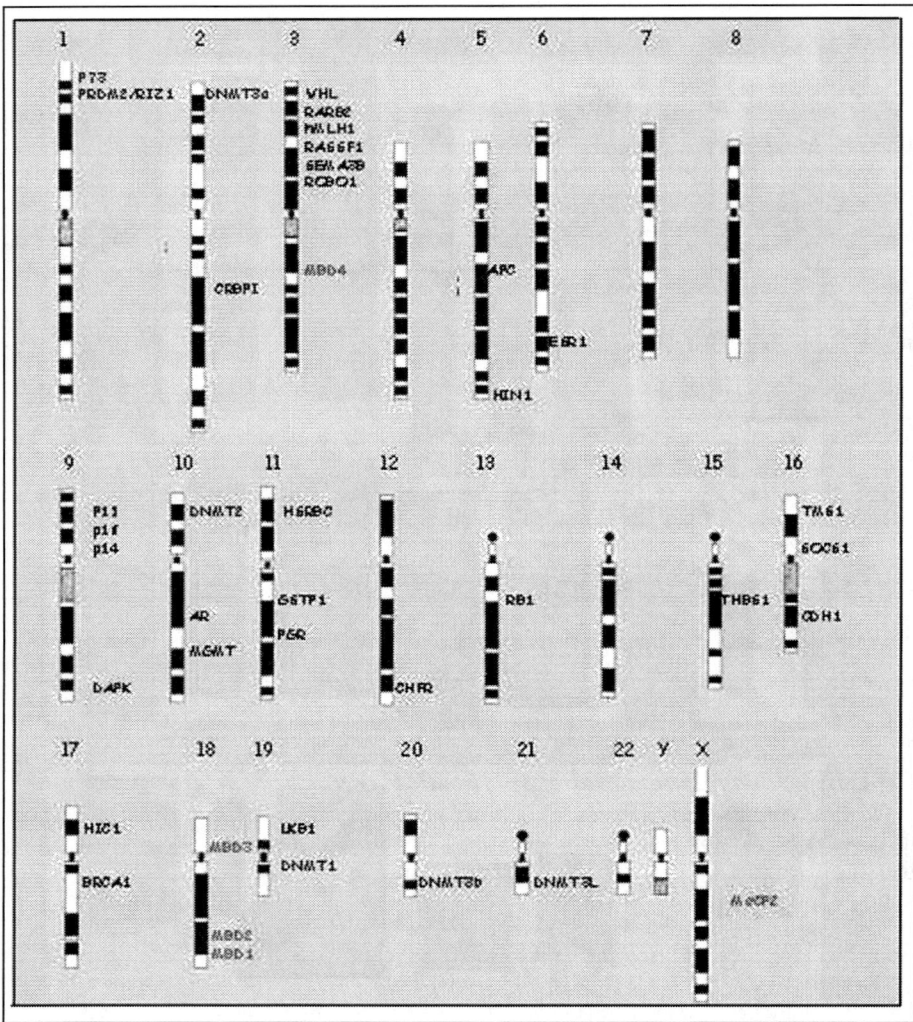

Figure 3. Hypermethylated genes in human cancers (black), MBDs (red) and DNMTs (blue). A color version of this figure is available online at http://www.Eurekah.com/.

high number of hypermethylated islands. This is the rate expected for events occurring randomly and being selected because they confer advantage to the cancer cell, excluding the existence of any significant methylator phenotype.

Epigenetics Hits in a Cancer Cell

The epigenetic balance of the cell suffers a dramatic alteration in cancer: transcriptional silencing of tumor suppressor genes by CpG island hypermethylation and histone deacetylation, global genomic hypomethylation and genetic defects in chromatin-related genes.

Not every gene is methylated in every tumor type. There is a delicate profile of hypermethylation that occurs in human tumors, but CpG island methylation affects all cellular pathways. The growing list of genes inactivated by promoter region hypermethylation provides an opportunity to examine the patterns of inactivation of such genes among different

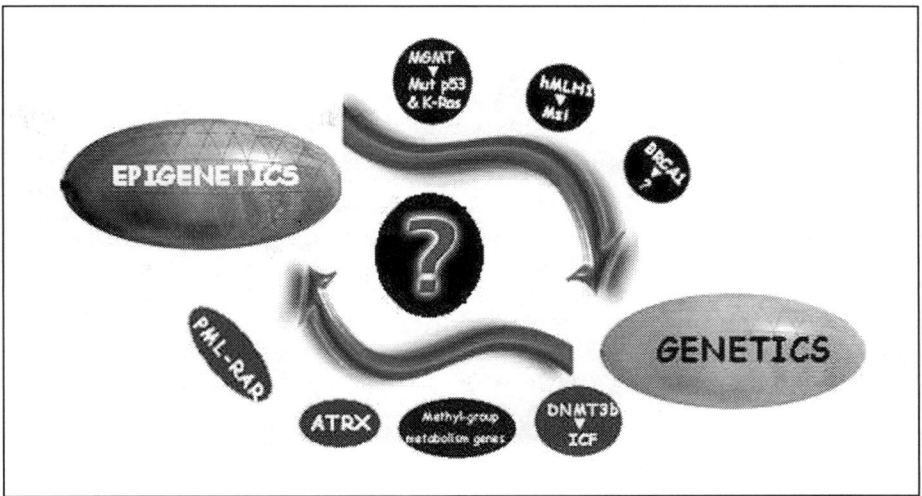

Figure 4. Epigenetics and genetics: friend or foe.

tumors. Usually one or more genes are hypermethylated in every tumor type. However, the profile of promoter hypermethylation for the genes differs for each cancer type providing a tumor-type and gene specific profile. In each case and tumor type, this epigenetic lesion occurs in the absence of a genetic lesion

If we look at our gene hypermethylation profile from the tumor type point of view, the picture is particularly interesting. Gastrointestinal tumors (colon and gastric) share a set of genes undergoing hypermethylation characterized by *p16INK4a, p14ARF, MGMT, APC* and *hMLH1*, while other aerodigestive tumor types, such as lung and head and neck, have a different pattern of hypermethylated genes including *DAPK, MGMT, p16INK4a* but not *hMLH1* or *p14ARF*. Similarly, breast and ovarian cancers be inclined to methylate certain genes including *BRCA1, GSTP1* and *p16INK4a*. This gene hypermethylation profile of human cancer is consistent with the data of particular "methylotypes" proposed for single tumor types.

In a cancer cell there is a clear distortion in the expression profiles and the presence of a dramatic change in the methylation patterns is one of the guilty parts. First, there is a disregulation in the methylating enzymes. Second, there is a global hypomethylation when compare to a normal cell; this is achieved due to a generalized demethylation in the CpGs dotted in the body of the genes and may be involved in causing global genomic fragility and reexpresion of inserted viral sequences. And third and finally, there are local and discrete regions normally devoid of methylation that suffer an intense hypermethylation.

Hypermethylation and Mutation: A Love and Hate Relation

When the first genetic mutation was discovered in a human cancer, the idea that a large number of genes would be found mutated in all tumors was everywhere. However, twenty years later only two genes, the oncogene K-ras and the tumor suppressor p53 have found to be consistently mutated in a high proportion of tumors. This concept of how to transform expectations to reality could also be applied to CpG island hypermethylation.

One of the most critical steps in giving CpG island methylation its accurate value is the fact that it should occur in the absence of gene mutations. Both events (genetic and epigenetic) abolish normal gene function (Fig. 4). The selective advantage of promoter hypermethylation in this context is provided by multiple examples. The cell cycle inhibitor p16^{INK4a} in one allele of the HCT-116 colorectal cancer cell line has a genetic mutation while the other is wild-type: p16INK4a hypermethylation occurs only on the wild-type allele, while the mutated allele is kept unmethylated.[12] Examples like that put CpG island hypermethylation evenly balanced with gene mutation for accomplishing selective gene inactivation.

A subset of human tumors display a bizarre genetic phenotype defined by the microsatellite instability (MSI) phenomena. MSI+ tumors are defined because they show aberrant insertions or deletions of mono- or dinucleotides repeats when the tumors are compared with their normal counterparts. The tumor types mainly involved in the disease are colorectal, endometrial and gastric carcinoma. In these HNPCC families the defect is attributed to germline mutations in the DNA mismatch-repair (MMR) genes, mainly *hMLH1* and hMSH2, while other components of the MMR pathway such as hMSH3, hMSH6, hPMS1 and hPMS2 seem to play a minor role in the disease. MSI positive tumors were also observed in spontaneous cases, however, MMR mutations were found in less than 10% of sporadic MSI+ tumors. The explanation of this data is that the presence of MSI is due to transcriptional inactivation of *hMLH1* by promoter hypermethylation.[13,14]

Other exciting example of how promoter hypermethylation affects the genome of the cancer cell is provided by the DNA repair gene O^6-methylguanine DNA methyltransferase (*MGMT*). The DNA repair gene *MGMT* is transcriptionally silenced by promoter hypermethylation in primary human tumors.[15] These tumors then accumulate a considerable number of G to A transition mutations affecting key genes such as *p53* and *K-ras*, in a similar way that loss of the *hMLH1* mismatch repair gene by methylation targets other genes.

Two more genes related to potential DNA lesions undergo inactivation by promoter hypermethylation, the glutathione S-transferase P1 (*GSTP1*) and the breast cancer familial gene *BRCA1*. Changes of *GSTP1* expression may prevent DNA damage, but its cause was imprecise until aberrant methylation of the *GSTP1* CpG island in prostate, breast and kidney carcinoma was reported.[16,17] The case of the tumor suppressor gene *BRCA1* gene, responsible for almost half of the cases of inherited breast cancer and ovarian cancer, is also relevant. *BRCA1* promoter hypermethylation leading to *BRCA1* loss of function is present in breast and ovarian primary tumors and cell lines.[18,11] What are the links between changes in *BRCA1* protein levels and DNA damage? Two hypothesis were defended: BRCA1 cooperates with the RNA helicase A and the histone deacetylase complex in the transcriptional regulation of DNA integrity maintenance genes and BRCA1 plays an important role in DNA repair forming supercomplexes with proteins like *ATM*, *RAD51* and *hMSH2*.

Now, let us check the other side of the coin. We could mention different examples where a link between a genetic modification and epigenetic fluctuations is established: ATRX, ICF syndrome, PML-RAR fusion protein and Methyl-group Metabolism genes. Mutations in *ATRX* give rise to characteristic developmental abnormalities including severe mental retardation, facial dysmorphism, urogenital abnormalities and alfa-thalasaemia. Mutations in *ATRX* give rise to changes in the pattern of methylation of several highly repeated sequences. ATRX is localized to pericentromeric heterochromatin and might exert chromatin-mediated effects in the nucleus and act as a transcriptional regulator through an effect on chromatin. A human genetic disorder (ICF syndrome) has been shown to be caused by mutations in the DNA methyltransferase 3B (DNMT3B) gene. A second human disorder (Rett syndrome) has been found to result from mutations in the MECP2 gene, which encodes a protein that binds to methylated DNA. Global genome demethylation caused by targeted mutations in the DNA methyltransferase-1 (Dnmt1) gene has shown that cytosine methylation plays essential roles in X-inactivation, genomic imprinting and genome stabilization. The leukemia-promoting PML-RAR fusion protein induces gene hypermethylation and silencing by recruiting DNA methyltransferases to target promoters and that hypermethylation contributes to its leukemogenic potential.[19,20] Retinoic acid treatment induces promoter demethylation, gene reexpression, and reversion of the transformed phenotype. Furthermore, germline variants in the methyl-group metabolism genes involved in the regeneration of the universal methyl-donor SAM (S-adenosyl-methionine) are also associated with different DNA methylation patterns in the cancer cell.[21] These results establish a mechanistic link between genetic and epigenetic changes during transformation and suggest that hypermethylation contributes to the early steps of carcinogenesis.

Why Do CpG Islands Become Hypermethylated?

The CpG island are usually unmethylated in all normal tissues and span the 5' end of genes. If transcription factors are available and the island remains in an unmethylated state with open chromatin configuration-associated with hyperacetylated histones, transcription will occur. Certain CpG islands are normally methylated: imprinted genes and genes of one X-chromosome in women. DNA methylation has also a role in repressing parasitic DNA sequences.

In the transformed or malignant cell certain CpG islands of tumor suppressor genes (real or putative) will become hypermethylated.[22,23] This is probably a progressive process, in contrast to the sudden appearance of a gene mutation. Perhaps several "steps" of disregulated methylation will be necessary to produce the dense hypermethylation necessary for transcriptionally silencing that particular promoter anchored in the CpG island. Two obvious theories can be postulated for this aberrant de *novo* methylation. First, the cancer methylation spreads from normal methylation-centers surrounding the methylation-free CpG island, for example from Alu regions.[24] Second, a basal status of methylation exists and certain single CpG dinucleotides in the island became methylated and subsequently this attracted more methylation. This process has a positive cooperative effect until hypermethylation is achieved. A model that combines prior gene silencing with "seeds" of methylation has been proposed for the GSTP1 in prostate cancer.[25] Both hypotheses are plausible and compatible. However, there is not definitive support for either.

Another question is why certain CpG islands become hypermethylated in cancer. It has been known for a long time that an overall increase in the enzymatic DNA methyltransferase activity occurs in tumors versus normal tissues (reviewed in ref. 23). This finding has been supported as a result of the molecular characterization of the genes encoding several DNA methyltranferases (DNMT1, DNMT3a, DNMT3b, DNMT3L and DNMT2), which has shown that the mRNA transcripts of DNMT1 (the classical methylation maintenance enzyme) and DNMT3b (the de novo methylation enzyme) are increased in several solid and hematological tumors.[26]

However, the most critical question is still unclear: why do certain CpG islands become hypermethylated while others remain unmethylated in a cancer cell? Certain CpG islands become hypermethylated rather than others because they confer a selective advantage for the survival of that particular cancer cell. For example BRCA1 undergoes promoter hypermethylation only in breast and ovarian tumors[3,18] because only in these tumors types does the lack of this transcript have important cellular consequences. This Darwinian concept is supported by the classical genetic studies of familial tumors: carriers of BRCA1 germline mutations develop predominantly breast and ovarian tumors and carriers of hMLH1 germline mutations mostly develop colorectal, gastric and endometrial tumors. There is a perfect match between the genetic and epigenetic worlds.

DNA Hypomethylation in the Context of CpG Island Hypermethylation

CpG islands become hypermethylated but the genome of the cancer cell undergoes a dramatic global hypomethylation: 20-60% less genomic 5mC resulting from the hypomethylation of the "body" of genes and repetitive DNA sequences. Global DNA hypomethylation may contribute to carcinogenesis causing chromosomal instability, reactivation of transposable elements and loss of imprinting.

The presence of alterations in the profile of DNA methylation in cancer was initially thought to be exclusively a global hypomethylation of the genome[27] (reviewed in ref. 28) that would possibly lead to massive overexpression of oncogenes whose CpG islands were normally hypermethylated. Nowadays, this is considered to be an unlikely or, at best, incomplete scenario. The idea that the genome of the cancer cell undergoes a reduction of its 5-methylcytosine content in comparison to the normal tissue from which it originated is essentially correct, and is also corroborated in a large survey of sporadic and inherited breast and colon tumors.[11]

The popularity of the concept of demethylation of oncogenes leading to their activation is in clear decadency. The first experiments supporting this hypothesis effectively demonstrated DNA hypomethylation, but as only certain methyl-sensitive restriction sites were used a significant amount of this "demethylation" was present in the "body" of the genes (internal exons and introns) rather than in the canonical CpG island. In fact, the vast majority of CpG islands are completely unmethylated in normal tissues (reviewed in ref. 1), with the logical exceptions of imprinted genes and X-chromosome genes in females.

How Does CpG Island Hypermethylation Lead to Transcriptional Gene Silencing?

Throughout the last twenty years research on cell signaling has carefully characterized the components involved in the transmission of signals. The same molecular dissection should now be applied to elucidate how CpG island hypermethylation leads to transcriptional gene silencing. Perhaps, each step of this chain is specific to each gene or group of genes. One clue to unscrambling the enigma was the discovery that DNA methylation results in the formation of nuclease-resistant chromatin and the subsequent repression of gene activity.[29]

Nowadays the most widely accepted explanation of events starts with the binding of certain methyl-binding proteins (MBDs) to the methylated CpG dinucleotides of the densely hypermethylated CpG island. The search for proteins with different binding properties for methylated and unmethylated DNA initially yielded two activities which were named MeCP1 and MeCP2, the first being a complex of proteins and the second a single polipeptide.[30] Further database searches revealed novel MBD-containing proteins, MBD1, MBD2, MBD3 and MBD4. A new question then arises: are there MBDs specific for subgroups of hypermethylated CpG island of tumor suppressor genes in cancer? Different methylation densities may attract different MBDs for example. Two recent reports have addressed this problem in one of the most interesting epigenetics spots in the human genome: the p15^{INK4b}/ p16^{INK4a}/p14^{ARF} locus in the 9p21 chromosomal region. These studies demonstrate that MeCP2[2] and MBD2[31] bind to the hypermethylated CpG islands of p14^{ARF} and p16^{INK4a}. If we improve the instrumental tools, it will signal that it is time to start mapping all the CpG island promoters of tumor suppressor genes for their MBD binding patterns.

Another critical result was the association of MeCP2 and histone deacetylase (HDAC) activity in repressing transcription.[32,33] The remaining MBDs have also proved to be members of similar HDAC complexes. Thus, the current model propose that MBDs recruit HDAC activities to methylated promoters which, in turn deacetylate histones, leading to a chromatin-repressed state of gene transcription. Considering the CpG islands that undergo hypermethylation in the cancer cell, the association of hypoacetylated histones H3 and H4 with a hypermethylated CpG island has now been demonstrated for the p16^{INK4a}, p14^{ARF}, BRCA1, COX-2 and TMS1 genes. Thus, CpG island hypermethylation and histone hypoacetylation seems to be firmly associated.

Methodologies

Study of CpG Island Methylation in Cancer Cells

The epigenetic alteration has become a centre of scientific attraction, especially due to its relation to gene silencing in disease. The first aim of the researcher should be the study of functional methylation, which is normally assumed to be dense CpG island hypermethylation associated with transcriptional silencing. The presence of 5-methylcytosine (mC) in the promoter of specific genes alters the binding of transcriptional factors and other proteins to DNA. It also attracts methyl-DNA-binding proteins and histone deacetylases that close the chromatin around the gene transcription start site. Both mechanisms block transcription and cause gene silencing.[34] Thus, methylation of C residues in genomic DNA plays a key role in the regulation of gene expression.[35]

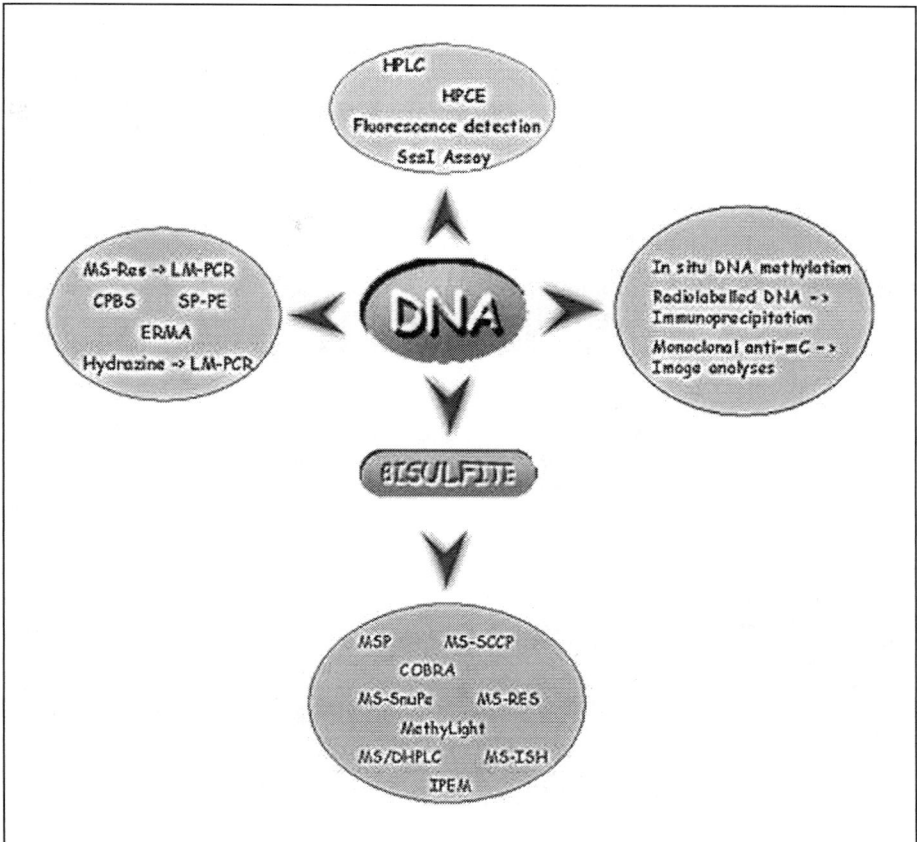

Figure 5. DNA methylation methods.

Quantitative and Qualitative Information on Genomic DNA Methylation

There is currently a wide range of methods designed to yield quantitative and qualitative information on genomic DNA methylation (Fig. 5). Particularly, optimisation of the methods based on bisulfite modification of DNA permits the analysis of limited CpGs in restriction enzyme sites (COBRA, SnuPE), the overall characterisation based on differential methylation states (MSP, MS-SSCP) and, allows very specific patterns of methylation to be revealed (bisulfite DNA sequencing).

To gain a deeper understanding of the DNA methylation patterns of the CpG islands the use of bisulfite-modified DNA is nowadays necessary.[9] Bisulfite converts unmethylated cytosine to uracil, while methylated cytosine does not react. This reaction constitutes the basis for differentiating between methylated and unmethylated DNA.

The Bisulfite Based Methods

Sequencing

In general, after denaturation and bisulfite modification, double-stranded DNA is obtained by primer extension and the fragment of interest is amplified by PCR.[9] Standard DNA sequencing of the PCR products may then detect methylcytosine. This approach has been helpful in the study of the DNA methylation of genes associated with cancer, such as APC[35] and Rb.[36]

Methylation-Specific PCR (MSP)

MSP is the most widely used technique for studying the methylation of CpG islands.[3,10] The differences between methylated and unmethylated alleles that arise from sodium bisulfite treatment are the basis of MSP. Primer design is a critical and complex component of the procedure. The great sensitivity of the method allows the methylation status of small samples of DNA, even those from paraffin-embedded or microdissected tissues. MSP has been widely proposed as a rapid and cost-effective clinical tool of use in the study CpG island hypermethylation in human cancer. For example, MSP has been successfully used to detect tumoral DNA in the serum of cancer patients.[15]

Other Bisulfite-Based Techniques

1. Combined bisulfite restriction analyses (COBRA)[37] constitute a highly specific approach releasing on the creation or modification of a target for restriction endonuclease after bisulfite treatment.
2. Methylation-sensitive Single nucleotide Primer Extension (Ms-SnuPE) employs bisulfite/PCR combined with single-nucleotide primer extension to analyze DNA methylation status quantitatively in a particular DNA region without using restriction enzymes.[38]
3. Methylation-Sensitive Single-Strand Conformation Analysis (MS-SSCP). Bisulfite modification of DNA generates sequence disparities between methylated and unmethylated alleles, which can be resolved by SSCP.[39]

Quantification of Global Methylation

Levels of methylcytosine occurrence in the genomic DNA can be measured by high-performance separation techniques or by enzymatic/chemical means. When separation devices are available, high-performance capillary electrophoresis (HPCE) may be the best choice since it is faster, cheaper and more sensitive than HPLC.[40,41] By means of labelled anti-methylcytosine antibodies, DNA methylation can be monitored in metaphase chromosomes, hetero/euchromatin and, most importantly, on a cell-by-cell basis within the same sample. The latter alternative, which generally yields qualitative results, is of great interest in cancer research as it can reveal methylation differences between normal and tumour tissues in the same sample.

HPLC-Based Methods

Relative mC contents of genomic DNA can be analysed by chemical hydrolysis to obtain the total base composition of the genome and subsequent fractionation and quantification of hydrolysis products using HPLC technologies. The degree of DNA methylation of several samples has been quantified by this method, but at least 2.5 μg DNA are generally required to quantify 5-methylcytosine with a low standard deviation for replicate samples. Sensitivity of the system can be increased with mass spectrometry detection, which has a detection limit 10^6 times the limit of absorption spectroscopy detectors.

HPCE-Based Methods

The development of high performance capillary electrophoretic (HPCE) techniques has given rise to an approach to research that has several advantages over other current methodologies used to quantify the extent of DNA methylation.[41,42] This method is faster than HPLC (taking less than 10 min per sample) and is also reasonably inexpensive since it does not require continuous running buffers and displays a great potential for fractionation (theoretically up to 10^6 plates). Approximately 1 methylcytosine in 200 cytosine residues can be detected by this method using 1 μg genomic DNA. To increase sensitivity, laser-induced fluorescence (LIF) and mass spectrometry detectors should be used.

Analyses of Genome-Wide Methylation by Chemical or Enzymatic Means

As previously stated, quantifying the degree of DNA methylation by HPLC or HPCE requires access to sophisticated equipment that is not always available. The radioactive labelling of CpG sites using the methyl-acceptor assay (reviewed in ref. 43) has been developed to address this problem but, among the technique's other drawbacks, it can only monitor CpG methylation changes, and so CpNpG methylation cannot be detected. This method uses bacterial *SssI* DNA methyltransferase to transfer tritium-labelled methyl groups from S-adenosylmethionine (SAM) (S-adenosyl-L-[*methyl*-³H]methionine) to unmethylated cytosines in CpG targets. The data obtained from a scintillation counter are used to calculate the number of methyl-groups incorporated in the DNA.

In Situ Hybridisation Methods for Studying Total Cytosine Methylation

Global DNA methylation can also be quantified by methylcytosine-specific antibodies.[44] An outstanding advantage of this approach is that it may be carried out on a cell-by-cell basis rather than in a heterogeneous population. Apart from classical immunoassay detection, approaches that involve quantifying the retention of radiolabelled DNA by polyclonal antibodies on nitrocellulose filters, immunoprecipitation, gel filtration and visualisation under electron microscopy, cytosine methylation can also be detected in metaphase chromosomes and in chromatin using monoclonal antibodies combined with fluorescence staining. An alternative to fluorescence detection is to connect a coloured enzyme-dependent reaction.

Mapping Methylcytosines by Nonbisulfite Methods

The most widely used methods for studying DNA methylation patterns of specific regions of DNA with no base modifications are based on the use of methylation-sensitive and insensitive restriction endonucleases (MS-REs).[45] One of the restriction enzymes of the isoschizomer pair is able to cut the DNA only when its target is unmethylated whereas the other is not sensitive to methylated cytosines. The most common isoschizomers used are the Hpa II/Msp I pair. Even though these pairs of enzymes can cleave hemimethylated DNA, they do not distinguish between cytosines methylated at different positions in the pyrimidinic ring.[46] However, there are several restriction enzymes that recognise the localisation of the methyl group.[47]

Nonbisulfite methods for the quantification of DNA methylation patterns are simple, rapid and can be used for any known-sequence genomic DNA region. These methods are extremely specific but their limitation to specific restriction sites reduces their value.

Finding New Hypermethylated "Hot-Spots"

Classical DNA methylation research concentrates on investigating the methylation status of cytosines occurring in known (or partially known) DNA sequences. However, alternative ways of investigating genome-wide methylation by searching for unidentified spots have been developed.

The Restriction Landmark Genomic Scanning (RLGS)[48] technique is one of the earliest ways reported for genome-wide methylation-scanning.[49] DNA is radioactively labelled at methylation-specific cleavage sites and then size-fractionated in one dimension. The digestion products are then digested with any restriction endonuclease that is specific for high-frequency targets. Fragments are then separated in the second dimension, yielding a number of scattered methylation-related "hot-spots".

Gonzalgo et al[38] described other suitable tool for screening the genome for regions displaying altered patterns of DNA methylation. The method, termed methylation-sensitive arbitrary primed PCR (AP-PCR), is a simple DNA fingerprinting technique that relies on arbitrarily primed PCR amplification followed by digestion with restriction isoschizomers. Another approach is "CpG island amplification" (MCA).[50] DNA is digested with restriction isoschizomers and restriction products are PCR-amplified after end-adaptor ligation. Even though methylated CpG islands are preferably amplified, cloning of truly CpG-rich DNA regions is frequently a laborious task. Another original approach to isolated methylated CpG-rich regions has recently been described.[51] This method employs affinity chromatography of a fragment of the methyl-CpG binding domain of MeCP2 to purify methylated CpG-rich fragments from mixtures obtained by digestion with methylation-specific restriction endonucleases. Chosen fragments are then cloned into a lamda Zap II vector and fragments that are mostly rich in CpG dinucleotides are isolated by segregation of partially melted molecules (SMP) in polyacrylamide gels containing a linear gradient of chemical denaturant.

Undoubtedly, one of the most effective means of genome-wide searching for CpG islands is the use of the novel CpG island arrays technology. Huang et al[52] proposed an array-based method, termed differential methylation hybridisation (DMH), which allows the simultaneous determination of the methylation rate of >276 CpG island loci. A modification of this method for the study of DNA methylation in cancer is the methylation-specific oligonucleotide (MSO) microarray.[53] After bisulfite treatment and PCR amplification, products are array hybridized. Methylation- specific microarray is designed to be able to detect methylation at specific nucleotide positions. Quantitative differences can be obtained by fluorescence detection.

Overall, DNA methylation can be studied using a great variety of experimental techniques, involving a multidisciplinary perspective on DNA methylation status. One of the approaches is the quantification of the overall degree of DNA methylation. This can be accomplished by high-performance separation techniques, by enzymatic/chemical means and even by in situ hybridisation using antibodies anti-methylcytosine. To analyse the DNA methylation status of a particular DNA sequence in depth, methylation-sensitive restriction endonucleases are commonly used. Beyond this, once bisulfite modification of the DNA has been accomplished, it is possible to obtain quantitative or semi-quantitative data regarding allele-specific methylation. Finally, a number of innovative techniques have been developed to investigate new methylation hot-spots within the whole genome of whose sequences we have no prior knowledge. Future steps towards automation and multi-assay arrays will allow large numbers of samples to be checked simultaneously.

Translational Studies of CpG Island Hypermethylation: From the Bench to the Bedside

Great expectations have been raised by the large amount of genetic information regarding cancer biology that has been gathered in the past two decades. CpG island hypermethylation of tumor suppressor genes may be a very valuable tool. One obvious advantage over genetic markers is that the detection of hypermethylation is a "positive" signal that can be accomplished in the context of a group of normal cells, while certain genetic changes such as homozygous deletions are not going to be detected in a background of normal DNA. Furthermore, while mutations occur at multiple sites and can be of very different types, promoter hypermethylation occurs within the same region of a given gene in each form of cancer, thus we do not need to test the methylation status first to assay the marker in serum or a distal site. Three major clinical areas can benefit from hypermethylation-based markers: detection, tumor behavior and treatment.

    a. Detection of cancer cells using CpG island hypermethylation as a marker. If we want to use these epigenetic markers, we will need to use quick, easy, nonradioactive and sensitive ways to detected hypermethylation in CpG islands of tumor suppressor genes, such as methylation-specific PCR technique. The detection of DNA hypermethylation in biological fluids of cancer patients (and even patients at risk of cancer) should lead to create consortiums of different institutions to develop comprehensive studies to validate the use of these markers in the clinical environment. We opened a new avenue of research in 1999 with the demonstration that it was possible to detect the presense of hypermethylated CpG islands of tumor suppressor genes in the serum DNA of cancer patients.[54]

b. CpG island hypermethylation as a marker for tumor behavior. There are two components: prognostic and predictive factors. Prognostic factors will give us information about the virulence of the tumors. For example, Death Associated Protein Kinase (DAPK) and p16[INK4a] hypermethylation has been linked to aggressive tumors in lung and colorectal cancer patients.[55,56] The second component is the group of factors that predict response to therapy. For example, the response to alkylating agents (BCNU and cyclophosphamide) is enhanced in those human primary tumors (gliomas and lymphomas, respectively) where the DNA repair gene MGMT is hypermethylated.[57,58]

c. CpG island hypermethylation as target for therapy. We have been able to reactivate hypermethylated genes in vitro. One obstacle to the transfer of this technique to human primary cancers is the lack of specificity of the drugs used. Demethylating agents such as 5-azacytidine or 5-aza-2-deoxycytidine (Decitabine)[23] inhibits the DNMTs and cause global hypomethylation, and we cannot reactivate exclusively the particular gene we would wish to. If we consider that only tumor suppressor genes are hypermethylated this would not be a great problem. However, we do not know if we have disrupted some essential methylation at certain sites, and global hypomethylation may be associated with even greater chromosomal instability. Another disadvantage is the toxicity to normal cells. However, these compounds and their derivatives have been used in the clinic with some therapeutic benefit, especially in hematopoietic malignancies.[59,60] The discovery that lower doses of 5-azacytidine associated with inhibitors of HDACs may also reactivate tumor suppressor genes was hopeful. Nevertheless, we are still left with the obstacle of nonspecificity.

Conclusions

Epigenetic changes have become established in recent years as being one of the most important molecular signatures of human tumors. The discovery of hypermethylation of the CpG islands of certain tumor suppressor genes in cancer links DNA methylation to the classic genetic lesions with the disruption of many cell pathways, from DNA repair to apoptosis, cell cycle and cell adherence. Promoter hypermethylation is now considered to be a bona-fide mechanism for gene inactivation.

The picture that has emerged in recent years has shown us that cancer is a poligenetic disease but also a poliepigenetic disease, where genes involved through multiple pathways from cell cycle to apoptosis, from cellular adhesion to hormonal response are inactivated by promoter hypermethylation. The patterns of epigenetic lesions are extremely specific in human cancer and reflect the idiosyncrasy of each cell type. The analysis of candidate genes can be seen as only a part of the methylation changes in cancer. First, there are certainly still numerous genes that undergo epigenetic inactivation waiting to be discovered. The completion of the human genome sequence and the use of several described techniques to find new genes with differential methylation will be extremely useful for this purpose. The spectrum of epigenetic alterations for a relatively small subset of genes provides a potentially powerful system of biomarkers for developing molecular detection strategies for virtually every form of human cancer.

Acknowledgements

We would like to thank Dr. Mario F. Fraga for his excelent help in the discussion of the DNA methylation technologies and Dr. Esteban Ballestar for his helpful chromatin comments.

References

1. Bird AP. CpG-rich islands and the function of DNA methylation. Nature 1986; 321:209-213.
2. Nguyen C, Liang G, Nguyen TT et al. Susceptibility of nonpromoter CpG islands to de novo methylation in normal and neoplastic cells. J Natl Cancer Inst 2001; 93:1465-72.
3. Esteller M, Corn PG, Baylin SB et al. A gene hypermethylation profile of human cancer. Cancer Res. 2001; 61:3225-3229.
4. Greger V, Passarge E, Hopping W et al Epigenetic changes may contribute to the formation and spontaneous regression of retinoblastoma. Hum Genet 1989; 83:155-8.

5. Herman JG, Latif F, Weng Y et al. Silencing of the VHL tumor-suppressor gene by DNA methy-lation in renal carcinoma. Proc Natl Acad Sci USA 1994; 91:9700-4.
6. Merlo A, Herman JG, Mao L et al. 5' CpG island methylation is associated with transcriptional silencing of the tumour suppressor p16/CDKN2/MTS1 in human cancers. Nat Med 1995; 1:686-692.
7. Herman JG, Merlo A, Mao L et al. Inactivation of the CDKN2/p16/MTS1 gene is frequently associated with aberrant DNA methylation in all common human cancers. Cancer Res 1995; 55:4525-30.
8. Gonzalez-Zulueta M, Bender CM, Yang AS et al. Methylation of the 5' CpG island of the p16/CDKN2 tumor suppressor gene in normal and transformed human tissues correlates with gene silencing.
9. Clark SJ, Harrison J, Paul CL et al. High sensitivity mapping of methylated cytosines. Nucleic Acids Res 1994; 22:2990-2997.
10. Herman JG, Graff JR, Myohanen S et al. Methylation-specific PCR: a novel PCR assay for methy-lation status of CpG islands. Proc Natl Acad Sci USA 1996; 93:9821-6.
11. Esteller M, Fraga MF, Guo M et al. DNA methylation patterns in hereditary human cancers mimic sporadic tumorigenesis. Hum Mol Genet 2001; 10:3001-7.
12. Myohanen SK, Baylin SB, Herman JG. Hypermethylation can selectively silence individual p16ink4A alleles in neoplasia. Cancer Res 1998; 58:591-3.
13. Herman JG, Umar A, Polyak K et al. Incidence and functional consequences of hMLH1 promoter hypermethylation in colorectal carcinoma. Proc Natl Acad Sci USA 1998; 95: 6870-68:5.
14. Esteller M, Levine R, Baylin SB et al. MLH1 promoter hypermethylation is associated with the microsatellite instability phenotype in sporadic endometrial carcinomas. Oncogene 1998; 17:2413-2417.
15. Esteller M, Hamilton SR, Burger PC et al. Inactivation of the DNA repair gene O6-methylguanine-DNA methyltransferase by promoter hypermethylation is a common event in primary human neoplasia. Cancer Res 1999; 59:793-797.
16. Lee WH, Morton RA, Epstein JI et al. Cytidine methylation of regulatory sequences near the pi-class glutathione S-transferase gene accompanies human prostatic carcinogenesis. Proc Natl Acad Sci USA 1994; 91,11733-7.
17. Esteller M, Corn PG, Urena JM et al. Inactivation of glutathione S-transferase P1 gene by pro-moter hypermethylation in human neoplasia. Cancer Res 1998; 58:4515-4518.
18. Esteller M, Silva JM, Dominguez G et al. Promoter hypermethylation and BRCA1 inactivation in sporadic breast and ovarian tumors. J Natl Cancer Inst 2000; 92:564-9.
19. Di Croce L, Raker VA, Corsaro M et al. Methyltransferase recruitment and DNA hypermethylation of target promoters by an oncogenic transcription factor. Science 2002; 295:1079-1082.
20. Esteller M, Fraga MF, Paz MF et al. Cancer epigenetics and methylation. Science 2002; 297:1807-1808.
21. Paz MF, Avila S, Fraga MF et al. Germ-line variants in methyl-group metabolism genes and sus-ceptibility to DNA methylation in normal tissues and human primary tumors. Cancer Res 2002; 62:4519-4524.
22. Jones PA, Laird PW. Cancer epigenetics comes of age. Nat Genet 1999; 21:163-7.
23. Baylin SB, Esteller M, Rountree MR et al. Aberrant patterns of DNA methylation, chromatin formation and gene expression in cancer. Hum Mol Genet 2001; 10:687-92.
24. Graff JR, Herman JG, Myohanen S et al. Mapping patterns of CpG island methylation in normal and neoplastic cells implicates both upstream and downstream regions in de novo methylation. J Biol Chem 1997; 272:22322-9.
25. Song JZ, Stirzaker C, Harrison J et al. Hypermethylation trigger of the glutathione-S-transferase gene (GSTP1) in prostate cancer cells. Oncogene 2002, 21, 1048-61.
26. Robertson KD, Uzvolgyi E, Liang G et al. The human DNA methyltransferases (DNMTs) 1, 3a and 3b: coordinate mRNA expression in normal tissues and overexpression in tumors. Nucleic Acids Res 1999; 27:2291-8.
27. Feinberg AP, Vogelstein B. Hypomethylation distinguishes genes of some human cancers from their normal counterparts. Nature 1983; 301:89-92.
28. Ehrlich M, DNA hypomethylation and cancer. In: Melanie Ehrlich, ed. DNA Alterations in Can-cer: Genetic and Epigenetic Changes. Natick: Eaton Publishing, 2000:273-291.
29. Keshet I, Lieman-Hurwitz J, Cedar H. DNA methylation affects the formation of active chroma-tin. Cell 1986; 44:535-543.
30. Lewis JD, Meehan RR, Henzel WJ et al. Purification, sequence, and cellular localization of a novel chromosomal protein that binds to methylated DNA. Cell 1992; 69:905-914.

31. Magdinier F, Wolffe AP. Selective association of the methyl-CpG binding protein MBD2 with the silent p14/p16 locus in human neoplasia. Proc Natl Acad Sci USA 2001; 98:4990-4995.
32. Nan X, Ng HH, Johnson CA et al. Transcriptional repression by the methyl-CpG-binding protein MeCP2 involves a histone deacetylase complex. Nature 1998; 393:386-389.
33. Jones PL, Veenstra GJ, Wade PA et al. Methylated DNA and MeCP2 recruit histone deacetylase to repress transcription. Nat Genet 1989; 19:187-191.
34. Ballestar E, Esteller M. The impact of chromatin in human cancer: linking DNA methylation to gene silencing. Carcinogenesis 2002; 23:1103-9.
35. Esteller M, Sparks A, Toyota M et al. Analysis of adenomatous polyposis coli promoter hypermethylation in human cancer. Cancer Res 2000; 60:4366-71.
36. Wolffe AP, Jones PL, Wade PA DN.A d emethylation. Proc Natl Acad Sci USA 1999; 96:5894-5896.
37. Stirzaker C, Millar DS, Paul PM et al. Extensive DNA methylation spanning the Rb promoter in retinoblastoma tumors. Cancer Res 1997; 57:2229-2237.
38. Xiong Z, Laird PW. COBRA: a sensitive and quantitative DNA methylation assay. Nucleic Acids Res 1997; 25:2532-2534.
39. Gonzalgo ML, Bender CM, You EH et al. Low frequency of p16/CDKN2A methylation in sporadic melanoma: comparative approaches for methylation analysis of primary tumors. Cancer Res 1997; 57:5336-5347.
40. Bianco T, Hussey D, Dobrovic A. Methylation-sensitive, single-strand conformation analysis (MS-SSCA): A rapid method to screen for and analyze methylation. Hum Mutat 1999; 14:289-293.
41. Fraga MF, Rodriguez R, Canal MJ. Rapid quantification of DNA methylation by high performance capillary electrophoresis. Electrophoresis 2000; 14:2990-4.
42. Fraga MF, Uriol E, Borja Diego L et al. High-performance capillary electrophoretic method for the quantification of 5-methyl 2'-deoxycytidine in genomic DNA: application to plant, animal and human cancer tissues. Electrophoresis 2002; 23:1677-1681.
43. Fraga MF, Esteller M. DNA methylation: a profile of methods and applications. Biotechniques 2002; 33:632-649.
44. Wu J, Issa JP, Herman J et al. Expression of an exogenous eukaryotic DNA methyltransferase gene induces transformation of NIH 3T3 cells. Proc Natl Acad Sci USA 1993; 90:8891-8895.
45. Miller OJ, Schnedl W, Allen J et al. 5-Methylcytosine localised in mammalian constitutive heterochromatin. Nature 1974; 251:636-637.
46. Cedar H, Solage A, Glaser G et al. Direct detection of methylated cytosine in DNA by use of the restriction enzyme MspI. Nucleic Acids Res 1979; 6:2125-2132.
47. Butkus V, Petrauskiene L, Maneliene Z et al. Cleavage of methylated CCCGGG sequences containing either N4-methylcytosine or 5-methylcytosine with MspI, HpaII, SmaI, XmaI and Cfr9I restriction endonucleases. Nucleic Acids Res 1987; 15:7091-7102.
48. McClelland M, Nelson M, Raschke E. Effect of site-specific modification on restriction endonucleases and DNA modification methyltransferases. Nucleic Acids Res 1994; 22:3640-3659.
49. Hatada I,Hayashizaki Y, Hirotsune S et al. A genomic scanning method for higher organisms using restriction sites as landmarks. Proc Natl Acad Sci USA 1991; 88:9523-9527.
50. Kawai J, Hirotsune S, Hirose K et al. Methylation profiles of genomic DNA of mouse developmental brain detected by restriction landmark genomic scanning (RLGS) method. Nucleic Acids Res 1993; 21:5604-5608.
51. Toyota M, Ahuja N, Ohe-Toyota M et al. CpG island methylator phenotype in colorectal cancer. Proc Natl Acad Sci USA 1999; 96:8681-6.
52. Shiraishi M, Chuu YH, Sekiya T. Isolation of DNA fragments associated with methylated CpG islands in human adenocarcinomas of the lung using a methylated DNA binding column and denaturing gradient gel electrophoresis. Proc Natl Acad Sci USA 1999; 96:2913-2918.
53 Huang TH, Perry MR, Laux DE. Methylation profiling of CpG islands in human breast cancer cells. Hum Mol Genet 1999; 8:459-470.
54. Gitan RS, Shi H, Chen CM et al. Methylation-specific oligonucleotide microarray: a new potential for high-throughput methylation analysis. Genome Res 2002; 12:158-164.
55. Esteller M, Sanchez-Cespedes M, Rosell R et al. Detection of aberrant promoter hypermethylation of tumor suppressor genes in serum DNA from nonsmall cell lung cancer patients. Cancer Res 1999; 59:67-70.
56. Tang X, Khuri FR, Lee JJ et al. Hypermethylation of the death-associated protein (DAP) kinase promoter and aggressiveness in stage I nonsmall-cell lung cancer. J Natl Cancer Inst 2000; 92:1511-6.
57. Esteller M, Gonzalez S, Risques RA et al. K-ras and p16 aberrations confer poor prognosis in human colorectal cancer. J Clin Oncol 2001;19:299-304.

58. Esteller M, Garcia-Foncillas J, Andion E et al. Inactivation of the DNA-repair gene MGMT and the clinical response of gliomas to alkylating agents. N Engl J Med 2000; 343:1350-1354.
59. Esteller M, Gaidano G, Goodman SN et al. Hypermethylation of the DNA repair gene O(6)-methylguanine DNA methyltransferase and survival of patients with diffuse large B-cell lymphoma. J Natl Cancer Inst 2002; 94:6-7.
60. Wijermans PW, Krulder JW, Huijgens PC et al. Continuous infusion of low-dose 5-Aza-2'-deoxycytidine in elderly patients with high-risk myelodysplastic syndrome. Leukemia 1997; 11:1-5.
61. Schwartsmann G, Fernandes MS, Schaan MD et al. Decitabine (5-Aza-2'-deoxycytidine; DAC) plus daunorubicin as a first line treatment in patients with acute myeloid leukemia: preliminary observations. Leukemia 1997; 11(Suppl 1):S28-31.

CHAPTER 7

The Loss of Methyl Groups in DNA of Tumor Cells and Tissues:
The Immunochemical Approach

Alain Niveleau, Chandrika Piyathilake, Adriana de Capoa, Claudio Grappelli, Jean-Marc Dumollard, Lucien Frappart and Emmanuel Drouet

Introduction

The existence of 5-methyldeoxycytidine (5-MeCyd) has been first demonstrated in 1958.[1] For several years the presence of this naturally modified base in DNA remained unexplained and its role was ignored until a relationship was established between the expression of ovalbumin and the methylation status of the gene coding for this protein in various tissues.[2] The gene was not methylated in the oviduct whereas it was methylated in the brain where ovalbumin was not expressed. This first observation prompted numerous studies that investigated the distribution and the role of this modified base in genomic DNA.[3-8] A further significant progress was accomplished when it was demonstrated that tumor genomes were globally less methylated than their normal counterparts.[9-14] This important observation was made using DNA that had been extracted from tissue samples and then submitted to extensive digestion by nucleases and reverse phase high pressure liquid chromatography analysis.[15] The development of gene-specific hybridization techniques together with the availability of an ever-increasing number of methylation-sensitive restriction enzymes[16-18] allowed the methylation status of numerous genes to be investigated, especially DNA derived from tumors.[19-21] This gene-specific analysis has revealed that in addition to genome-wide hypomethylation,[22-29] local hypermethylated sites were identified in tumors[30-31] in concert with an increased DNA methyltransferase activity.[32-36] For several years most studies in the field aimed at the detection of altered methylation patterns of oncogenes[37-41] or tumor suppressor genes.[42-55]

The very first evidence for the presence of 5-MeCyd-rich regions in genomic DNA was provided by the pioneering work of Erlanger and Beiser[56] who took advantage of the specificity of the immunostaining techniques and who were able to obtain polyclonal antibodies specifically directed against the methyl group on carbon 5 of the pyrimidine ring. This remarkable achievement gave rise to an important set of results illustrating the uneven distribution of 5-MeCyd along metaphase chromosomes. Visualizing clusters of 5-MeCyd with antibodies in situ showed for the first time that pericentromeric regions were strongly methylated, and that this feature was repeatedly observed for numerous animal tissues.[57-59] The high specificity of the immunological tools and the possibility of obtaining a source of immunoglobulins with constant characteristics prompted us to raise monoclonal antibodies against several modified bases of tRNA and to develop an ELISA test that could serve as an alternative to the HPLC analysis described by Waalkes et al.[60] These authors showed that the urinary excretion of modified bases of tRNA was increased in cancer patients as compared to healthy individuals and that these tRNA breakdown products could be considered as reliable tumor markers.[61-62] In

DNA Methylation and Cancer Therapy, edited by Moshe Szyf. ©2005 Eurekah.com and Kluwer Academic/Plenum Publishers.

mice, the excretion of these breakdown products increases before the tumor is detectable by palpation,[63] thereby becoming an early signal of tumorigenesis. Monitoring rates of excretion of these modified nucleosides during therapy could provide an insight into the evolution of the pathology. An antibody specific for 5-methylcytosine (5-MeCyd) was among the set of mono- clonal antibodies that we obtained.[64] Early experiments utilizing this antibody revealed that labelling metaphase chromosomes with our 5-MeCyd monoclonal antibodies provided similar staining results[65] to the ones described by Erlanger and coworkers. Modifying the technique enabled us to unmask new 5-MeCyd-rich sites in chromosomes.[66-67] We then detected abnor- mal methylation patterns in heterochromatin regions of metaphase chromosomes prepared from lymphocytes of ICF patients.[68] During these investigations we observed that interphase nuclei lying besides metaphase chromosomes in the same spread were stained. The labelling was intense and the spatial distribution of the fluorescent signal was heterogeneous within each nucleus. This observation prompted us to develop protocols to label the nuclear compartment of either isolated cells or tissue sections.

Besides the frequently observed cancer-associated regional hypermethylation, the preva- lence of global DNA hypomethylation in many types of human cancer[69-70] suggests that this alteration plays a significant and fundamental role in tumorigenesis. The most likely mecha- nisms through which global DNA hypomethylation may participate in neoplastic transforma- tion are activation of oncogenes that are normally silenced by methylation[81] and induction of genomic instability[71-77] that results in abnormal chromosomal structures.[78-80]

A causal role for cancer-associated DNA hypomethylation in oncogenic transformation or tumor progression has been supported by results derived from several lines of animal experi- ments and cell culture studies. Inhibitors of DNA methylation were shown to be oncogenic in rodents.[82-83] The treatment of low-metastatic human or mouse cell lines with methylation in- hibitors was shown to induce their conversion to highly metastatic cells.[84-85] Based on this data we reasoned that a method that allowed assessing the global methylation status of DNA, on a cell by cell basis, while preserving the nuclear morphology and the tissue architecture, features that are important for pathology, would be of primary diagnostic importance. We developed and applied this methodology to study DNA methylation at the cytological level in several types of human tumors. Some of the results obtained by these studies are presented here.

Results

Chronic Lymphocytic Leukaemia B Cells

A sequential immunolabelling method that can distinguish between T and B cells on a single slide[86] was used to assess the methylation status of eu- and hetero- chromatin in inter- phase nuclei from CLL patients who had not received any therapy. In accordance with previous observations[87,88] significant differences were measured between control and CLL nuclei with regards to the following parameters:

1. The total number of heterochromatic (condensed) regions within each sample
2. The average number of heterochromatic regions per nucleus
3. The mean area of the condensed regions per nucleus
4. The mean optical density (OD) of condensed regions in each nucleus
5. The mean OD of nuclear euchromatin

These results are illustrated by Figure 1. Peripheral blood lymphocytes from a healthy indi- vidual can be seen on Figure 1A-C. The nuclear compartment appears as a heterogeneously stained structure with dark regions. As shown previously,[87] these regions correspond respec- tively to eu-chromatin and hetero-chromatin visualized by Giemsa staining and their surface and optical density can be evaluated by image analysis. Lymphocytes from a CLL patient are shown in Figure 1D. This example illustrates that a low staining intensity with 5-MeCyd is observed in both euchromatin and heterochromatin of leukemic cells when compared to nor- mal cells. Figure 2 summarizes the differences measured between normal and CLL cells (statis- tical significance Student's t-test for unpaired data: p<0.001). They confirm data on DNA

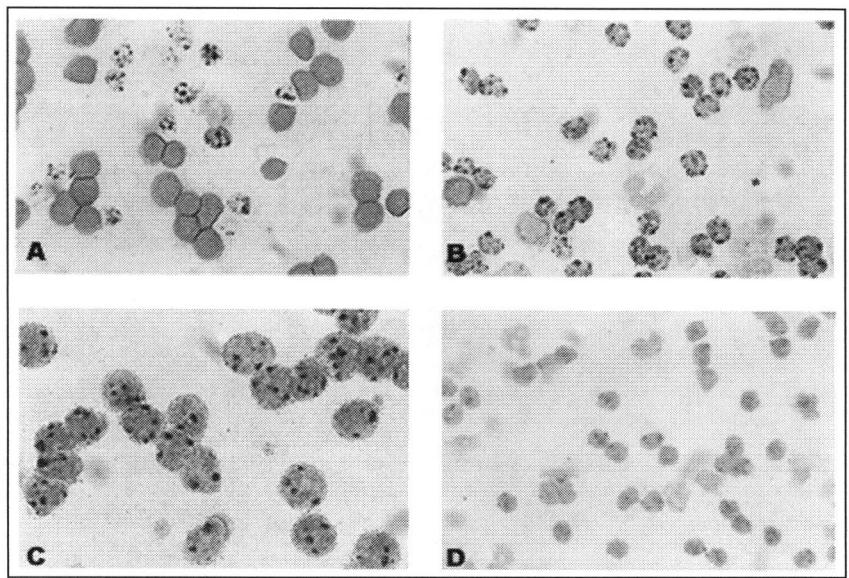

Figure 1. Immunolabeling of peripheral blood lymphocytes. Immunolabeled lymphocytes from healthy individuals are shown in A-C. (objective 100x). Lymphocytes from a CLL patient are shown in D. In A, T lymphocytes that have been labelled with an anti-CD3-peroxidase conjugate (Dako, Glostrup, Denmark) appear coated with a brown di-aminobenzidine (DAB) precipitate that keeps anti-5MeCyd antibodies from reaching the nucleus. Thus, only B lymphocytes nuclei are accessible and display the blue staining due to 4-chloro-1-alphanaphtol (4CIN) used as a substrate for peroxidase. Inversely, in B, B lymphocytes were labelled with anti-CD22-peroxydase conjugates (Ortho Diagnostics, Raritan, NJ) and revealed (brown color) by DAB-hydrogen peroxide, whereas T lymphocytes nuclei are accessible by the anti-5MeCyd antibodies and appear stained in blue by 4CIN.

Control and pathological slides were fixed in cold absolute methanol-glacial acetic acid (3:1) for 15 min. After aging at room temperature for 2 weeks in a dry atmosphere, control, dysplastic and tumor slides were submitted to ultra-violet irradiation as previously described.[87] Indirect immunostaining was performed with anti 5-MeCyd monoclonal antibodies previously obtained and characterised.[64] In all experiments the cells to be analyzed were selected by three experienced independent observers for being well spread and not overlapping. Digitalized images were acquired with a Leica Diaplan microscope (obj 40x) equipped with a b/w VC-44 CCD camera (Mannheim,Germany) and analyzed by Image Pro-Plus 3.1 software (Media Cybernetics, Milano). The number, size and optical density (OD) of the immunolabelled heterochromatic areas were measured in controls and pathological samples. To avoid interferences affecting the OD of the heterochromatin in cytospin preparations from all samples cell density was adjusted to obtain a monolayer of single cell thickness. Occasionally superimposed cells were not included in the quantitative analysis. A color version of this figure can be viewed at www.Eurekah.com.

hypomethylation of CLL populations that was demonstrated by other authors using either methylation sensitive restriction enzymes analysis[23] or HPLC. Our results demonstrate that the levels of DNA methylation of normal and leukemic lymphocytes can be quantified on an individual cell and point out a significant difference in the methylation status of eu- and hetero-chromatin in both CLL and normal cells.

Lung

The methylation status of DNA can be assessed by the radio-labelled methyl incorporation (RMI) assay. This assay measures methylation of DNA indirectly by determining its capacity to be methylated in vitro with *Sss* I methylase in the presence of S-adenosylmethionine that is used as the methyl donor.[89] The level of incorporation 3-H-labeled methyl residues correlates

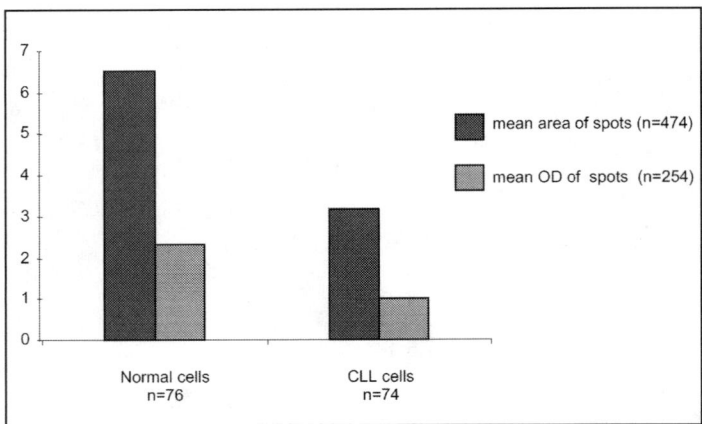

Figure 2. Differences between normal cells and CLL cells. The surface (μm^2) and density (on a 0 to 7 scale) of each heterochromatin region was assessed with Image Pro Plus 3.1 in samples from 2 healthy individuals and 2 B-cell chronic lymphocytic leukemia patients. The statistical analysis of data was performed by Student's t-test for unpaired data.

with the frequency of unmethylated CpGs in the DNA. However, since any given tissue sample contains a mixed population of cells (fibroblasts, lymphocytes, etc., in addition to cancer and normal epithelial cells), the RMI assay provides only an average value of the different populations of cells in a sample. This value varies with the heterogeneity of the sample. In our hands, the RMI assay exhibits day-to-day variability. However, the relative methylation values for groups of samples processed on the same day are comparable, especially when matched cancer and normal samples are used.

In our recent studies,[89] we have evaluated the global methylation status of DNA of specific types of cells involved in carcinogenesis using the immunochemical approach. Sections were labelled with anti-5MeCyd antibodies,[64] and no labelling was observed outside of the nuclei. This observation indicates that the antibodies reach the nucleus and that they bind methylated DNA. As with any nuclear marker, there may be a steric hindrance to the binding of methyl-CpG by an antibody due to the large size of an immunoglobulin relative to the size of the methylated CpG doublets. Although we have used the appropriate antigen retrieval technique for this antibody, endogenous DNA binding proteins in the nucleus might interfere with binding. The ultimate test of the validity of the antibody is establishing a correlation between the intensity of labelling with anti-5MeCyd antibodies and other diagnostic and prognostic factors. As described in the studies that follow, we have shown in our primary studies that this correlation exists. Since the commonly used RMI assay also has some deficiencies, as noted above, it is important to evaluate global DNA methylation by more than one approach.

To address these issues in lung carcinogenesis, we evaluated the status of global DNA methylation using the anti-5-MeCyd monoclonal antibody, in randomly selected lung specimens of sixty cigarette smokers who developed squamous cell carcinoma (SCC) and thirty cigarette smokers who did not. The racial representation in this study was mainly Caucasians.[90] In this study, 5-MeCyd immunostaining scores of normal bronchial epithelial cells in noncancer specimens were not significantly different from 5-MeCyd scores of uninvolved bronchial epithelial cells associated with SCC (p = 0.67). Scores of epithelial hyperplastic lesions of noncancer samples, however, were significantly higher compared to hyperplastic lesions associated with SCC (p = 0.02). Scores of normal bronchial epithelial cells in noncancer specimens were significantly higher compared to scores of both SCC-associated epithelial hyperplasia and SCC (p < 0.0001 and 0.0002, respectively). While 5-MeCyd scores were not significantly different between SCC-associated uninvolved bronchial epithelial cells and epithelial hyperplasia, they were significantly different between epithelial hyperplasia and SCC and also between uninvolved bronchial epithelial cells and SCC (Wilcoxon Sign Rank test p values 0.49, 0.01 and 0.0005 respectively). These observations suggest that altered global DNA methylation is potentially an important epigenetic predictor of susceptibility for lung cancer.

Since we observed a large variation in global DNA methylation among subjects, similarly to previous reports in breast carcinomas,[91] we calculated the ratio between 5-MeCyd scores of SCC and matched uninvolved bronchial mucosa for each subject. A lower ratio represents a more marked hypomethylation in SCC compared with adjacent uninvolved tissues. Forty-four of 60 SCCs had scores for both SCC and uninvolved bronchial mucosa and hence were available to calculate this ratio. The ratio was significantly lower with advanced stage and size of the tumor. Although the SCC/U ratio was 3 fold lower in subjects diagnosed with distant metastasis, this difference did not reach statistical significance, probably because a large majority of subjects presented with no distant metastasis at the time of surgery. The SCC/U ratio appeared to be unrelated to nodal status and grade of differentiation of the tumor. These results suggest that altered global DNA methylation is important in the progression of SCCs of the lung.

Race and Age Dependent Alterations in Global Methylation of DNA in Squamous Cell Carcinoma of the Lung

In a recent study we investigated the influence of race and age-dependent alterations in global DNA methylation on the development and progression of SCCs of the lung. Global methylation status was evaluated in SCC and in the associated uninvolved bronchial mucosa and epithelial hyperplasia of 53 Caucasians and 23 African Americans using the 5-MeCyd antibody described above. A low score indicates global hypomethylation of DNA. 5-MeCyd scores of SCC (0.59 ± 0.06) were significantly lower compared to 5-MeCyd scores of uninvolved bronchial mucosa (UBM) (0.87 ± 0.07) and epithelial hyperplasia (EH) 0.82 ± 0.07) in Caucasians ($p< 0.05$). In African Americans, 5-MeCyd scores of SCC (0.55 ± 0.09) were not significantly different from scores of UBM (0.60 ± 0.09) and EH (0.54 ± 0.14), suggesting an involvement of methylation in the development of SCCs in Caucasians, but not in African Americans. Figures 3 and 4 illustrate the pattern of DNA methylation in different types of cells in Caucasians and African Americans respectively.

In this study, we also demonstrate that 5MeCyd scores of cancer cells in Caucasians are lower in younger (< 65-years) subjects compared to older (> 65-years) subjects. Since cancers in younger subjects tend to be more aggressive than cancers in older subjects, these observations suggest that hypomethylation may have contributed to the aggressiveness of cancers of younger Caucasians. Hypomethylation of SCCs in white men is associated with shorter survival from

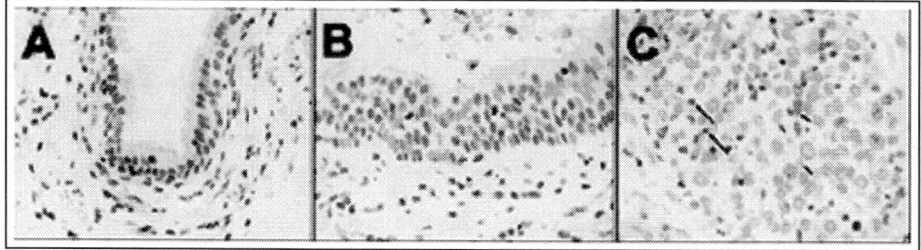

Figure 3. Immunohistochemical staining of normal and cancer tissues in a SCC specimen from a Caucasian subject, with anti-5-MeCyd antibodies in histologically normal uninvolved bronchial epithelium (A), in epithelial hyperplasia (B) and SCC (C), (magnification X 400) 4 μm thick tissue sections were processed as described in ref. 89. Three observers scored the immunostaining as regards the grading of the intensity (on a scale going from 0=no staining up to 4=intense staining) and the percentage of cells for each intensity level.

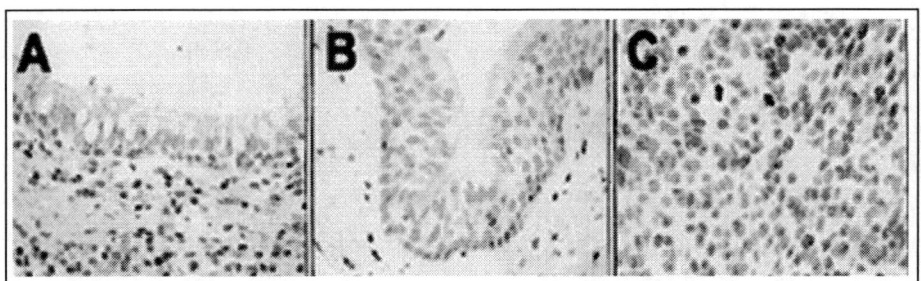

Figure 4. Immunohistochemical staining of normal and cancer tissues in a SCC specimen from an African American subject, with anti-5-MeCyd antibodiesin histologically normal uninvolved bronchial epithelium (A), epithelial hyperplasia (B) and SCC (C) (magnification X 400). Same technique as in Figure 5.

the disease. These preliminary results suggest that the methylation status of DNA may affect the development, aggressiveness and prognosis of SCCs in Caucasians. These observations suggest that careful attention should be given to racial distribution of study populations in investigations of DNA methylation.[92]

In summary, we wish to state that evaluation of global DNA methylation by immunohistochemistry allows a detailed evaluation of the pattern of methylation in the process of carcinogenesis and should have important applications in future studies of cancer.

Colon

An average global hypomethylation of 8% to 10% has been observed in colon adenomas or adenocarcinomas and a strong correlation between genomic instability and the DNA methylation status was demonstrated in colorectal cancer cells.[13,14,22,24,25,71-73]

Figure 5 is given as an example illustrating the difference in the staining intensity between normal mucosa (Fig. 5A,C) and the adenocarcinomatous zone (Fig. 5B,D) in paired samples from one of the patients. The immunostaining is localized in the nuclei of cells. The staining pattern of pleomorphic nuclei in the neoplastic area is obviously different from the one observed for the normal counterpart. The morphologically altered nuclei display densely labelled spots within faintly labelled areas whereas normal nuclei are darker and uniformly stained. This visually-detected difference in the staining intensity between the two types of cells is confirmed by image analysis as shown in Table 1 in which the average integrated optical density of the nuclei in neoplastic and normal tissue are reported for each patient, demonstrating a constant and significantly lower intensity of staining for the former type of cells (Student t test: p <0.05). The possibility that this difference could be due to a difference in the accessibility of DNA can be ruled out for the following reasons:

1. 5-MeCyd-rich regions are mainly present in the compacted heterochromatin compartment[86,88] that should then be less accessible and less labelled than the open euchromatin. Obviously an exact opposite pattern of staining is observed.
2. In a work published elsewhere[93] mouse embryos labelled with anti-DNA antibodies gave rise to the same signal in the paternal and maternal pronuclei whereas the anti-5-MeCyd signal was markedly reduced in the female genome. Therefore the differences observed in the present work cannot be attributed to a problem of accessibility.

As shown in Table 1 the average integrated optical density of the nuclei in the neoplastic area was 0,826 ±0,097 and that of the normal tissue was 0,961 ±0,099 (Student t test p<0.05).

To verify further the specific binding of the monoclonal antibody to DNA, we analysed deproteinized DNA with our antibodies. DNA samples were extracted from normal and from malignant tissues and processed as described by H. Sano et al.[94-95] The immunoblotting generated signals such as those shown in Figure 6. The staining patterns demonstrate that

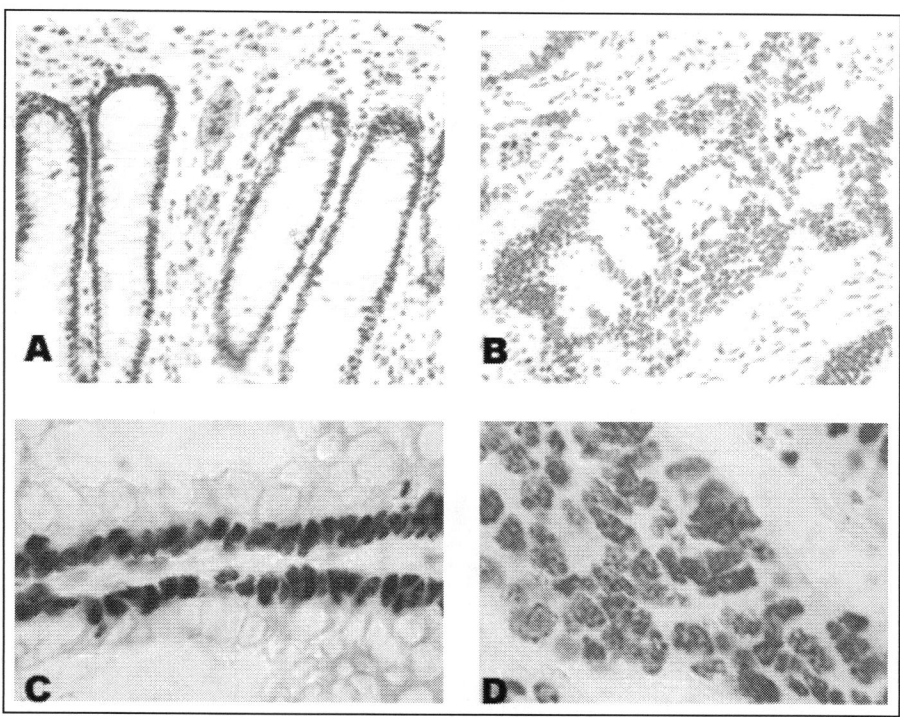

Figure 5. Indirect immunoperoxydase labeling of normal and adenocarcinoma human tissue sections with anti-5-MeCyd antibodies. Example of dense nuclear labelling observed in normal colon tissue (A,C) and heterogenous moderately stained nuclei in well differentiated colorectal adenocarcinoma (B,D). After antigen retrieval in a microwave oven, sections were treated as described in ref. 92. The optical density of the nuclei of epithelial cells was measured in 20 microscopic fields belonging to the malignant portion of the sample and in 20 fields of the normal tissue, for each patient, with a Leica Quantimet Analysis System.

smears of various shapes and intensities can be observed before and after digestion, the longest smear being obtained as expected with Mspl. These results are comparable to those obtained by other authors who used radiolabelled probes for hybridisation.[74,96] When applied to clinical samples, as expected, nearly identical indexes were obtained for each sample digested with Mspl which is not sensitive to methylation, (r-tumor/r-control=1.09), whereas a lower sensitivity, was observed for DNA from normal tissues digested with HpaII than for HpaII digested DNA from malignant lesions (r-tumor/r-control=1.194) confirming the hypomethylated status of the latter. Three other paired DNA samples, which were obtained from patients that were not included in the immunohistochemical study gave similar results under the same conditions (Table 2).

An image analysis of the immunostained samples demonstrates a 16% lower intensity of staining in the malignant portions of the sample, when compared to their nonmalignant counterpart. Feinberg et al[14] detected a range of 8% to 10% DNA hypomethylation in colon adenocarcinomas when they analysed by nuclease digestion and HPLC (high pressure liquid chromatography) DNA extracted from the whole tissue samples. In difference from these previous analyses, the immunohistochemical method used here looks at individual nuclei of altered epithelial cells. This may account for the difference between our results and the ones obtained by other laboratories. Our data is in accordance however with data acquired through analyses performed with biochemical methods.[13-15]

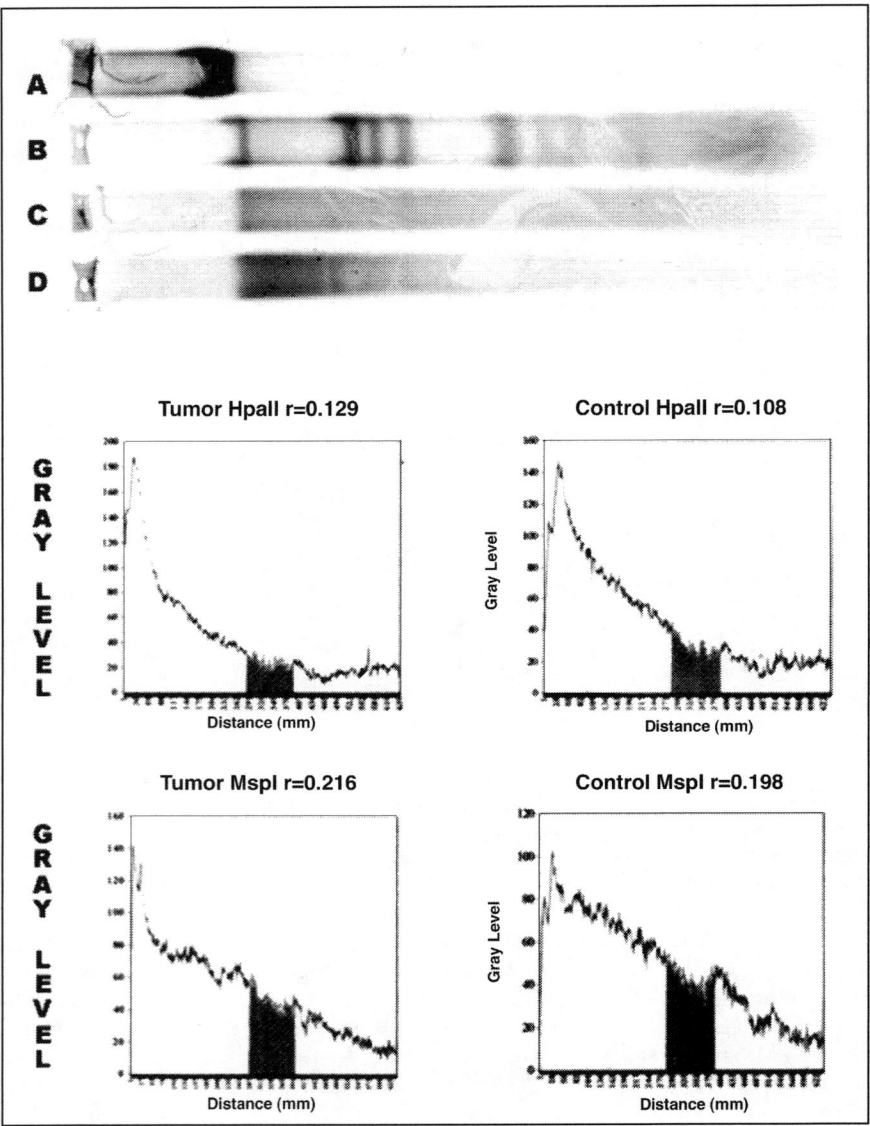

Figure 6. Immunoblotting of genomic DNAs. DNA was extracted from four pairs of normal and malignant tissues with phenol-chloroform, digested for 16 hours at 37°C with MspI or HpaII restriction enzymes (Roche, Meylan, France) at 10 units/µg of DNA. DNA samples were loaded on 1% agarose gels. After electrophoresis, DNA fragments were transferred under alkaline conditions onto Hybond N membranes (Amersham). Images of immunoblots were recorded with a Kodak DCS 200 digital camera. Gray level-based intensity measurements were performed along each lane with the SigmaScan/Image software (Jandel Scientific GmbH, Erkrath, Germany). Examples of profiles acquired for two DNA samples extracted from normal and tumor tissues are shown below the Southwestern blots. The shaded area indicates the portion of the lane corresponding to satellite DNA. Density profiles obtained after scanning the membrane allowed to calculate ratios between the signal recorded in the low molecular weight zone corresponding to satellite DNA as described in reference 96 and the signal recorded in the totality of the lane. The ratios r between the surface of this portion and the surface of the entire lane are indicative of the sensitivity of each sample to digestion. A) non digested DNA. B) DNA digested with Eco R1. C) DNA digested with MspI. D) DNA digested with HpaII.

Table 1. Immunohistochemistry of 5-MeCyd in colon adenocarcinoma tissue sections

Case Number	Location	Age	Stage (Dukes)	Differ-entiation	OD Tumor	OD Normal	Difference
1	Right Colon	84	D	WD	0.796 (0.039)	1.014 ± 0.04	21.4%
2	Right Colon	63	D	WD	0.766 (0.013)	0.997 ± 0.026	23.2%
3	Right Colon	75	D	MD	0.816 (0.041)	1.025 ± 0.035	20.0%
4	Right Colon	85	C	MD	0.773 (0.019)	1.021 ± 0.046	24.3%
5	Right Colon	69	D	WD	0.792 (0.041)	1.030 ± 0.038	23.0%
6	Left Colon	50	B	MD	0.764 (0.033)	1.019 ± 0.048	25.1%
7	Left Colon	58	C	WD	0.739 (0.051)	1.021 ± 0.031	27.6%
8	Left Colon	64	C	PD	0.751 (0.039)	1.018 ± 0.021	26.2%
9	Left Colon	60	B	WD	0.723 (0.026)	1.026 ± 0.026	29.5%
10	Left Colon	77	B	WD	0.837 (0.037)	1.010 ± 0.032	17.1%
11	Left Colon	74	A	WD	0.784 (0.043)	1.006 ± 0.024	22.1%
12	Left Colon	91	B	WD	0.818 (0.055)	1.027 ± 0.039	20.4%
13	Left Colon	62	C	MD	0.787 (0.048)	1.009 ± 0.032	22.0%

Nuclear densities after immunostaining with anti-5MeCyd antibodies in colon biopsies. Paraffin-embedded sections from paired tissues were processed as described in ref. 100. The values indicate the average optical density (mean ± 2SD) for each sample. OD, optical density; WD, well-differentiated; MD, moderately differentiated; PD, poorly differentiated. Dukes classification:. The classification was established according to Dukes CE J Path Bact 1932; 35:323-344.

Table 2. Assessment of global DNA methylation indexes by Southwestern blotting

Case Number	Localization	Age	Dukes Stage	Differentiation	Southwestern Blot Index
8	left colon	64	C	PD	21.3
14	left colon	60	A	WD	14.8
15	right colon	55	D	MD	19.4
16	left colon	76	B	WD	15.2

Samples from patient n° 8 were obtained from the same patient as in Table 1. The three other pairs were from other patients not included in the set of samples studied by immunohistochemistry.

Post-Transplantation Lymphomatous Diseases

In immunocompromised individuals, B cell proliferation is strongly associated with high loads of Epstein-Barr viral particles in the peripheral blood. ZEBRA is a viral protein involved in the switch between latency and the lytic cycle. We have previously demonstrated the following points.

1. High EBV serum loads and elevated titers of anti-ZEBRA antibodies are observed in patients with EBV-harboring tumor cells of Hodgkin's disease.[97]
2. Global DNA hypomethylation is detected in EBV-transformed interphase nuclei.[98]
3. The methylation status of genomic DNA can be assessed in human tumor cells and tissues by immunochemistry on a cell-by cell basis.[86,90]

We used the immunochemistry approach to study the following questions:

1. What is the effect of hypomethylation induced by 5-azadeoxycytidine on the expression of ZEBRA in cells latently infected with EBV?

2. Determine the putative expression of ZEBRA in paraffin-embedded tissue section from patients with post-transplantation lymphoproliferative diseases
3. Determine the global methylation status of DNA in these malignant tissues compared with normal samples.

Expression of ZEBRA in EBV-Immortalized Cells and Human Lymphomatous Tissue

Viral replication does not occur in EBV immortalized AKATA cells. However, 5-azadeoxycytidine induces release of viral particles to the medium that is preceded by synthesis of ZEBRA. This effect is illustrated by Figure 7. While no labelling is detected in nontreated cells (Fig. 7A) or mock-treated cells (Fig. 7B), the expression of ZEBRA is induced 24 hours after adding the hypomethylating agent to the culture medium (Fig. 7C,D). In the B-95-8 cell line ZEBRA is constitutively expressed as shown in Figure 7E,F. The induction of ZEBRA expression by the DNA methylation inhibitor was confirmed by a Western blotting shown in Figure 8.

The ZEBRA viral transactivator is also expressed in a significant proportion of tissue samples (6/12) obtained from the same group of PTLD patients as illustrated in Figure 9 using the same immunodetection protocol.

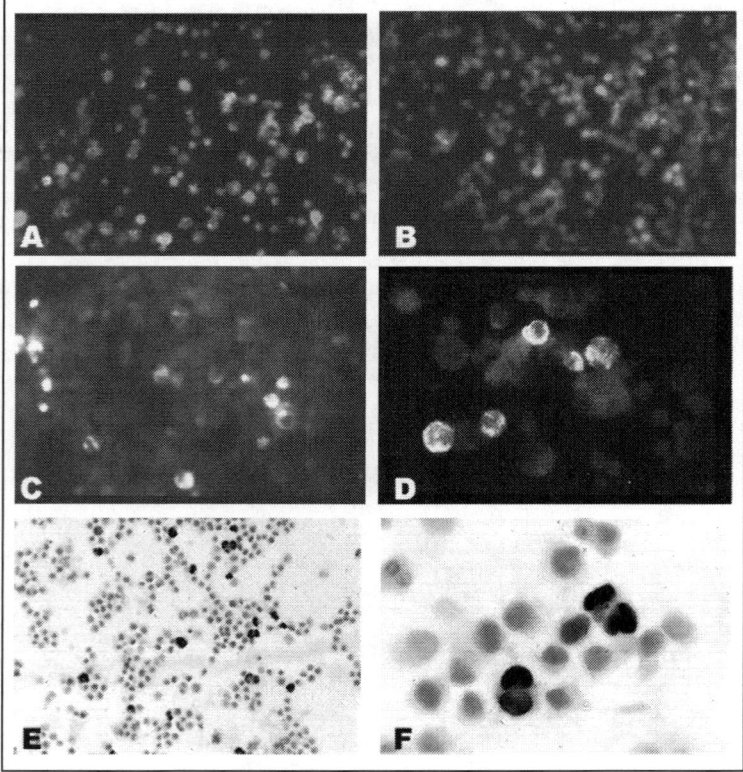

Figure 7. Immunodetection of ZEBRA in EBV-immortalized cells. The same monoclonal antibody raised against a peptide fragment of ZEBRA was used to detect the expression of the viral transactivator in two cell lines. AKATA nonproducing cells were treated with 5 azacytidine and the expression of ZEBRA was detected with a FITC-anti-peptide conjugate. B95-8 constitutively-producing cells were labelled with the anti-ZEBRA antibody. The binding was revealed with the UltraVision Detection System anti-mouse HRP/DAB kit from (Lab Vision Corp, Fremont, Ca). A) nontreated AKATA cells (obj 20x). B) mock-treated AKATA cells (obj 20x). C) AKATA cells treated with 5 μM 5-AzadCfor 48h (obj 20x). D) AKATA cells treated with 5 μM 5-AzadC for 48h (obj 32x). E,F) nontreated B95-8 cells (obj 9.5x and 40x).

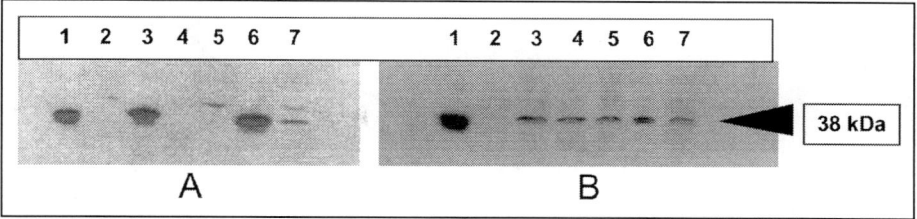

Figure 8. Expression of ZEBRA detected by Western blotting (ECL detection). A) lanes 1 and 6: purified protein; lane 3 extract from B 95-8 cells; lane 2 and 5 extract from nontreated AKATA cells; lane 4 extract from AKATA cells treated with 5 μM 5-AzadC for 24h; lane 7 extract from AKATA cells treated with 5 μM 5-AzadC for 48h. B) lane 1 purified protein; lane 2 mock treated Akata cells; lane 3 extract from AKATA cells treated with 3 μM 5-AzadC for 48h; lane 4 extract from AKATA cells treated with 6 μM 5-AzadC for 48h; lane 5 extract from AKATA cells treated with 9 μM 5-AzadC for 48h; lane 6 extract from AKATA cells treated with 15 μM 5-AzadC for 48h; lane 7 extract from AKATA cells treated with 30 μM 5-AzadC for 48h.

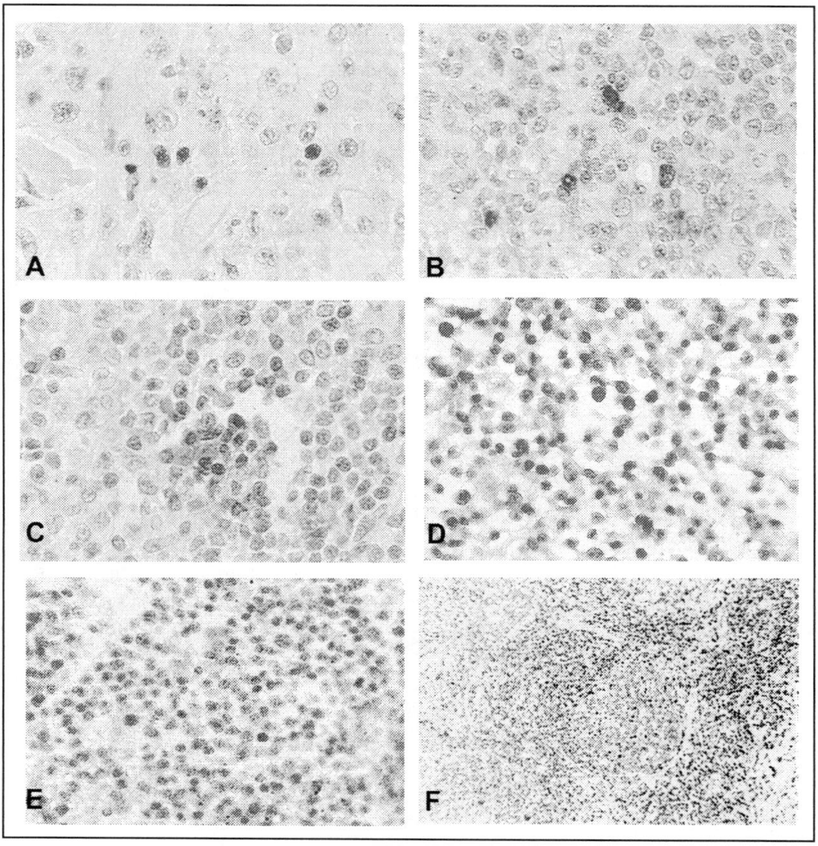

Figure 9. Immunodetection of the presence of ZEBRA in post-transplantation lymphomatous tissues. Tissue sections from six different patients were labelled with the same anti-ZEBRA monoclonal antibody as the one used in the previous experiments. The binding of this antibody was detected with second antibody conjugated with peroxidase in the presence of hydrogen peroxide and diamino-benzidine (A-C,F) or Vector Red (D,E).

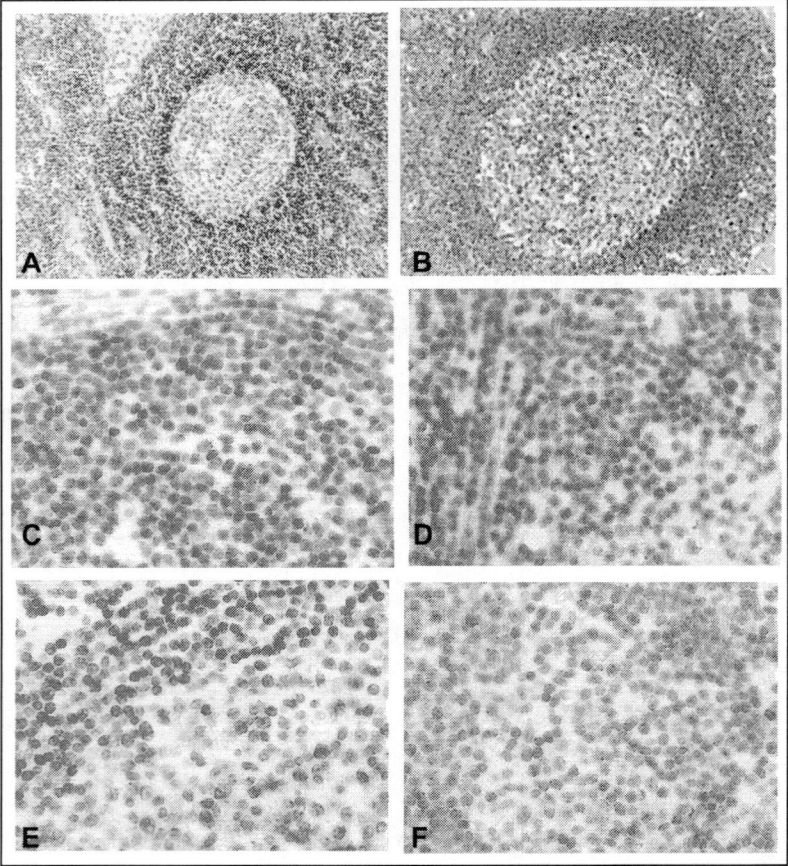

Figure 10. DNA methylation status in normal lymphoid tissues. Normal lymphoid tissues were reacted with the anti-5MeCyd monoclonal antibody under the same conditions as for colon tissues. A) Giemsa staining (obj 10x). B) Anti-5MeCyd immunolabeling (obj 10x). Same technique as for Figure 3: obj 40x.

Immunolabeling of 5-MeCyd

Monoclonal 5-MeCyd antibodies were used to determine the distribution of methylated DNA in tumors versus normal samples. Characteristic patterns of immunostaining can be seen in Figures 10 and 11. Sections of normal reactive lymph nodes stained with Giemsa or labelled with anti-5MeCyd antibodies are shown in Figure 10A,B respectively. Cells present in the mantle zone display a dense staining pattern as illustrated by Figure 10C,D, whereas cells occupying the germinal centre are significantly paler as seen in Figure 10E,F. A panel of images recorded with pathological samples illustrates the various aspects observed. As illustrated by Figure 11, not only is the spatial distribution of cells markedly disturbed, but the intensity of labelling in each sample is also highly heterogeneous and significantly reduced, compared to normal tissues.

The results presented above were quantified by image analysis of immunolabelled tissues and were summarized in Tables 3 and 4. The data presented in the tables indicates that while the variation in optical densities recorded in normal tissues is limited to 2.9% it varies between 4.6% and 7% in malignant zones suggesting a significant disruption of the methylation pattern in tumor samples.

Table 3. Immunohistolabeling of 5-MeCyd in normal reactive lymph nodes

Sample	Gray Level (0-256)		Difference (%)
	Mantle	Germinal Center	
1	92.6	90.1	- 2.6
2	101.8	99.1	- 2.6
3	101.5	98.6	- 2.9

Sections were processed as for colon tissues. Pictures were acquired with a Leica DC 50 digital camera. The optical density of nuclei was assessed with the Scion Image Analysis System (Scion Corp. Md USA).

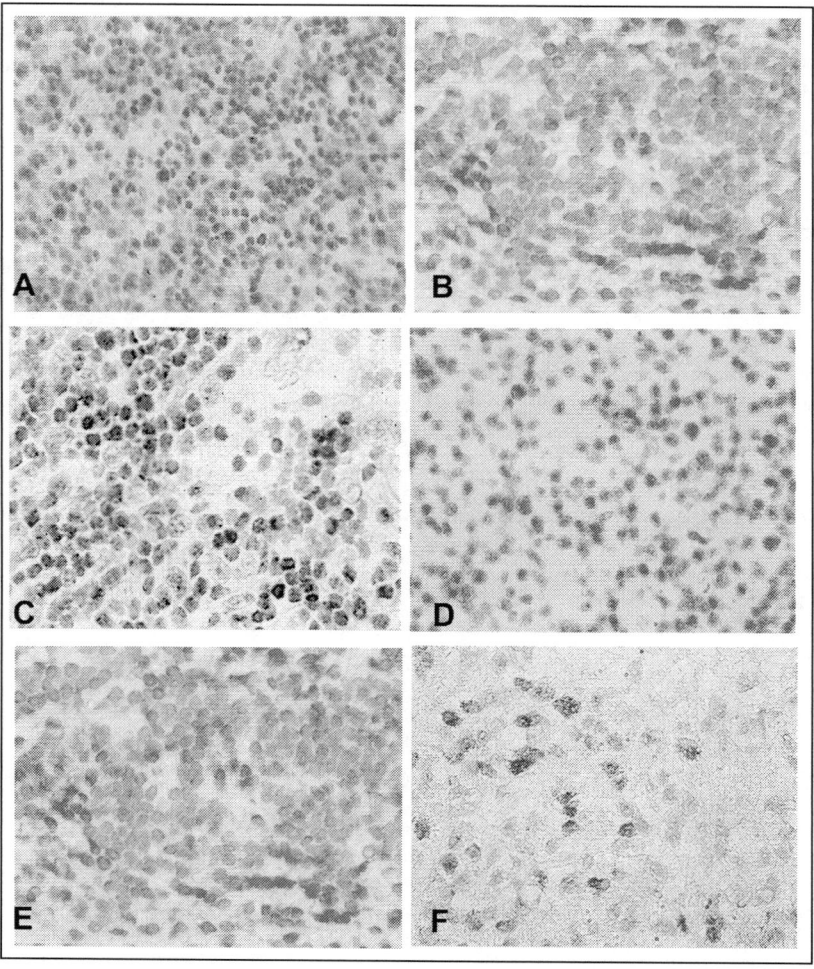

Figure 11. DNA methylation status of malignant lymphoid tissues. Same immunolabeling method as in Figure 10. All pictures were taken att: obj 40 x.

Table 4. Immunohistolabeling of 5-MeCyd in lymphomatous tissues

Sample	Grading	Gray Level (0-256)		Difference (%)
		Normal Zone	Malignant Zone	
1	low	99.2	94.5	-4.7
2	low	97.9	92.5	-5.4
3	low	98.2	92.1	-6.2
4	low	94.7	88.6	-4.7
5	intermediate	101.8	94.6	-6.4
6	high	98.2	93.5	-7.0

Tissue sections were processed and the optical density of nuclei was assessed as for normal tissues.

It is plausible that the global hypermethylation observed in the tumor samples not only participates in the establishment of the malignant phenotype as already observed for other malignancies but also unlocks the silencing of expression of viral genes such as ZEBRA. Induction of such antigens thereby enhances the replication and dissemination of new viral particles able to further infect and transform B cells in immunocompromised patients.

Breast Tissues

As is the case for numerous other tumors as discussed above, breast cancer is characterized by an overall hypomethylation of DNA[26,91,96] and by concomitant alterations of the methylation pattern of specific genes[51,52,99] relative to normal tissue. Using immunostaining with 5-MeCyd monoclonal antibodies we observed significant differences in intensity of staining between normal or benign and malignant lesions Immunostaining experiments demonstrated that normal cells-containing regions in breast biopsies obtained from nontreated patients were intensely labelled when compared to adjacent regions containing malignant cells. Figure 12A,B illustrates the staining pattern observed with a benign lesion (well differentiated grade I according to the Scarff, Bloom and Richardson classification). The low magnification picture (Fig. 12A) shows that the size and optical densities of most nuclei are homogeneous. The tissue architecture is characteristic of a normal glandular structure with numerous acini. At a higher magnification (Fig. 12B) one can observe a dense labelling of both myoepithelial and epithelia cells.

Figure 12C,D illustrates the coexistence, in the same tissue section, of an infiltrating ductal carcinoma (poorly differentiated grade III) with normal tissue. Both the tissue organization and the staining intensities are strikingly different.

The same kind of difference can be observed between adjacent zones in another sample (IDC grade II moderately differentiated) at low magnification (Fig. 12E). In the same tissue sample a different lesion appears heterogeneously labelled as illustrated by Figure 12F at a higher magnification.

Figure 13 illustrates immunostaining patterns of samples that were processed separately: Although the coexistence of normal and malignant lesions in the same tissue section could not be observed, the difference between grades is easily detected. In Figure 13A,B we stained well-differentiated grade I samples. Figure 13C,D illustrates the staining of moderately differentiated IDC grade II. The greatest heterogeneity was observed for poorly differentiated grade III IDC, as illustrated by Figure 13E,F.

Immunostaining experiments performed on 24 informative samples (6 benign and 18 malignant lesions) were quantified by image analysis and the results were summarized in Table 5.

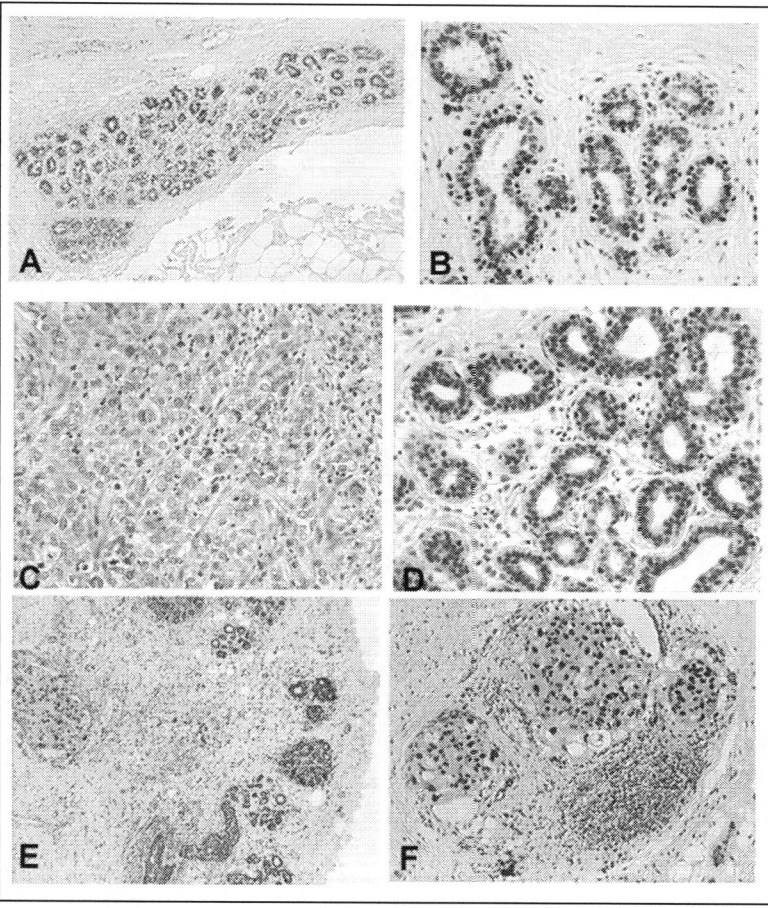

Figure 12. DNA methylation status of benign and malignant lesions coexisting in the same section. Immuno-staining with anti-5MeCyd antibodies was performed as for other tissues. A) (obj10x) and B) (obj 32x) well differentiated IDC SBR grade I. B,C) (obj 32x) IDC SBR grade III poorly differentiated. C) malignant lesion; D) normal tissue. E) (obj 10x) and F) (obj 20x) moderately differentiated IDC SBR grade II.

The results illustrate that nuclei belonging to malignant lesions consistently display a lower optical density than nuclei present in normal zones. This feature is verified not only in samples collected from separate patients but also in adjacent fields belonging to the same section, which excludes a possible inter-specimens bias due to thickness variation or uneven processing of samples. We observed a larger difference in methylation levels between normal and tumor breast tissue than what we observed in the colon (Fig. 5). We believe that this discrepancy could be explained by technical differences in the processing of these samples. First, the breast tissue samples were not fixed with paraformaldehyde like the colon or lung tissues but they were processed with Bouin's fixative (a mixture of formaldehyde and picric acid). It is well known that treating cells and tissues with acidic solutions removes histones and DNA-binding proteins, thereby increasing the accessibility of DNA to the antibodies. We have shown that isolated cells in suspension can be treated with acid to unmask DNA while preserving the nuclear architecture so as to perform flow cytometry analysis.[100] One can assume that fixation

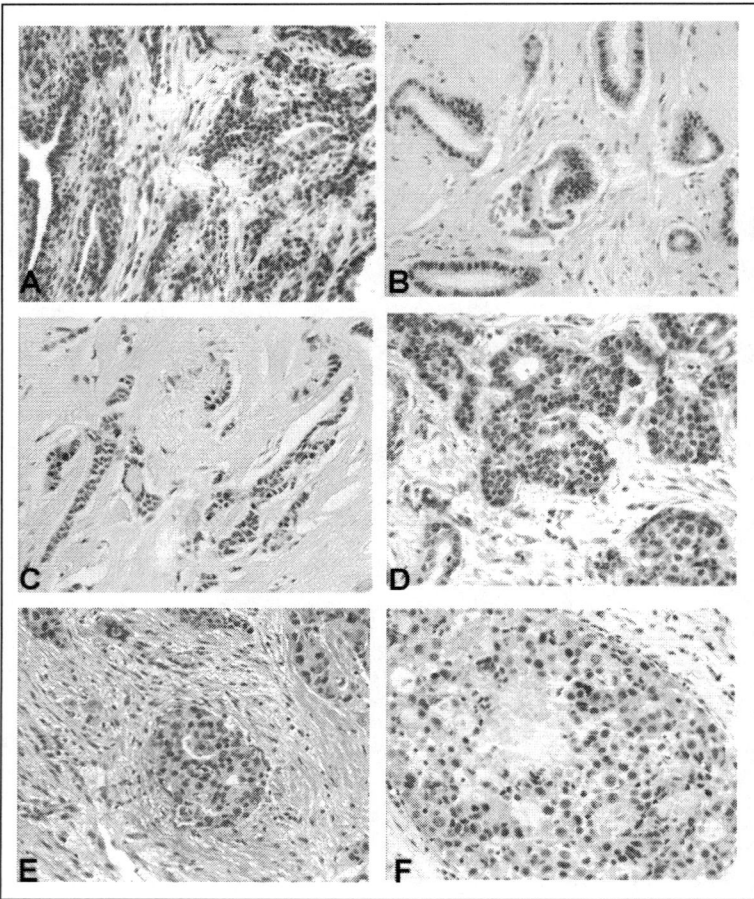

Figure 13. DNA methylation status in benign and malignant lesions observed in separate sections. A,B) (obj 32x) well differentiated IDC SBR grade I. C,D) (obj 32x) moderately differentiated IDC SBR grade II. E,F) (obj 32x) poorly differentiated IDC SBR grade III.

of paraffin-embedded biopsies with Bouin's mixture is more efficient than acid treatment of tissues previously fixed with paraformaldehyde. Second, it was not necessary to use the microwaves antigen retrieval procedure to achieve immunolabeling of breast derived specimens. In mammary tumors, healthy tissue can often be found nearby the malignant lesion.

In summary, the results reported here indicate that besides standardized morphological criteria such as anisocytosis, anisocaryosis, mitoses and differentiation, a difference in the optical density measured with anti-5MeCyd antibodies between individual cells within the same tissue section and also between sections from different samples can be used in tumor pathology and identification and staging of malignancy. Although the limited number of patients in each group precludes generalized conclusions, there is a clear statistically significant difference between normal and malignant tissues that can justify adding the 5-MeCyd immunostaining pattern as an additional pathological marker of malignancy.

Table 5. Immunohistolabeling of 5-MeCyd-rich regions in breast tissues

Patient N°	Type	Stage (S.B.R.)	Differenciation	OD Tumor	OD Normal	T/N
A. Separately analyzed benign lesions						
12496	benign	1	WD	-	134 ± 10.9	
11645	benign	1	WD	-	134 ± 15.2	
9010	benign	1	WD	-	121 ± 10.6	
8196	benign	1	WD	-	148 ± 5.7	
8171	benign	1	WD	-	156 ± 7.0	
7360	benign	1	WD	-	92.3 ± 8.1	
B. Separately analyzed malignant tissue samples						
2410	IDC	3	MD	95.2 ± 16.1	-	
7105	-	-	LD	78.9 ± 10.3	-	
7360	IDC	2	WD	93.2 ± 8.1		
9031	IDC	2	LD	94.7 ± 6.6	-	
11466	IDC	2	LD	80.5 ± 8.4	-	
12386	IDC	2	MD	95.6 ± 10.4	-	
C. Analysis of adjacent fields including benign and malignant tissues						
569	IDC	-	LD	70.7 ± 1.9	126.1 ± 0.3	0.56
2164	Special	-	MD	96.1 ± 11.6	173.8 ± 4.9	0.55
2169	Special	-	MD	55.1 ± 9.8	122.6 ± 7.1	0.45
2242	IDC	2	WD	108 ± 8.5	171.8 ± 5.6	0.63
6887	IDC	2	MD	94,2 ± 3.36	133.6 ± 2,7	0.71
6974	IDC	2	MD	104.1 ± 13.2	147.5 ± 11.4	0.71
7358	IDC	2	LD	79.5 ± 11.5	139.0 ± 2.4	0.57
7388	IDC	-	MD	117.9 ± 6.5	166.2 ± 1.5	0.71
8030	IDC	3	LD	67.9 ± 7.5	113.3 ± 1.7	0.60
8123	IDC	3	LD	84.9 ± 7.0	130.7 ± 2.2	0.65
9621	Special	-	MD	75.9 ± 11.2	147.1 ± 9.1	0.52
9875	IDC	-	LD	68.8 ± 2.4	103.6 ± 2.9	0.66

SBR: Scarff, Bloom and Richardson classification. Immunolabeling of nuclei in human breast tissue sections. Tissue sections were labelled with anti-5MeCyd antibodies and peroxydase-anti-mouse conjugates. The values indicate the average optical density (mean ± 2SD) for each sample. OD: optical density; Gradings correspond to the Scarff, Bloom and Richardson classification (S.B.R.). IDC: infiltrating ductal carcinoma.

Conclusions

The number of publications studying DNA methylation has increased exponentially since a relationship was established between tumorigenesis and genome-wide hypomethylation two decades ago. Following the very first results which measured global methylation by chromatographical approaches, numerous sophisticated methods were developed to investigate the methylation status of individual genes[101-105] and to highlight the correlation between alterations of the DNA methylation pattern and tumorigenesis[106-110] especially in relation to the silencing of tumor suppressor genes following the hypermethylation of their promoter.[111-113]

The demonstration that DNA methyltransferase activity is associated with the replication machinery[114] and that the acetylation/deacetylation of histones correlates with alterations in DNA methylation patterns highlights the significance of the relationship between DNA methylation and chromatin organization and carcinogenesis.[115-118] The immunochemical approach that we developed had illustrated this particular relationship between DNA methylation and chromatin compaction.[88] Our results show that a low immunohistochemical signal with anti 5-meCyd antibodies is associated with a change in chromatin condensation in tumor cells and tissues. Considering that the percentage of cytosines that are modified by methylation reaches about 5% in a normal diploid cell, a 5 to 10% global hypomethylation corresponds toa loss of between $0.5.10^6$ and 3.10^6 methyl groups. These groups are distributed aong the promoters of silent genes but also along noncoding regions of the genome such as satellite DNA. It therefore stands to reason that such an alteration will profoundly disturb the specific binding of proteins involved in the maintenance of nuclear structures and will provoke an instability of the genome. Therefore it is not surprising that this profound disruption of chromatin organization induced by modifications of the methylation pattern can be visualized by our immunolabelling approach.

Global DNA hypomethylation is an early event in colorectal tumorigenesis and the progression of the disease is associated with an increase in DNA methyltransferase activity and with the appearance of hypermethylated sites. These biochemical changes are paralleled by modifications in the histopathological patterns of colorectal tumors that include successive steps progressing from adenomas to carcinomas. An increase in epithelial dysplasia and cell dedifferentiation are considered as markers of multistep carcinogenesis. Global DNA hypomethylation is detected at the early adenoma stage. Global hypomethylation can be considered as a general feature of tumorigenesis since it is shown here to be a hallmark of other kinds of tumors. The immunohistochemical method described here for determining the state of methylation of cells in situ allows the pathologist to collect important data on the DNA methylation status of various regions in the biopsy, on a cell by cell basis, while preserving tissue architecture and cellular morphology.

Acknowledgements

Authors are grateful to Dr. J. Taylor and Dr. D.V. Spagnolo, Clinical Director at the The Western Australian Centre for Pathology and Medical Research, Nedlands, 6009 Western Australia, for the gift of tissue samples from PTLD patients.

This work was supported by Ligue Nationale Contre le Cancer, Fondation pour la Recherche Médicale and Fondation de France (grants to A.N.) Ministero dell'Università e della Ricerca Tecnologica (MURST) to A. dC, Galileo France-Italy exchange program (grants to A.N. and A. dC). National Cancer Institute (grant to C.P. n° KO7 CA 70160).

References

1. Hotchkiss RD. The quantitative separation of purines, pyrimidines and nucleosides by paper chromatography. J Biol Chem 1948; 175:315-332.
2. Mandel JL, Chambon P. DNA methylation: Organ specific variations in tne methylation pattern within and around ovalbumin and other chicken genes. Nucleic Acids Res 1979; 20:2081-2103.
3. Razin A, Riggs AD. DNA methylation and gene function. Science 1980; 210:604-610.
4. Ehrlich M, Wang RY. 5-Methylcytosine in eukaryotic DNA. Science 1981; 212:1350-1357.
5. Ehrlich M, Gama-Sosa MA, Huang L et al. Amount and distribution of 5-methyl-cytosine in human DNA from different type of tissues or cells. Nucl Acids Res 1982; 10:2709-2721.
6. Bird A. CpG-rich islands and fuction of DNA methylation. Nature 1986; 321:209-213.
7. Hsieh CL. Dependence of transcriptional repression on CpG methylation density. Mol Cell Biol 1994; 14:5487-5494.
8. Muiznieks I, Doerfler W. The impact of 5'-CG-3' methylation on the activity of different eukaryotic promoters: A comparative study. FEBS Lett 1994; 344:251-254.
9. Lapeyre JN, Becker FF. 5-methylcytosine content of nuclear DNA during chemical hepatocarcinogenesis and in carcinomas which result. Biochem Biophys Res Commun 1979; 87:698-705.

10. Gama-Sosa MA, Slagel VA, Trewyn RW et al. The 5-methylcytosine content of DNA from human tumors. Nucleic Acids Res 1983; 11:6883-6894.

11. Diala ES, Cheah MS, Rowitch D et al. Extent of DNA methylation in human tumor cells. J Natl Cancer Inst 1983; 71:755-764.

12. Feinberg AP, Vogelstein B. Hypomethylation distinguishes genes of some human cancers from their normal counterparts. Nature 1983; 301:89-92.

13. Goelz SE, Vogelstein B, Hamilton SR et al. Hypomethylation of DNA from benign and malignant human colon neoplasms. Science 1985; 228:187-190.

14. Feinberg AP, Gehrke CW, Kuo KC et al. Reduced genomic 5-methylcytosine in human colonic neoplasia. Cancer Res 1988; 48:1159-1161.

15. Gehrke CW, Zumwalt RW, MC Cune RA et al. Quantitative high-performance liquid chromotography analysis of modified nucleosides in physiological fluids, tRNA and DNA. Recent Results Cancer Res 1983; 84:344–359.

16. Bird A, Southern EM. Use of restriction enzymes to study eukaryotic DNA methylation: I. The methylation pattern in ribosomal DNA from Xenopus Laevis. J Mol Biol 1978; 118:27-47.

17. Cedar H, Solage A, Glaser G et al. Direct detection of methylated cytosine in DNA by use of the restriction enzyme MspI. Nucleic Acids Res 1979; 6:2125-2132.

18. Singer-Sam J, Lebon JM, Tanguay RL et al. A quantitative HpaII-PCR assay to measure methylation of DNA from a small number of cells. Nucl Acids Res 1990; 18:687-692.

19. Baylin SB, Esteller M, Rountree MR et al. Aberrant patterns of DNA methylation, chromatin formation and gene expression in cancer. Hum Mol Genet 2001; 10:687-692.

20. Counts JL, Goodman JI. Alterations in DNA methylation may play a variety of roles in carcinogenesis. Cell 1981; 83:13-15.

21. Jones PA. DNA methylation errors and cancer. Cancer Res 1996; 56:2463-2467.

22. Fearon ER, Jones PA. Progressing toward a molecular description of colorectal cancer development. FASEB J 1992; 6:2783-2790.

23. Wahlfors J, Hiltunen H, Heinonen K et al. Genomic hypomethylation in human chronic lymphocytic leukemia. Blood 1992; 80:2074-2080.

24. Cravo M, Pinto R, Fidalgo P et al. Global DNA hypomethylation occurs in the early stages of intestinal type gastric carcinoma. Gut 1996; 39:434-438.

25. Chen RZ, Petterson U, Beard C et al. DNA hypomethylation leads to elevated mutation rates. Nature 1998; 395:89-93.

26. Narayan A, Ji W, Zhang XY et al. Hypomethylation of pericentromeric DNA in breast adenocarcinomas. Int J Cancer 1998; 77:833-838.

27. Qu G, Dubeau L, Narayan A et al. Satellite DNA hypomethylation vs overall genomic hypomethylation in ovarian epithelial tumors of different malignant potentials. Mutation Research 1999; 423:91-101.

28. Stern LL, Mason JB, Selhub J et al. Genomic DNA hypomethylation, a characteristic of most cancers, is present in peripheral leukocytes of individuals who are homozygous for the C677T polymorphism in the methylenetetrahydrofolate reductase gene. Cancer Epidemiol Biomarkers Prev 2000; 9:849-53.

29. Lin CH, Hsieh SY, Sheen IS et al. Genome-wide hypomethylation in hepatocellular carcinogenesis. Cancer Res 2001; 61:4238-43.

30. Baylin SB, Fearon ER, Vogelstein B et al. Hypermethylation of the 5' region of the calcitonin gene is a property of human lymphoid and acute myeloid malignancies. Blood 1987; 70:412-417.

31. Makos M, Nelkin BD, Lerman MI et al. Distinct hypermethylation patterns occur at altered chromosome loci in human lung and colon cancer. Proc Nalt Acad Sci USA 1992; 89:1929–1933.

32. el-Deiry WS, Nelkin BD, Celano P et al. High expression of the DNA methyltransferase gene characterizes human neoplastic cells and progression stages of colon cancer. Proc Natl Acad Sci USA 1991; 88:3470-3474.

33. Issa J-P, Vertino PM, Wu J. Increased cytosine DNA-methyltransferase activity during colon cancer progression. J Natl Cancer Inst 1993; 85:1235-1240.

34. Melki JR, Warnecke P, Vincent PC et al. Increased DNA methyltransferase expression in leukaemia. Leukemia 1998; 12:311-316.

35. Jakob CA, Guldenschuh I, Hürlimann R et al. 5'cytosine DNA-methyltransferase mRNA levels in hereditary colon carcinomas. Virchows Arc 1999; 434:57-62.

36. Chen B, Liu X, Savell VH et al. Increased DNA methyltransferase expression in rhabdomyosarcomas. Int J Cancer 1999; 83:10-14.

37. Feinberg AP, Vogelstein B. Hypomethylation of ras oncogenes in primary human cancers. Biochem Biophys Res Com 1983; 111:47-54.

38. Cheah MS, Wallace CD, Hoffman RM. Hypomethylation of DNA in cancer cells: A site-specific change in the c-myc oncogene. J Natl Cancer Inst 1984; 73:1057-1065.

39. Sharrard RM, Royds JA, Rogers S et al. Patterns of methylation of the c-myc gene in human colorectal cancer progression. Br J Cancer 1992; 65:667-672.

40. Nakayama M, Wada M, Harada T et al. Hypomethylation status of CpG sites at the promoter region, and overexpression of the human MDR1 gene in acute myeloid leukemias. Blood 1998; 2:4296-4307.

41. Tao L, Yang S, Xie M et al. Hypomethylation and overexpression of c-jun and c-myc protooncogenes and increased DNA methyltransferase activity in dichloroacetic and trichloroacetic acid-promoted mouse liver tumors. Cancer Letters 2000; 158:185-193.

42. Greger V, Debus N, Lohman D et al. Frequency and parental origin of hypermethylated Rb1 alleles in retinoblastoma. Human Genet 1994; 94:491-496.

43. Greenblatt MS, Bennett WP, Hollstein M et al. Mutations in the p53 tumor suppressor gene: clues to cancer etiology and molecular pathogenesis. Cancer Res 1994; 54:4855-4878.

44. Magewu AN, Jones PA. Ubiquitous and tenacious methylation of the CpG site in codon 248 of the p53 gene may explain ist frequent appearance as a mutational hot spot in human cancer. Mol Cell Biol 1994; 14:4225-4232.

45. Gonzalez-Zuleta M, Bender CM, Yang AS et al. Methylation of the 5' CpG island of the p16/CDKN2 tumor-suppressor gene in normal and transformed human tissues correlates with gene silencing. Cancer Res 1995; 55:4531-4535.

46. Herman JG, Merlo A, Mao L et al. Inactivation of the CDKN2/p16/MTS1 gene is frequently associated with aberrant DNA methylation in all human common cancers. Cancer Res 1995; 55:4525-4530.

47. Little M, Wainwright B. Methylation and p16: Suppressing the suppressor. Nature Med 1995; 1:633-634.

48. Merlo A, Herman JG, Mao L et al. 5' CpG island methylation is associated with transcriptional silencing of the tumor suppressor p16/CDKN2/MTS1 in human cancers. Nature Med 1995; 1:686-692.

49. Dodge JE, List AF, Futscher BW. Selective variegated methylation of the p15 CpG island in acute myeloid leukaemia. Int J Cancer 1998; 78:561-567.

50. Esteller M, Sanchez-Cespedes M, Rosell R et al. Detection of aberrant promoter hypermethylation of tumor suppressor genes in serum DNA from nonsmall cell lung cancer patients. Cancer Res 1999; 59:67-70.

51. Dammann R, Yang G, Pfeifer GP. Hypermethylation of the cpG island of Ras association domain family 1A(RASSF1A), a putative tumor suppressor gene from the 3p21.3 locus, occurs in a large percentage of human breast cancers. Cancer Res 2001; 61:3105-3109.

52. Baldwin RL, Nemeth E, Tran H et al. BRCA1 promoter region hypermethylation in ovarian carcinoma: A population-based study. Cancer Res 2000; 60:5329-5333.

53. Sanchez-Cespedes M, Esteller M, Wu L et al. Gene promoter hypermethylation in tumors and serum of head and neck cancer patients. Cancer Res 2000; 60:892-895.

54. Esteller M. Epigenetic lesions causing genetic lesions in human cancer: Promoter hypermethylation of DNA repair genes. Eur J Cancer 2000; 36:2294-2300.

55. Lehmann U, Hasemeier B, Lilischkis R et al. Quantitative analysis of promoter hypermethylation in laser-microdissected archival specimens. Lab Invest 2001; 1:635-638.

56. Erlanger BF, Beiser SM. Antibodies specific for ribonucleosides and ribonucleotides and their reaction with DNA. Proc Natl Acad Sci USA 1964; 52:68-74.

57. Beiser SM, Erlanger BF, Tanenbaum SW. Conjugated and synthetic antigens: Preparation of purine- and pyrimidine-protein conjugates. In: Williams CA, Chase MW, eds. Methods in Immunology and Immunochemistry. New York: Academic Press, 1967:180-185.

58. Schrek RR, Erlanger BF, Miller OJ. The use of antinucleoside antibodies to probe the organization of chromosomes denatured by ultraviolet organization. Exp Cell Res 1974; 88:31-39.

59. Miller OJ, Schnedl W, Allen J et al. 5-methylcytosine localised in mammalian heterochromatin. Nature 1974; 251:636-637.

60. Waalkes TP, Gehrke CW, Bleyer WA et al. The urinary excretion of nucleosides of tRNA in patients with advanced cancer. Cancer 1975; 36:390-397.

61. Borek E, Baliga BS, Gehrke CW et al. High turnover of tRNA in tumor tissue. Cancer Res 1977; 37:398-402.

62. Speer J, Gehrke CW, Kuo KC et al. tRNA breakdown products as markers for cancer. Cancer 1979; 44:2120-2123.

63. Thomale J, Nass G. Elevated urinary excretion of RNA catabolites as an early signal of tumor development in mice. Cancer Lett 1982; 15:149-159.

64. Reynaud C, Bruno C, Boullanger P et al. Monitoring of urinary excretion of modified nucleosides in cancer patients using a set of six monoclonal antibodies. Cancer Lett 1991; 61:255-262.

65. Dante R, Baldini A, Miller DA et al. Methylation of the 5' flanking sequences of the ribosomal DNA in human cell lines and in a human-hamster hybrid cell line. J Cell Biochem 1992; 50:357-362.

66. Montpellier C, Bourgeois C, Kokalj-Vokac N et al. Detection of methylcytosine rich heterochromatin on banded chromosomes. Application to cells with various status of DNA methylation. Cancer Genet Cytogenet 1994; 78:87-93.

67. Barbin A, Montpellier C, Kokalj-Vokac N et al. In situ DNA methylation analysis of human tumor cells: Mapping of methylcytosine-rich bands. Human Genetics 1994; 94:684-692.

68. Miniou P, Jeanpierre M, Blanquet V et al. Abnormal methylation pattern in constitutive and facultative (X inactive chromosome) heterochromatin of ICF patients. Human Mol Genetics 1994; 3:2093-2102.

69. Counts J, Goodman JI. Hypomethylation of DNA: An epigenetic mechanism involved in tumor promotion. Mol Carcinog 1994; 11:185-188.

70. Schmutte C, Yang AS, TuDung T et al. Mechanisms for the involvement of DNA methylation in colon carcinogenesis. Cancer Res 1996; 56:2375-2381.

71. Lengauer C, Kinzler KW, Vogelstein B. DNA methylation and genetic instability in colorectal cancer cells. Proc Natl Acad Sci USA 1997; 94:2545-2550.

72. Ahuja N, Mohan AL, Li Q et al. Association between CpG island methylation and microsatellite instability in colorectal cancer. Cancer Res 1997; 57:3370-3374.

73. Steinbeck RG. Chromosome division figures reveal genomic instability in tumorigenesis of human colon mucosa. Brit J Cancer 1998; 77:1027-1034.

74. Cunningham JM, Christensen ER, Tester DJ et al. Hypermethylation of the hMLH1 promoter in colon cancer with microsatellite instability. Cancer Res 1998; 58:3455-3460.

75. Xu G, Bestor TH, Bourc'his D et al. Chromosome instability and immunodeficiency syndrome caused by mutations in a DNA methyltransferase gene. Nature 1999; 402:187-191.

76. Shannon B, Kay P, House A et al. Hypermethylation of the Myf-3 gene in colorectal cancers: Associations with pathological features and with microsatellite instability. Int J Cancer 1999; 84:109-113.

77. Vilain A, Bernardino J, Gerbault-Seureau M et al. DNA methylation and chromosome instability in lymphoblastoid cell lines. Cytogenet Cell Genet 2000; 90:93-101.

78. Lewis J, Bird A. DNA methylation and chromatin structure. FEBS Lett 1991; 285:155-159.

79. Laird PW, Jaenisch R. The role of DNA methylation in cancer Genetics and epigenetics. Annu Rev Genet 1996; 30:441-464.

80. Laird PW. Oncogenic mechanisms mediated by DNA methylation. Molecular Medicine today 1997;223-229.

81. Weissbach A, Ward C, Bolden A. Eukaryotic DNA methylation and gene expression. Curr Topics Cell Regul 1989; 30:1-21.

82. Carr BI, Reilly JG, Smith SS et al. The tumorigenicity of 5-azacytidine in the male Fischer rat. Carcinogenesis 1984; 5:1583-1590.

83. Cavaliere A, Bufalari A, Vitali R. 5-Azacytidine carcinogenesis in BALB/c mice. Cancer Lett 1987; 37:51-58.

84. Kerbel RS, Frost P, Liteplo R et al. Possible epigenetic mechanisms of tumor progression: Induction of high-frequency heritable but phenotypically unstable changes in the tumorigenic and metastatic properties of tumor cell populations by 5-azacytidine treatment. J Cell Physiol 1984; 3:87-97.

85. Ormerod EJ, Everett CA, Hart IR. Enhanced experimental metastatic capacity of a human tumor cell line following treatment with 5-azacytidine. Cancer Res 1986; 46:884-890.

86. de Capoa A, Grappelli C, Febbo FR et al. Methylation levels of normal and chronic lymphocytic leukemia B lymphocytes: Computer-assisted quantitative analysis of anti-5-methylcytosine antibody binding to individual nuclei. Cytometry 1999; 36:157-159.

87. de Capoa A, Di Leandro M, Grappelli C et al. Methylation status of individual interphase nuclei in human cultured cells: A semiquantitative computerised analysis. Cytometry 1998; 31:85-92.

88. de Capoa A, Romana Febbo F, Giovanelli F et al. Reduced levels of Poly(ADP)-ribosylation result in chromatin compaction and hypermethylation as shown by a cell-by-cell computer-assited quantitative analysis. FASEB J 1999; 13:89-93.

89. Piyathilake CJ, Johanning GL, Frost AR et al. Immunohistochemical evaluation of global DNA methylation: Comparison wit in vitro radiolabeled methyl incorporation assay. Biotechnic and Histochemistry 2000; 75:251-258.

90. Piyathilake CJ, Frost AR, Bell WC et al. Altered global methylation of DNA: An epigenetic differ-ence in susceptibility for lung cancer is associated with its progression. Human Pathol 2001; 32:856-62.
91. Soares J, Pinto AE, Cunha CV et al. Global DNA hypomethylation in breast carcinoma: Correla-tion with prognostic factors and tumor progression. Cancer 1999; 85:112-118.
92. Piyathilake CJ, Macaluso M, Henao O et al. Race and age dependant differences in global methy-lation of DNA. FASEB J 2001; 15:494-12.
93. Mayer W, Niveleau A, Walter J et al. Demethylation of the zygotic paternal genome. Nature 2000; 403:501-502.
94. Sano H, Royer HD, Sager R. Identification of 5-methylcytosine in DNA fragments immobilized on nitrocellulose paper. Proc Natl Acad Sci USA 1980; 77:3581-3585.
95. Sano H, Imokawa M, Sager R. Detection of heavy methylation in human repetitive DNA subsets by a monoclonal antibody against 5-methylcytosine. Biochim Biophys Acta 1988; 951:157-165.
96. Bernardino J, Roux C, Almeida A et al. DNA hypomethylation in breast cancer: An independent parameter of tumor progression? Cancer Genet Cytogenet1997; 97:83-89.
97. Drouet E, Brousset P, Fares F et al. High Epstein-Barr virus (EBV) serum load and elevated titers of anti-zebra antibodies in patients with EBV-harboring tumor cells of Hodgkin's disease. J Med Virol 1999; 57:383-389.
98. Yang X, Yan L, Davidson NE. DNA methylation in breast cancer. Endocr Relat Cancer 2001; 8:115-127.
99. Habib M, Fares F, Bella C et al. DNA global hypomethylation in EBV-transformed interphase nuclei. Exp Cell Res 1999; 249:46-53.
100. Hernandez-Blazquez FJ, Habib M, Dumollard JM et al. Evaluation of global DNA hypomethylation in human colon cancer tissues by immunohistochemistry and image analysis. Gut 2000; 47:689-693.
101. Hakkarainen M, Wahlfors J, Myohannen S et al. Hypermethylation of calcitonin gene regulatory sequences in human breast cancer as revealed by genomic sequencing. Int J Cancer Pred Oncol 1996; 69:471-474.
102. Huang TH, Perry MR, Laux DE. Methylation profiling of CpG islands in human breast cancer cells. Human Mol Genet 1999; 8:459-470.
103. Wang K, Gan L, Jeffery E et al. Monitoring gene expression profile changes in ovarian carcinomas using cDNA microarray. Gene 1999; 229:101-108.
104. Costello JF, Fruhwald MC, Smiraglia DJ et al. Aberrant CpG-island methylation has nonrandom and tumor-type-specific patterns. Nat Genet 2000; 24:132-138.
105. Yan PS, Chen CM, Shi H et al. Dissecting complex epigenetic alterations in breast cancer using CpG island microarrays. Cancer Res 2001; 61:8375-8380.
106. Tycko B, Ashkenas J. Epigenetics and its role in disease. J Clin Invest 2000; 105:245-246.
107. Robertson KD, Wolffe AP. DNA methylation in health and disease. Nat Rev Genet 2000; 1:11-19.
108. Warnecke PM, Bestor TH. Cytosine methylation and human cancer. Curr Opin Oncol 2000; 12:68-73.
109. Esteller M, Fraga MF, Guo M et al. DNA methylation patterns in hereditary human cancers mimic sporadic tumorigenesis. Hum Mol Genet 2001; 10:3001-3007.
110. Feinberg AP. Cancer epigenetics takes center stage. Proc Natl Acad Sci USA 2001; 98:392-394.
111. Robertson KD. DNA methylation, methyltransferases and cancer. Oncogene 2001; 20:3139-3155.
112. Herman JG, Baylin SB. Promoter-region hypermethylation and gene silencing in human cancer. Curr Top Microbiol Immunol 2000; 249:35-54.
113. Esteller M, Corn PG, Baylin SB et al. A gene hypermethylation profile of human cancer.Cancer Res 2001; 61:3225-3229.
114. Leonhardt H, Page AW, Weier HU et al. A targeting sequence directs DNA methyltransferase to sites of DNA replication in mammalian nuclei. Cell 1992; 71:865-873.
115. Schubeler D, Lorincz MC, Cimbora DM et al. Genomic targeting of methylated DNA: Influence of methylation on transcription, replication, chromatin structure, and histone acetylation. Mol Cell Biol 1992; 20:9103-9112.
116. Leonhardt H, Cardoso MC. DNA methylation, nuclear structure gene expression and cancer. J.Cell Biochem 2000; 35(Suppl):78-83.
117. Baylin SB, Esteller M, Rountree MR et al. Aberrant patterns of DNA methylation, chromatin formation and gene expression in cancer. Hum Mol Genet 2001; 10:687-92.
118. Esteller M, Herman JG. Cancer as an epigenetic disease: DNA methylation and chromatin alter-ations in human tumors. J Pathol 2002; 196:1-7.

CHAPTER 8

Identifying Clinicopathological Association of DNA Hypermethylation in Cancers Using CpG Island Microarrays

Susan H. Wei, Timothy T.-C. Yip, Chuan-Mu Chen
and Tim H.-M. Huang

Abstract

Hypermethylation of promoter CpG islands has been associated with gene silencing in cancer. Increasingly, these CpG islands have potential clinical utility as molecular markers for cancer diagnosis. Here we describe a microarray-based technique, called differential methylation hybridization (DMH), for simultaneous screening of methylation alteration across thousands of CpG island loci in one tumor sample at a time. We also describe a second approach, called methylation target array (MTA), for detecting methylation alteration of a single CpG island locus across hundreds of tumor DNA samples. The DMH and MTA assays are complementary to each other in that DMH allows for rapid identification of multiple loci hypermethylated in tumor genomes while MTA can rapidly assess the utility of these loci as markers for clinical diagnosis. Furthermore, the use of clustering algorithms to analyze the array data of multiple CpG island loci can identify an association of DNA hypermethylation with specific clinicopathological features of tumors.

Introduction

Cancer is a heterogeneous group of diseases with a wide spectrum of molecular alterations and clinicopathological manifestations. One hallmark of cancer is the accumulation of aberrantly methylated CpG dinucleotides located in promoter CpG islands of genes.[1-3] Through chromatin restructuring at the methylated sites, the expression of genes may be silenced leading to a loss of control of cell growth and subsequent tumor development.[4] Numerous genes of tumor suppressors, cyclins, and other biomodulators playing pivotal roles in tumorigenesis are reportedly modified in this fashion,[5] and may serve as epigenetic markers for cancer diagnosis. For instance, hypermethylation of the *DAPK1* promoter is correlated with poor survival in stage I nonsmall-cell lung cancer,[6] *hMLH1* promoter hypermethylation and gene down-regulation results in drug-resistance in ovarian cancer cells,[7] *CDKN2A* promoter hypermethylation is correlated with the progression of adult T-cell leukemia,[8] and *MGMT* promoter hypermethylation is useful for predicting the responsiveness of gliomas to alkylating agents[9] (see Table 1 for additional examples, see refs. 8-22).

CpG island hypermethylation in cancers is increasingly shown to be complex and to affect multiple loci concurrently in the tumor genome.[23-25] Moreover, this type of alteration is not random and frequently is cancer-specific.[26] While there remain benefits in candidate gene approaches for searching single CpG islands hypermethylated in cancer cells, the development

DNA Methylation and Cancer Therapy, edited by Moshe Szyf. ©2005 Eurekah.com and Kluwer Academic/Plenum Publishers.

Table 1. Clinicopathological associations of aberrant DNA methylation

Genes	Cancers – Associated Diseases	Correlations and Associations	Refs.
MGMT	Gliomas	Methylation correlated with tumor regression and prolonged overall and disease-free survival in patients treated with alkylating agent.	9
DAPK	NSCLC	Association of hypermethylation with advanced clinical stage ($P = 0.003$), tumor size increase ($P = 0.009$) and lymph node involvement ($P = 0.04$); odds ratios increase with clinical stages. Association with poor prognosis in stage I patients ($P < 0.001$).	6 10
	H&NCa	Hypermethylation correlated with lymph node involvement and advanced clinical stages.	11
P21	ALL	Methylation correlated significantly with decreased disease free and overall survival at 7 & 9 years.	12
BRCA1	Breast cancer	Methylation correlated with decreased ER ($P = 0. 016$) and p27 ($P = 0.018$) expression i ncreased p21 expression ($P = 0.011$) and methylated tumor is usually high grade.	13
E-Cad and *ER*	Breast cancer	Coincident methylation of both genes increased significantly from ~20% in CIS to ~50% metastatic lesions.	14
HMLH1	Ovarian cancer	Methylation associated with cisplatinum resistance.	7
P16	NSCLC	1. Hypermethylation associated with pack-years smoked ($P = 0.007$), smoking duration ($P = 0.0009$) and negatively with the time sincequitting smoking ($P = 0.03$). 2. In stage 1 adenocarcinoma, an independent risk factor predicting shorter post-surgery survival ($P = 0.03$).	15
	Lung SCC	Frequency of hypermethylation increased from basal cell hyperplasis (17%) to squamous metaplasia (24%) to CIS (50%)	16
	ATL	Hypermethylation more frequently found in more malignant lymphoma (73%) and acute (47%) ATL types than the less malignant chronic (17%) & smoldering (17%) types.	8
HIC1	ALL, CML, NHL	*HCL1* gene methylation was found in 100% recurrent ALL and 100% blast crisis, but not at initial diagnosis of CML.	17
APC, *P16*	Barrett's esophagus	Hypermethylation of both genes found as early as in Barrett's metaplasia and displasia	18,19
Multiple genes	Bladder cancer	Hypermethylation correlated with patients' survival, tumor grade, growth pattern, muscle invasion, tumor stage and DNA ploidy pattern.	20
Multiple genes	Ulcerative colitis	Dramatic increase of methylation from non-dysplastic colitis to high grade dysplasia.	21
Multiple genes	Cervical cancer	Methylation extent increased from non-dyplastic low grade CIN to high grade & finally to invasive carcinoma.	22

ALL – acute lymphoblastic leukemia; APC – adenomatous polyposis coli gene; ATL - adult T cell leukemia; *BRCA1* – mutation in breast cancer 1 gene; CIN – cervical intraepithelial neoplasia; CIS – carcinoma in situ; CML – chronic myelogenous leukemia; *DAPK* – death-associated protein kinase; E-cad – E-cadherin; ER – estrogen receptor; *GSTP1* – glutathione S-transferase P1; *HIC1* – hypermethylated in cancer 1 gene; *hMLH1* – human Mut L homologue 1 DNA repair gene; H&NCa – head & neck cancer; p16[INK4a/CDKN2A] – a multiple tumor suppressor gene 1 (*MTS1*) & also called cyclin dependent kinase inhibitor 2A; p21[CIP/WAF/SDI1] – a tumor suppressor gene & also called cyclin dependent kinase inhibitor 1; *MGMT* – O6-methylguanine methyl transferase; NHL- non-Hodgkin lymphoma; NSCLC – non small cell lung carcinoma; lung SCC – squamous cell carcinoma in lung.

of efficient techniques to scan methylation throughout the whole genome would greatly enhance the identification of markers useful for tumor classification. Several methylation-scanning techniques have been developed including, but not limited to, arbitrarily primed-PCR,[27] MethyLight,[28] methylated CpG island amplification,[29] landmark restriction genomic scanning,[30] and oligonucleotide microarray[31] (see also ref. 32 for overview).

In the following sections, we describe two array-based techniques, differential methylation hybridization (DMH)[33] and methylation target array (MTA),[34] for high-throughput methylation screening. We first discuss the principles of DMH and MTA and then their applications in detecting methylation alterations in various cancers. In its use of probes and targets for hybridization, DMH is similar in development to the cDNA microarray. The "probes" for DMH are panels of CpG island tags arrayed on the stationary matrix and hybridized with the combined normal and tumor DNAs in the soluble phase, referred to as "targets." Conversely, MTA uses "targets" affixed on nylon membrane and hybridized with cancer-related CpG island "probes" one at a time.

Genomic Targets for DMH and MTA

To prepare targets for DMH or MTA analyses, genomic DNA is digested with a 4-base endonuclease, such as *Mse*I (T↓TAA), *Tsp509*I (↓AATT), *Nla*III (CATG↓), or *Bfa*I (C↓TAG). These endonucleases restrict bulk DNA into small fragments (<200 bp), but their recognition sites rarely occur in GC-rich regions and thus most CpG islands remain intact after the restriction (see Fig. 1).[35] The combined utility of these 4 endonucleases is estimated to cover the whole repertoire of methylated CpG islands in the genome. The digested DNA fragments (0.2 to 2 kb in length), which are enriched for CpG islands, are next ligated to end-linkers and then restricted with 4-base methylation-sensitive endonucleases *Bst*UI and *Hpa*II. An initial analysis indicates that approximately 85-90 % of the *Mse*I-digested, GC-rich fragments contain the recognition sites for either *Bst*UI or *Hpa*II. Fragments with hypermethylated *Bst*UI or *Hpa*II sites in the tumor sample resist the digestion and are amplified by PCR using the flanking linker-primers. In the normal control, the same allelic fragments are usually unmethylated and are digested away and thus cannot be amplified. Between 15-25 cycles of amplification are used for target preparation. These low cycles of amplification are essential to prevent overabundance of some repeat sequences not digested away by the methylation-sensitive restriction. We have also found that under this amplification condition, the pool of hypermethylated single- or low-copy CpG islands is sufficiently magnified and consequently enhances the methylation differential for comparison between the tumor and normal control genomes.

DMH and Its Applications

The Principle—Probes Affixed and Targets Mobile

CpG island clones used for generating the probes were derived from a genomic library, CGI, available from the Human Genome Mapping Project Resource Centre (http://www.hgmp.mrc. ac.uk).[36] After subjecting CGI to a round of screening to eliminate repetitive elements, a total of 8,000 CGI clones not hybridized with repetitive Cot-1 DNA probe were selected and arranged onto a series of 96-well plates. These CGI inserts contained in plasmid vectors were amplified by colony-PCR, and PCR products were deposited as microdots (~0.05 µl; 0.1 µg/µl) on a 4.5 x 1.6 cm² polylysine-coated surface that covalently binds DNA to a microscope slide. After postprocessing to remove unbound DNA, the CpG island fragments left bound to slide are ready for hybridization with fluorescently labeled targets.[33] Alternatively, these CpG island probes can be arrayed on nylon membrane for hybridization with radiolabeled targets.[37]

Using the microarray setup as an example, DNA targets of normal control are coupled with the green-fluorescent Cy3 dye while tumor targets are coupled with the red-fluorescent Cy5 dye (see Section II). As the two DNA samples are combined for hybridization, allelic DNA

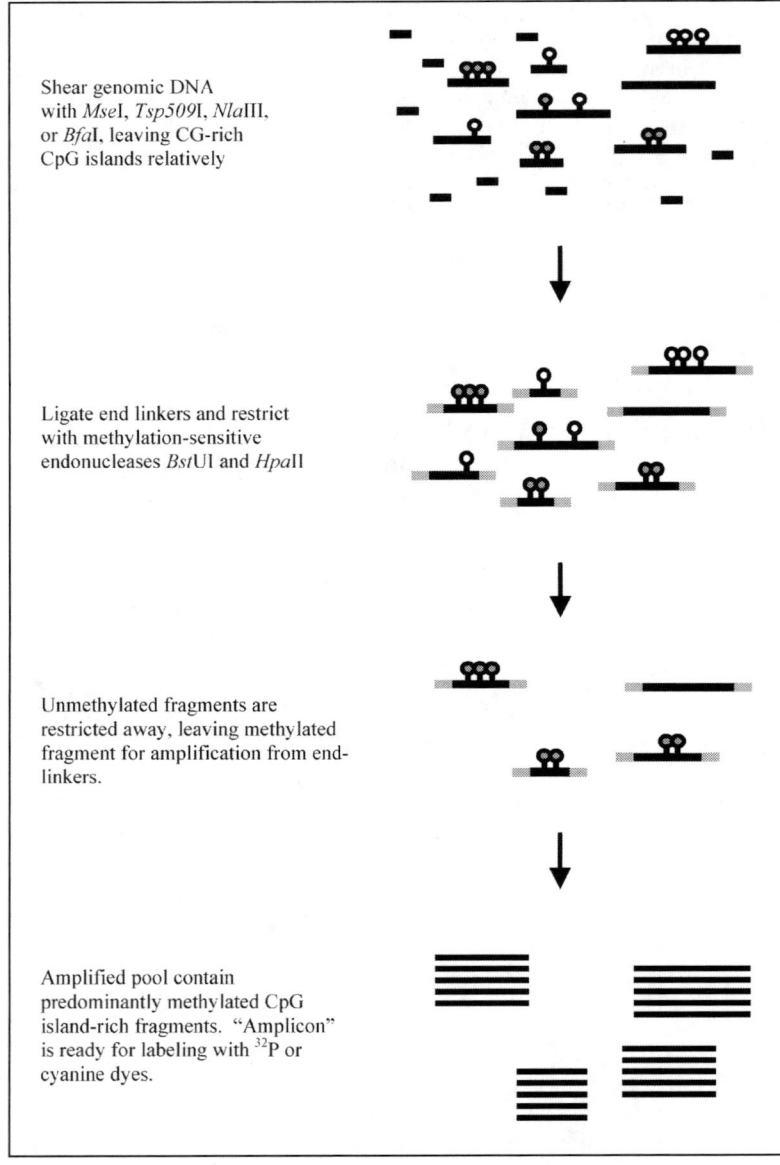

Shear genomic DNA
with *Mse*I, *Tsp509*I, *Nla*III,
or *Bfa*I, leaving CG-rich
CpG islands relatively

Ligate end linkers and restrict
with methylation-sensitive
endonucleases *Bst*UI and *Hpa*II

Unmethylated fragments are
restricted away, leaving methylated
fragment for amplification from end-
linkers.

Amplified pool contain
predominantly methylated CpG
island-rich fragments. "Amplicon"
is ready for labeling with ^{32}P or
cyanine dyes.

Figure 1. Schematic flowchart for methylation target preparation. Tumor and normal amplicon targets are similarly processed in the differential methylation hybridization (DMH) or methylation target array (MTA) assay.

fragments bind competitively to the CpG probes affixed on the glass slide. When an equal abundance of normal and tumor DNAs bind to a probe, the hybridized DNA is visually read as a yellow spot upon scanning and imaging, indicating the CpG island locus is methylated in both tumor and normal genomes. The presence of a red spot indicates a gain of CpG methylation in the tumor, but not in the normal, genome. Conversely, the presence of a green spot indicates a hypomethylation event in the tumor.

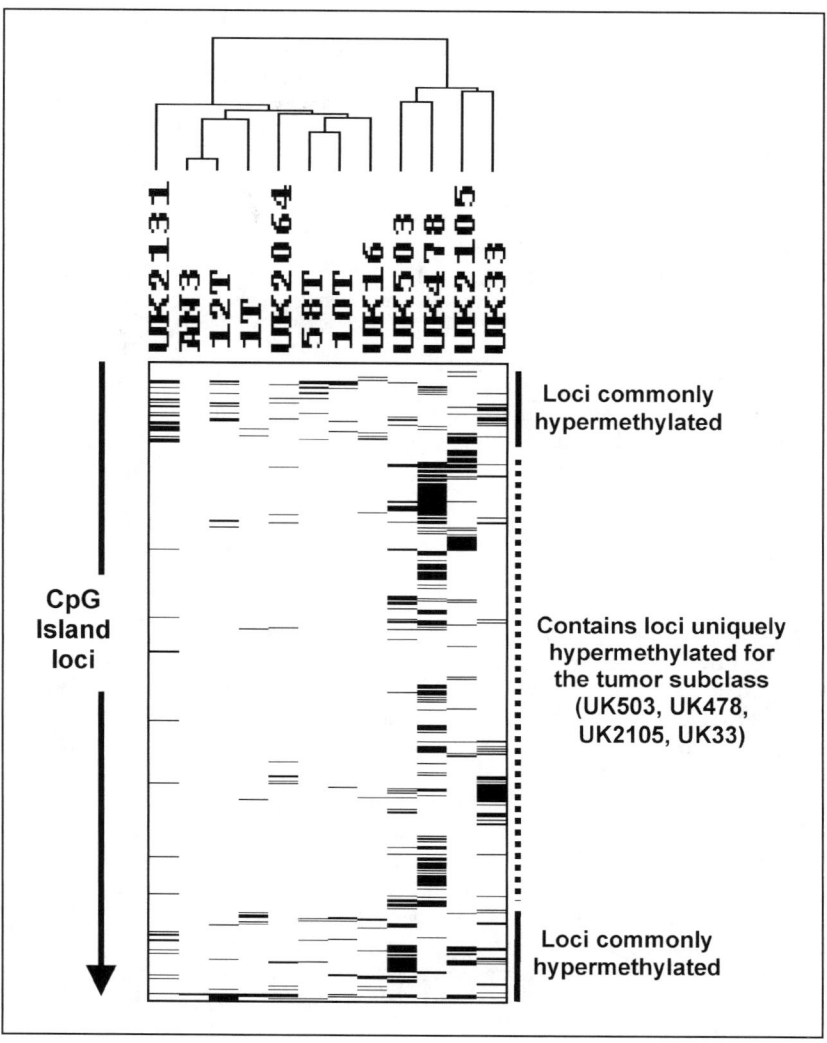

Figure 2. A representative hierarchical clustering of ovarian tumors. The panel depicts ~600 CpG island loci hypermethylated in at least one tumor. Tumors that carry similarly methylated loci, giving rise to their profiles, would tend to be clustered together. Loci commonly hypermethylated in all tumors as well as those hypermethylated in a subclass are detected (see an examples in ref. 46).

DMH has the capacity to generate thousands of data points in each experiment. Several bioinformatics tools, such as hierarchical clustering, self-organizing map, principal component analysis, and multidimensional-scaling, are available to manage and analyze data.[38,39]

Figure 2 shows a preliminary analysis of 12 ovarian carcinomas using hierarchical clustering, which groups together tumors of similar methylation profiles and identifies loci that are commonly hypermethylated in this tumor group as well as those that are unique to subgroups.

Points to Consider

The strength of DMH lies in its capacity to conduct a high-throughput scanning across thousands of CpG island loci in a timely manner. When applied in conjunction with clustering algorithms for analyzing many tumors, DMH becomes a powerful tool for delineating tumor relationships within the cohort and has the great potential for discovering molecular correlates for specific tumor subtypes. DMH is also flexible in that investigators can readily reassemble a smaller panel of CpG island probes for a more focused analysis in routine clinical samples.

DMH is limited in assessing the methylation details of individual CpG dinucleotides spanning a CpG island locus, however. As indicated earlier, this type of approach generates a DNA methylation profile of multiple CpG island loci for a tumor and is distinct from other approaches such as bisulfite sequencing,[40] which produces methylation patterns of several linked CpG dinucleotides within an interrogating CpG island locus.[41] Nevertheless, DMH assay is conducive to examining extensive CpG island hypermethylation in the genome, particularly for detecting densely hypermethylated gene promoters that may have a greater repressive effect on transcription.

As with other microarray techniques, DMH may produce false methylation findings.[33] One reason is the detection of genetic abnormalities due to the lack of methylation-sensitive sites in some CpG island loci. In this case, the observed differential hybridization signals in DMH could be attributed to gain or loss of copy-number of such loci in the tumor sample. As indicated earlier, almost all of the interrogating CpG island loci contain either the methylation-sensitive *Hpa*II or *Bst*UI sites. The combined use of *Hpa*II and *Bst*UI restrictions in the preparation of methylation targets certainly minimizes the detection of genetic alterations in the DMH assay. Therefore, this genetic event does not affect the overall evaluation of methylation profiles in tumor genomes.

Applications to Identify Clinicopathological Association

In a DMH study, the extent of hypermethylation averaged 1% of the ~8,000 CpG island loci examined in a panel of breast tumors,[33] which was consistent with earlier estimates for this cancer.[23,42] DNA hypermethylation is usually associated with more advanced disease. Poorly differentiated tumors were found to exhibit more hypermethylation than their moderately or well-differentiated counterparts (P = 0.041).[42] Furthermore, hierarchical clustering identified 3 subgroups of breast tumors based on their estrogen-receptor (ER)/progesterone-receptor (PR) status.[33] One group was ER/PR-positive and exhibited less hypermethylation while two distinct epigenetic subgroups were ER/PR-positive, showing poorer response to hormone therapy. Both commonly hypermethylated CpG islands as well as unique loci specific to the distinctions described above were found. These include 5' regulatory regions of genes, exonic regions, and a few intergenic sequences. Interestingly, a detailed investigation of one gene, *glypican 3* (*GPC3*), indicated that its promoter hypermethylation resulted in transcriptional silencing in vitro and was associated with ER/PR-negative breast tumors (P = 0.005).[33] GPC3 is an X-linked gene found to be mutated in the Simpson-Golabi-Behmel syndrome and reportedly, a candidate tumor suppressor for multiple embryonic neoplasms and pre and postnatal overgrowth.[43] This novel finding suggests that silencing of *GPC3* via DNA hypermethylation may play a role in breast cancer development.

In a second DMH study, a higher degree of hypermethylation across the ~8,000 CpG islands was detected, ranging from 0.9 to 4.5%, in a group of stages III and IV ovarian tumors.[44,45] Computation analyses, including hierarchical clustering, identified two subgroups of tumors with distinct methylation profiles in a select group of 182 CpG islands; tumor from the first group showed high levels of current methylation in this CpG island group, while the second group exhibited less methylation in these loci. Based on patient's response to chemotherapy as defined by the months of progression-free survival (PFS), these groups differed markedly from each other.[45] The first group, whose tumors were more hypermethylated, has a median PFS of 6 months whereas the second group whose tumors were less hypermethylated, has a median PFS of 15 months (P<0.001, log rank test). In clinical settings, these selected CpG island loci are therefore useful as DNA biomarkers for predicting chemotherapy outcome in ovarian cancer patients.

In a third study, DMH has identified a small group of colorectal tumors with high degrees of concurrent methylation,[46] consistent with a previous observation describing the so called CpG island methylator phenotype in this cancer.[47] Although no other clear correlations were found in this initial study, it demonstrated, nevertheless, that DMH was useful for assessing methylation of CpG islands across the genome and for classifying tumors.

MTA and Its Applications

The Principle—Targets Affixed and Probes Mobile

MTA was devised for analyzing extensive promoter hypermethylation for many tumors in a single experiment, similar in concept to the tissue microarray.[48] To increase throughput, methylation targets are prepared as described earlier, but in a 96-well format (Fig. 3). Genomic DNA is first restricted with *Mse*I, *Tsp509*I, *Nla*III, or *Bfa*I and interrogated for hypermethylation by restricting with methylation-sensitive *Bst*UI and *Hpa*II, prior to PCR amplification. Contrary to

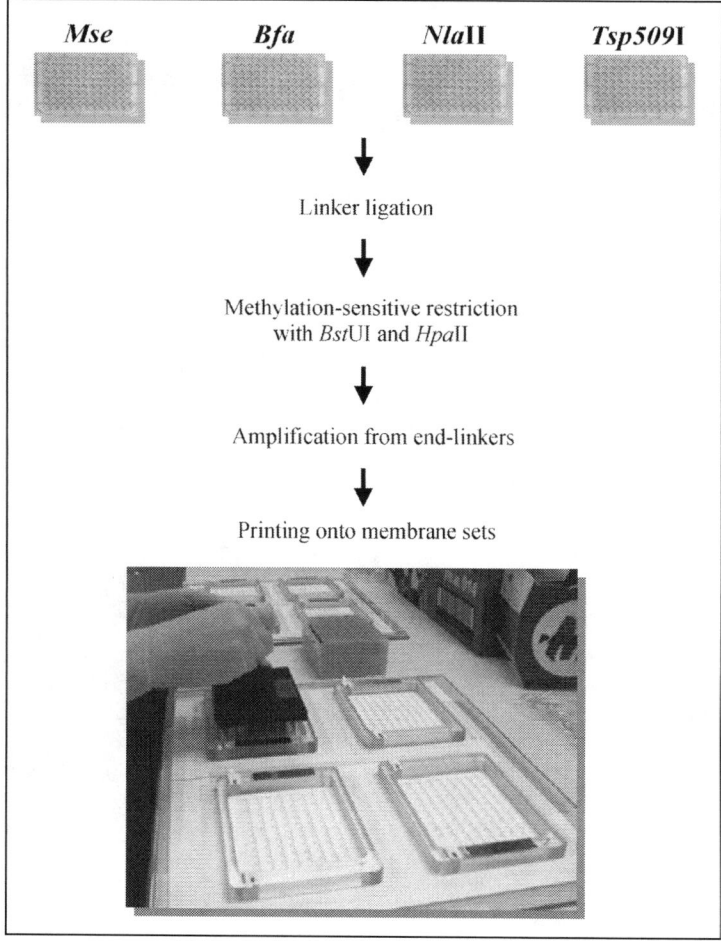

Figure 3. Preparation of targets for methylation target array. Tumor and normal amplicon targets are prepared in 96-well formats and affixed onto separate sets of membrane filters for probing with ^{32}P-labeled CpG island probes.

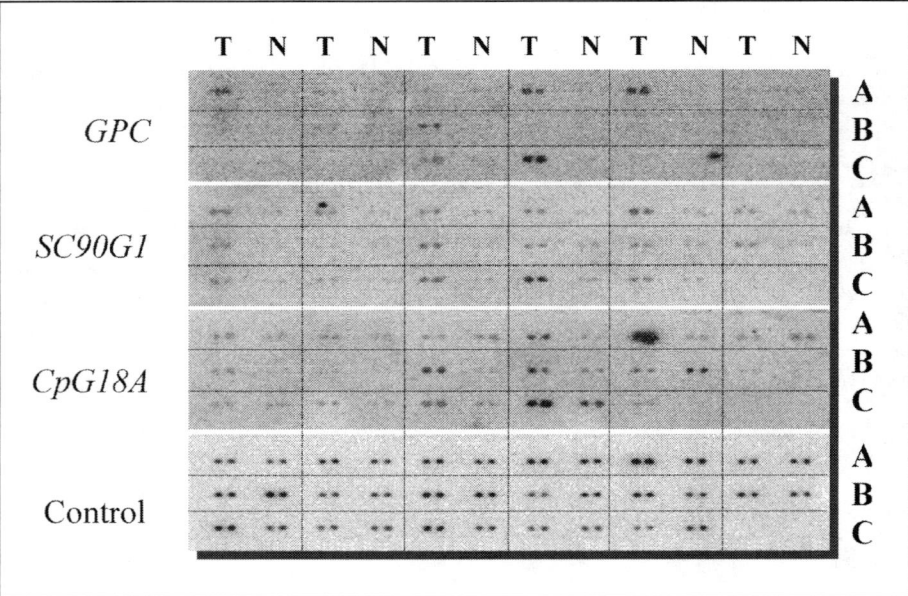

Figure 4. Representative results of methylation target array. Each tumor (T) target was spotted in duplicate on a membrane with its paired normal (N) control. Repetitive Cot-1 DNA, which is present in both tumor and control targets, was used as a control probe.

DMH in which the targets are in solution, the MTA targets are arrayed in duplicate onto nylon membrane and hybridized with a ^{32}P-labeled CpG island probe.

Figure 4 shows representative results of 18 paired tumor and normal targets. The hybridization intensity for a spot in the duplicate target is individually quantified and normalized, and positive hypermethylation is scored when the calculated intensities of tumor relative to normal control "spots" are above a set threshold.

Applications

In a pilot study, a total of 468 methylation targets prepared from 93 breast tumors, 20 normal controls and 4 breast cancer cell lines were arrayed onto nylon membranes. These MTA membranes were hybridized individually with probes derived from promoter CpG islands of 6 known tumor-suppressor genes (*GPC3, uPA, HOXA5, MUC2, p16*, and *BRCA2*).[34] Hypermethylation was found at the levels of 63%, 38%, 33%, 27%, 26%, and 0%, respectively, in the breast tumors examined, and the findings were independently confirmed in the *GPC3* and *p16* genes by Southern blot analysis and methylation-specific PCR. This initial study demonstrated that MTA represents a powerful, candidate gene approach for rapid methylation analysis of a single CpG island across hundreds of tissue genomes, and for generating methylation profiling of tumors when multiple promoter CpG islands are tested. Analyzed in conjunction with patients' clinical data, methylation profiles may be rapidly dissected to discover significant clinicopathological associations. A broader investigation with additional gene promoters is expected to unveil tumor subgroups comprised of different sets of hypermethylated loci.

Concluding Remarks

Promoter hypermethylation is an important epigenetic mechanism underlying the down-regulation of many cancer-related genes. The notion that accumulations of this heritable modification play a role in promoting tumor development suggests the clinical use of these molecular flags to monitor cancer progression, before the appearance of any diagnosable symptoms. DMH and MTA are high-throughput approaches complementary to one another for exploring methylation in the genome and for analyzing hypermethylation in large-scale tumor samples. In addition to tumor profiling, the DMH assay can also identify new genes that are aberrantly hypermethylated, and the MTA assay can rapidly assess the clinical utility of these molecular markers. Alternatively, using prescreened subpanels of the 8,000 CpG island clones, DMH may also lend itself to the translational arena for large-scale validation studies using specialized microarrays.

DNA methylation is reversible where demethylating agents are available for use in synergism with chemotherapy. Indeed, translational cancer research is dynamically positioned now for new strategies in therapeutics.[49] Along with several established assays that are highly sensitive for detecting DNA methylation at defined CpG dinucleotides, epigenomic-based research at the level of large-scale analyses such as the ones described here is possible for routine diagnosis of cancer.

References

1. Jones PA, Laird PW. Cancer epigenetics comes of age. Nat Genet 1999; 21:163-167.
2. Baylin SB, Esteller M, Rountree MR et al. Aberrant patterns of DNA methylation, chromatin formation and gene expression in cancer. Hum Mol Genet 2001; 10:687-692.
3. Feinberg AP. Cancer epigenetics takes center stage. Proc Natl Acad Sci USA 2001; 98:392-394.
4. Robertson KD. DNA methylation, methyltransferases, and cancer. Oncogene 2001; 20:3139-3155.
5. Costello JF, Plass C. Methylation matters. J Med Genet 2001; 38:285-303.
6. Kim DH, Nelson HH, Wiencke JK et al. Promoter methylation of DAP-kinase: association with advanced stage in nonsmall cell lung cancer. Oncogene 2001; 20:1765-70.
7. Strathdee G, MacKean MJ, Illand M et al. A role for methylation of the hMLH1 promoter in loss of hMLH1 expression and drug resistance in ovarian cancer. Oncogene 1999: 18: 2335-2341.
8. Nosaka K, Maeda M, Tamiya S et al. Increasing methylation of the CDKN2A gene is associated with the progression of adult T-cell leukemia. Cancer Res 2000; 60:1043-8.
9. Esteller M, Garcia-Foncillas J, Andion E et al. Inactivation of the DNA repair gene MGMT and the clinical response of gliomas to alkylating agents. N Eng J Med 2000; 343:1350-4.
10. Tang X, Khuri FR, Lee JJ et al. Hypermethylation of the death-associated protein (DAP) kinase promoter and aggressiveness in stage I nonsmall-cell lung. J Natl Cancer Inst 2000; 92:1511-6.
11. Sanchez-Cespedes M, Esteller M, Wu L et al. Gene promoter hypermethylation in tumors and serum of head and neck cancer patients. Cancer Res 2001; 60:892-5.
12. Roman-Gomez J, Castillejo JA, Jimenez A et al. 5' CpG island hypermethylation is associated with transcriptional silencing of the p21[CIP/WAF1/SDI1] gene and confers poor prognosis in acute lymphoblastic leukemia. Blood 2002; 99:2291-2296.
13. Niwa Y, Oyama T, Nakajima T. BRCA1 expression status in relation to DNA methylation of the BRCA1 promoter region in sporadic breast cancers. Jpn J Cancer Res 2000; 91:519-526.
14. Nass SJ, Herman JG, Gabrielson E et al. Aberrant methylation of the estrogen receptor and E-caderin 5' CpG islands increases with malignant progression in human breast cancer. Cancer Res 2000; 60:4346-4368.
15. Kim DH, Nelson HH, Wiencke JK et al. p16(INK4a) and histology-specific methylation of CpG islands by exposure to tobacco smoke in nonsmall cell lung cancer. Cancer Res 2001; 61:3419-3424.
16. Belinsky SA, Nikulua KJ, Palmisano WA et al. Aberrant methylation of p16INK4a is an early event in lung cancer and a potential biomarker for early diagnosis. Proc Natl Acad Sci USA 1998; 95:11891-11896.
17. Issa JPJ, Zehnbauer BA, Kaufmann SH et al. HIC1 hypermethylation is a late event in hematopoietic neoplasms. Cancer Res 1997; 57:1678-1681.
18. Eads CA, Lord RV, Kurumboor SK et al. Fields of aberrant CpG island hypermethylation in Barrett's esophagus and associated adenocarcinoma. Cancer Res 2000; 60:5021-5026.
19. Eads CA, Lord RV, Wickramasinghe K et al. Epigenetic patterns in the progression of esophageal adenocarcinoma. Cancer Res 2001; 61:3410-3418.
20. Maruyama R, Toyooka S, Toyooka KO et al. Aberrant promoter methylation profile of bladder cancer and its relationship to clinicopathological features. Cancer Res 2001; 61:8659-8663.
21. Issa JPJ, Ahuja N, Toyota M et al. Acclerated age-related CpG island methylation in ulcerative colitis. Cancer Res 2001; 61:3573-3577.
22. Virmani AK, Muller C, Rathi A et al. Aberrant methylation during cervical carcinogenesis. Clin Cancer Res 2001; 7:584-589.

23. Esteller M, Corn PG, Baylin SB et al. A gene hypermethylation profile in human cancer. Cancer Res 2001; 61:3225-3229.
24. Strathdee G, Appleton K, Illand M et al. Primary ovarian carcinomas display multiple methylator phenotypes involving known tumor suppressor genes. Am J Pathol 2001; 158:1121-1127.
25. Huang T H-M, Perry MR, Laux DE. Methylation profiling of CpG islands in human breast cancer cells. Hum Mol Genet 1999; 8:459-470.
26. Costello JF, Fruhwald MC, Smiraglia DJ et al. Aberrant CpG-island methylation has nonrandom and tumor-type-specific patterns. Nat Genet 2000; 25:132-138.
27. Gonzalgo ML, Liang G, Spruck CH et al. Identification and characterization of differentially methylated regions of genomic DNA by methylation-sensitive arbitrarily primed PCR. Cancer Res 1997; 57:594-599.
28. Eads CA, Danenberg KD, Kawakami K et al. MethyLight: a high-throughput assay to measure DNA methylation. Nucleic Acids Res 2000; 28:e32.
29. Toyota M, Ho C, Ahuja N et al. Identification of differentially methylated sequences in colorectal cancer by methylated CpG island amplification. Cancer Res 1999; 59:2307-2312.
30. Costello J, Plass C, Cavenee WK. Restriction landmark genomic scanning. In: Mills KI, Ransahoye BH, eds. DNA Methylation Protocols. Totowa: Humana Press, 2002:53-70.
31. Adorjan P, Distler J, Lipscher E et al. Tumor class prediction and discovery by microarray-based DNA methylation analysis. Nucleic Acid Res 2002; 30:e21.
32. DNA Methylation Protocols. Mills KI, Ransahoye BH, eds. Totowa: Humana Press, 2002.
33. Yan PS, Chen C-M, Shi H et al. Dissecting complex epigenetic alterations in breast cancer using CpG island microarrays. Cancer Res 2001; 61:8375-8380.
34. Chen C-M, Chen H-L, Yan PS et al. Methylation "tissue" array: a novel technique for simultaneous analysis of extensive DNA hypermethylation in multiple breast tumors. Proc Am Asso Cancer Res 2002; 43:1124a.
35. Shiraishi M, Lerman L, Seyika T. Preferential isolation of DNA fragments associated with CpG islands. Proc Natl Acad Sci USA 1995; 92:4229-4233.
36. Cross SH, Charlton JA, Nan X et al. Purification of CpG islands using a methylated DNA binding column. Nat Genet 1994; 6:236-244.
37. Yan PS, Wei SH, Huang TM-H. Differential methylation hybridization using CpG island arrays. In: Mills KI, Ransahoye BH, eds. DNA Methylation Protocols Totowa: Humana Press, 2002:87-100.
38. Eisen MB, Spellman PT, Brown PO et al. Cluster analysis and display of genome-wide expression patterns. Proc Natl Acad Sci USA 1998; 95:14863-14868.
39. Tamayo P, Slonim D, Mesirov J et al. Interpreting patterns of gene expression with self-organizing maps: methods and application to hematopoietic differentiation. Proc Natl Acad Sci USA 1999; 96:2907-2912.
40. Warnecke PM, Mann JR, Frommer M et al. Bisulfite sequencing in preimplantation embryos: DNA methylation profile of the upstream region of the mouse imprinted H19 gene. Genomics 1998; 51:182-190.
41. Huang TH-M, Plass C, Liang G et al. Epi meets genomics: Technologies for finding and reading the 5th base. In: Beck & Olek, eds. The Epigenome—Molecular Seek and Hide. Verlag - VCH Verlag Gm5H, 2002 (submitted).
42. Yan PS, Perry MR, Laux DE et al. CpG island arrays: an application toward deciphering epigenetic signatures of breast cancer. Clin Cancer Res 2000; 1432:1432-1438.
43. Veugelers M, Cat BD, Muyldermans SY et al. Mutational analysis of the GPC3/CPC4 glypican gene cluster on Xq26 in patients with Simpson-Golabi-Behmel syndrome: identification of loss-of-function mutations in the GPC3 gene. Hum Molec Genet 2000; 9:1321-1328.
44. Ahluwalia A, Yan P, Hurteau JA et al. DNA methylation and ovarian cancer. I. Analysis of CpG island hypermethylation in human ovarian cancer using differential methylation hybridization. Gynecol Oncol 2001; 82:261-268.
45. Wei SH, Chen C-M, Strathdee G et al. Methylation microarray analysis of late-stage ovarian carcinomas distinguishes progression-free survival in patients and identifies candidate epigenetic markers. Clin Cancer Res 2002; in press.
46. Yan PS, Efferth T, Chen H-L et al. Use of CpG island microarrays to identify colorectal tumors with a high degree of concurrent methylation. Methods, in press.
47. Toyota M, Ahuja N, Ohe-Toyota M et al. JP. CpG island methylator phenotype in colorectal cancer. Proc Natl Acad Sci USA 1999; 96-8681-8686.
48. Kononen J, Bubendorf L, Kallioniemi A et al. Tissue microarrays for high-throughput molecular profiling of tumor specimens. Nat Med 1998; 4:844-847.
49. Widschwendter M, Jones PA. The potential prognostic, predictive, and therapeutic values of DNA methylation in cancer. Clin Cancer Res 2002; 8:17-21.

CHAPTER 9

Methylation Analysis in Cancer:
(Epi)Genomic Fast Track from Discovery to Clinical Routine

Carolina Haefliger, Sabine Maier and Alexander Olek

Abstract

Aberrant DNA methylation is an early and common event in human cancers. Methylation acts as an epigenetic regulator of gene expression and is involved in cancer development as well as resistance to drug treatments. Specific methylation patterns have been shown for different cancer types and there is evidence that methylation can be used as a diagnostic tool. Several methods have been developed to study methylation on a genome wide basis. However they are labor intensive and can assess only a limited number of tissues at a time preventing the assessment of these genes in larger populations. Methylation microarrays now offer the possibility to validate these candidate genes statistically filling the gap between genome wide discovery methods and single gene assays which could be adjusted to routine clinical use. Here we show how all these methods can be combined to broaden our knowledge regarding DNA methylation and transform some of this information into powerful diagnostic tests.

DNA Methylation and Carcinogenesis

Mammalian cells have the capacity to epigenetically modify their genomes by the addition of a 5'methyl group to cytosines within the context of CpG dinucleotides.[1] This form of methylation has a nonrandom distribution throughout the genome. The CpGs occur in clusters, called CpG islands, in the 5' promoter region of genes and the vast majority of the genome remains CpG dinucleotide poor.[2] These islands can also be located in the first exon or in regions towards the 3'end of the genes.[3] CpG islands are unmethylated in the germline and in most somatic tissues. Correlation between DNA methylation and gene expression has been known for more than 20 years, although the exact mechanisms of how this happens are still unclear.[4] Many of the proteins involved in gene expression, promoter methylation, histone acetylation and chromatin structure have been shown to interact with each other (details in refs. 5-7).

Methylation changes contribute to carcinogenesis by several mechanisms. First, aberrant DNA methylation patterns have been clearly correlated to repression or overexpression of several cancer related genes.[8] Although hypermethylation of promoter regions of specific tumor suppressor genes is the main alteration, general hypomethylation is a common early feature of tumors, and correlates with an increased expression of several oncogenes.[9,10] Second, methylated CpGs are hot spots for mutations and can contribute to up to 30% of all point mutations in the germ line.[11] The increase in the mutation rate may be due to several factors including the differential repair efficiency of deamination of methylcytosines (that results in thymines) and

DNA Methylation and Cancer Therapy, edited by Moshe Szyf. ©2005 Eurekah.com and Kluwer Academic/Plenum Publishers.

cytosines (that form uracils), the first error being more difficult to detect.[12] Other factors that can play a role are the rate of spontaneous deamination of 5 methylcytosines and the rate of cell division.[13,14] Other ways to contribute to carcinogenesis may be alterations in the methylation machinery although this has not yet been shown to be linked to tumorigenesis.[15] Moreover an enhanced inactivation, via promoter hypermethylation, of certain tumor suppressor genes, in relationship with environmental carcinogens such as smoking and LUV has been demonstrated in several publications.[16,17]

During carcinogenesis, de novo methylation of certain genes has been widely demonstrated.[8] The exact mechanisms responsible for de novo methylation are not known. There is evidence that the abnormal methylation of individual CpG islands may reflect a widespread loss of protection against methylation.[10] DNA methyltransferases, such as DMNT3b, may be responsible of this de novo methylation and, together with DNMT1, cooperatively maintain methylation in cancer cells, as shown by Rhee et al.[18] CpG islands located in exons are more susceptible to de novo hypermethylation than promoter regions and they may be the first to hypermethylate and, from there methylation spreads to promoter regions. These cells would then acquire a growth advantage, via the spread of methylation and promoter inactivation, that may contribute to cancer development.[19]

Methylation Profile in Human Cancer

The systematic detection of methylation patterns is a powerful tool for human cancer profiling as studies of colon, breast, lung and several other tumor types have shown.[20] In this essay, colorectal and prostate cancer are taken as examples to show the implications of methylation in tumorigenesis.

With regard to sporadic colorectal cancer (CRC) development, Toyota et al describe a distinct pathway, termed CpG island methylator phenotype (CIMP).[21] They identified two types of methylation in colonic cells. Type A methylation is age-related, tissue specific and not cancer-related. The other phenotype, type C, is cancer specific and associated with transcriptional silencing of tumor suppressor genes such as *p16* and *hMLH1* (from the mismatch repair family) and is related to higher microsatellite instability (MSI high). They suggest that CIMP may have profound pathophysiological consequences in cancer development through the inactivation of tumor suppressor genes, metastasis supressor genes and angiogenesis inhibitors and that this defect could be due to a loss of protection against methylation. Further studies showed that methylation of the proximal region of *hMLH1* promoter, but not of the distal one, correlated with lack of expression and high MSI.[22] Furthermore, aberrant methylation of the *p16* gene is associated with higher grades of Dukes classification and may be a marker for poor prognosis. These markers can be detected in the serum of patients using highly sensitive detection methods.[23,24]

Differences in gene expression, and in the extent of CpG island methylation, enable the differentiation between MSI high hereditary and sporadic CRCs as well as between adenomas with different predisposition to develop invasive carcinoma.[25,26] Taken collectively, these characteristic differences in the type and extent of hypermethylation may show diverging developmental pathways and could become useful markers of category and progression.

Several specific genetic alterations have been identified in prostate cancer, including *ras* oncogene activation and inactivation of tumor suppressor genes as *p53*, *Rb* and *CDKN2a*. However, mutations in these genes are rare or are a late event in carcinogenesis such as *PTEN* inactivation .[27,28] On the other hand, promoter hypermethylation of *GSTP1* gene seems to be one of the most frequent alterations in prostate carcinogenesis.[29,30] This gene belongs to the GTS family of enzymes that catalyze intracellular detoxification reactions by conjugating glutathione with electrophilic compounds such as carcinogens and exogenous drugs.[31] *GSTP1* acts as a tumor suppressor gene as the loss of its expression leads to increased susceptibility of prostatic cells to the carcinogenic effects of these substances. Methylation of this gene is associated with loss of expression in the majority of prostate cancers and also in prostatic intraepithelial neoplasia (PIN) but not in normal tissue or in benign prostatic hyperplasia (BPH), suggesting that it is an early event in the development of malignancy.[32,33] The detection of *GSTP1* methylation changes in several body fluids such as plasma, serum, urine with or without previous prostatic massage, biopsy washings and ejaculate could provide an alternative to the currently used PSA (Prostatic Specific Antigen) assay for early detection.[34-36]

Technological Approaches for DNA Methylation Study: High Throughput Analysis

During the past decade, several technological approaches have been developed to enable different aspects of methylation analysis.

Genome Wide Screening

To screen for novel CpG islands throughout the whole genome several discovery techniques are available. All are based on methylation-specific restriction enzymes that cut only if the sequence is not methylated. Different methods are applied to identify differentially methylated sites, as Restriction Landmark Genomic Scanning (RLGS), which allows the simultaneous assessment of the methylation status and copy number of several thousand CpG islands.[37,38] This method enables the elucidation of different and characteristic methylation patterns within and between tumor types and the estimation of the overall influence of CpG island methylation on the cancer cell genome.[39,40] Other methods include Methylation-Sensitive Arbitrarily Primed PCR (AP-PCR), Methylated CpG island amplification (MCA) and differential methylation hybridization (DMH).[41-43] Although these techniques are all very important for the discovery of novel differentially methylated sites, the CpG sites identified with each of them is limited by the recognition sites of the restriction enzyme used. Except DMH, they are all difficult to automatize.

These discovery technologies are a first, unbiased approach to methylation in cancer and identify large number of candidate genes. These will include novel genes, other genes whose correlation with cancer has only been identified in a limited number of publications and also those with a well characterized methylation status. As the number of samples assessed by discovery methods is low, these genes frequently contain coincidental findings which need to be separated from the relevant ones by analysis of larger population.

Chip Technology

In order to approach studies with higher number of samples to assess the methylation status of many genes in parallel a high through put approach is required. The aim of chip studies is to filter the hundreds of genes identified with Genome wide screening, in order to identify the subset of genes whose correlation to a particular cancer is statistically significant.

We have recently shown that a microarray-based approach is a powerful tool to analyze hundreds of CpG sites in a large number of samples in parallel.[44] In order to distinguish methylated from unmethylated cytosines by a hybridization procedure, total DNA from the samples of interest are first treated with bisulphite. This procedure converts all the unmethylated cytosines to uracils whereas methylated cytosines remain unconverted.[45] Regions of interest are then amplified by PCR using fluorescently labeled primers. In the amplified nucleic acids the originally unmethylated CpG dinucleotides are converted to TG dinucleotides and the methylated ones are conserved. The PCR primers are designed complementary to regions which do not contain CpGs. Hence, in one reaction methylated and unmethylated genes are amplified without bias. All PCR products of each individual sample are combined and hybridized to glass slides. The slides have a pair of oligonucleotides for the analysis of each CpG site in question immobilized to the surface in the form of a microarray. The detection oligonucleotides are designed to hybridize specifically to the sequence surrounding the originally methylated or unmethylated CpG of interest (Fig. 1). Hybridization conditions are selected to allow detection of a single nucleotide difference between TG and

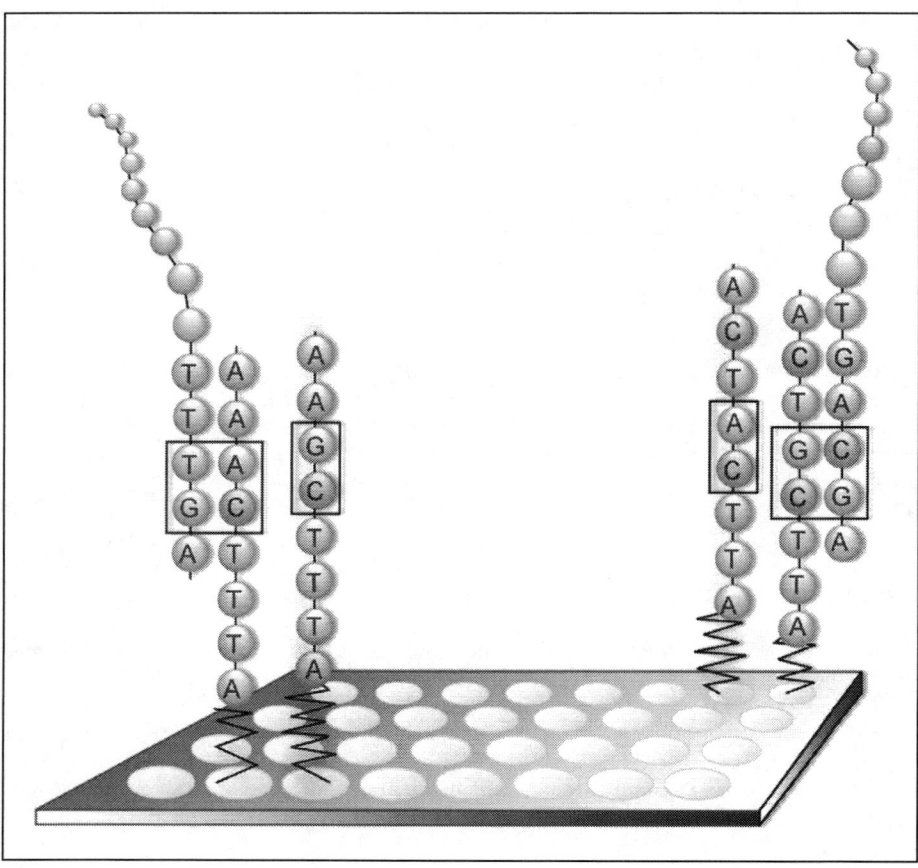

Figure 1. Microarray slide containing oligonucleotides for the methylated and unmethylated CpGs. Hybridization with sample DNA.

CG variants. Then, the ratio between the two signals is calculated based on the comparison of intensity of the fluorescent signals. The degree of methylation of a single position can be quantified by internal calibration of the microarray.

The data of the microarray hybridization is then analyzed by complex statistical tools such as support vector machines (SVM). First all CpGs are ranked for a given separation task by the significance of the difference between the two class means (e.g., between normal and neoplastic tissue). The significance of each CpG is estimated by a two sample t-test and the SVM is then trained upon the most significant CpG positions.[46] The optimal number of CpGs to distinguish between two particular classes depends on the complexity of the separation task; a proper feature selection improves the class prediction and provides useful information about the complexity of the problem.

Combination of microarray technology and bioinformatics analysis is highly effective in predicting known and discovering novel tumor classes, as established by Adorjan et al, 2002 in several different tissues. In this study we analyzed normal and neoplastic tissue from prostate and kidney, as well as T and B cells from normal and T-ALL and B-ALL (Acute Lymphocytic Leukemia) patients. Classification results are comparable to mRNA techniques regarding accuracy, safety and reproducibility. Advantages include the possibility of screening

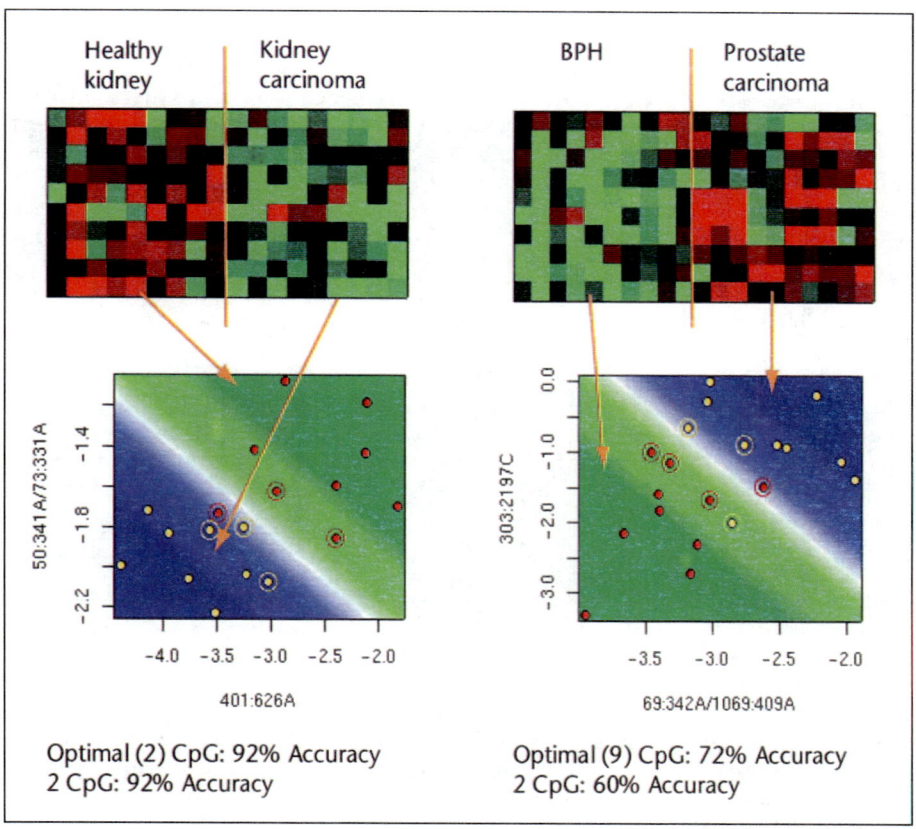

Figure 2. Above: Methylation patterns in kidney carcinoma and controls and benign prostate hyperplasia and prostate carcinoma as the log ratio of the CG and TG signal intensities. Red represents hypermethylation and green hypomethylation; columns are individual samples and rows are single CpGs. The most significant CpGs for differential diagnosis are shown in these graphics. Below: Class prediction of the same samples. SVM were trained with the most significant CpGs for discrimination of the two classes. Circled points are the supported vectors that defined the line between both class areas (green for normal or BPH and blue for cancer). Modified from Nuc Aid Res 2002, Vol 30, No 30 e21.

large populations and paraffin embedded samples; applications currently not suitable for mRNA analysis due to its relative instability in comparison with DNA. Moreover, mRNA expression profiling signal intensities are dependent on the absolute and relative amounts of different RNA species therefor comparison between independent samples is difficult. In contrast, in the methylation microarray approach the CT:CG ratio is used as an internal calibration, hence the amount of probe does not influence results.[47] (Fig. 2).

Compared to methylation-sensitive restriction enzyme based technologies, microarrays are less labor intensive, they can be easily automated and do not have the limitation of methylation sensitive restriction sites or the occurrence of false positives due to incomplete digestion. Also many samples can be analyze in parallel and less DNA is required.[44] By employing automatization, assessment of the methylation patterns of different cancer types will become a powerful tool for diagnostic and prognostic purposes in greater populations.

Sensitive Detection

Cancer patients have elevated amounts of free DNA in serum and other body fluids. These DNA is thought to be released from apoptotic or dead cancer cells. We and others have shown that aberrant methylation patterns in the tumor can be detected in several body fluids including serum, plasma, urine, ejaculate and in secretions such as sputum. In order to detect these methylation markers in small amounts of DNA, assays such as Methylation-specific PCR (MSP) have been developed which has been successful, for example, in screening of *the GPST1* gene in prostate cancer patients and for the *MGMT* gene in sputum of squamous cell carcinoma patients.[34,48]

A second generation of sensitive detection methods includes MethyLight, which combines MSP with fluorescent-based real-time PCR (TaqMan technology). This high throughput assay is highly specific, sensitive and reproducible and has several applications. These include, rapid screening for methylation state of a particular locus and determination of the relative prevalence of a specific DNA methylation pattern.[49] More recently, we have developed a third generation of sensitive assays with technological modifications, called HeavyMethyl.[52]

These methods can be useful as the validation step for markers, obtained by chip analysis, in large populations and will, in the near future, provide the platform for quick, simple assays for routine screening purposes.

Future Perspectives

DNA methylation patterns, although known for more than two decades, were neglected as there were no suitable technologies for their analysis. With the development of methylation microarrays, large number of samples can be analyzed making this technology useful for clinical and research applications.

Methylation can be used as a discovery tool in oncology to detect genes that cannot be found with other methods, e.g., genes with gradual expression alteration or those difficult to detect with expression analysis. Arrays can also be useful to select the best methylation markers of a specific tumor type. Then these markers can be transferred to a sensitive detection platform for screening of larger populations.

Moreover microarrays can be designed in such a way that they target genes from specific pathways that could be involved in drug response modulation. With these panels one would be able to identify markers of response or resistance to a drug in retrospective sample collections or during clinical trials.

References

1. Bird A. The essentials of DNA methylation. Cell 1992; 70:5-8.
2. Cooper DN, Krawczak M. Cytosine methylation and the fate of CpG dinucleotides in vertebrate genomes. Hum Genet 1989; 83:181-188.
3. Larsen F, Gundersen G, Lopez R et al. CpG islands as gene markers in the human genome. Genomics 1992; 13:1095-1107.
4. Taylor SM, Jones PA. Multiple new phenotypes induced in 10T1/2 and 3T3 cells treated with 5-azacytidine. Cell 1979; 17:771-779.
5. Yievin A, Razin A. Gene methylation patterns and expression. In: Jost JP, Saluz HP, eds. DNA methylation: Molecular and Biological Significance. Basel: Birkhauser Verlag, 1993:523-568.
6. Nguyen CT, Gonzales FA, Jones PA et al. Altered chromatin structure associated with methylation-induced gene silencing in cancer cells: correlation of accessibility, methylation, MeCP2 binding and acetylation. Nuc Acids Res 2001; 29(22):4598-4606.
7. Jones P, Veentra GJ, Wade PA et al. Methylated DNA and MePC2 recruit histone deacetylase to repress transcription. Nature Genet 1998:19:187-191.
8. Jones PA. DNA methylation errors and cancer. Cancer Res 1996; 56:2463-2467.
9. Feinberg AP, Vogelstein B. Hypomethylation distinguishes genes of some human cancers from their normal counterparts. Nature 1983; 301:8:9-92.
10. Jones PA, Laird PW. Cancer epigenetics comes of age. Nature Genet 1999; 21:163-167.
11. Cooper DN, Youssoufian H. The CpG dinucleotide and human genetic disease. Hum Genet 1988; 78:1:51-155.

12. Jiricny J. Mismatch repair and cancer. Cancer Surveys: Genetic instability in Cancer. Imperial Cancer Research Fund 1996; 28:47-68.

13. Shen JC, Rideout WM, Jones PA. The rate of hydrolitic deamination of 5-methylcytosine in double stranded DNA. Nucleic Acids Res 1994; 22:9:72-976.

14. Lieb M, Rehmat S. 5-methylcytosine is not a mutagen hot spot in non dividing Escherichia Coli. Proc Natl Acad Sci USA 1997; 94:940-945.

15. Shen JC, Rideout WM, Jones PA. High frequency mutagenesis by a DNA methyltransferase. Cell 1992; 71:1073-1080.

16. Soria J-C, Rodriguez M, Liu DD et al. Aberrant promoter methylation of multiple genes in bronchial brush samples form former cigarette smokers. Cancer Res 2002; 62:351-355.

17. Tommasi S, Denissenko MF, Pfeifer GP. Sunlight induces pyrimdine dimers preferentially at 5-methylcytosine bases. Cancer Res 1997; 57:4727-4730.

18. Rhee I, Bachman KE, Park BH et al. DNMT1 and DNMT3b cooperate to silence genes in human cancer cells. Nature 2002; 416:552-556.

19. Nguyen C, Liang G, Nguyen TT et al. Susceptibility of non promoter CpG islands to de novo methylation in normal and neoplastic cells. J Nat Cancer Inst 2001; 93(19):1465-1472.

20. Esteller M, corn PG, Baylin SB et al. A gene hypermethylation profile of human cancer. Cancer Res 2001; 61:3225-3229.

21. Toyota M, Ahuja N, Ohe-Toyota M et al. CpG island methylator phenotype in colorectal cancer. Proc Natl Acad Sci USA 1999; 96:8681-8686.

22. Deng G, Peng E, Gum J et al. Methylation of hMLH1 promoter correlates with silencing with a region-specific manner in colorectal cancer. Br J Cancer 2002; 86(4):574-579.

23. Yi J, Wang ZW, Cang H et al. P16 gene methylation in colorectal cancers associated with Duke's staging. World J Gastroenterol 2001; 7(5):722-725.

24. Zou HZ, Yu BM, Wang ZW et al. Clin Cancer Res 2002; 8(1):188-191.

25. Yamamoto H Min Y, Itoh F, Imsumran A et al. Differential involvement of the hypermethylator phenotype in hereditary and sporadic colorectal cancers with high frequency microsatellite instability. Genes Chromosomes Cancer 2002; 33(3):322-325.

26. Rashid A, Shen L, Morris JS et al. CpG island methylation in colorectal adenomas. Am J Pathol 2001; 159 (3):1129-1135.

27. Isaacs WB, Isaacs JT. Molecular genetics of prostate cancer progression. In: Ravanagh D, Liebel SA, Schar HI, Lange P, eds. Principles and practice of genotourinary oncology. Philadelphia, New York: Lippincott-Raven Publishers, 1996:403-408.

28. Cairns P, Okami K, Halachmi S et al. Frequent inactivation of PTEN/MMAC1 in primary prostate cancer. Cancer Res 1997; 57:4997-5000.

29. Lee WH, Isaacs WB, Bova GS et al. CG island methylation changes near the GSTP1 gene in prostatic carcinoma cells detected using the polymerase chain reaction: a new prostate biomarker. Cancer Epidemiol Biomark Prev 1997; 6:443-450.

30. Millar DS, Ow KK, Paul CL et al. Detailed methylation analysis of the glutathione S-transferase π (GSTP1) gene in prostate cancer. Oncogene 1999; 18:1313-1324.

31. Mannrvik B, Alin P, Guthenberg G et al. Identification of three classes of cytosolic glutathione transferase common to several mammalian species: correlation between structural data and enzymatic properties. Proc Natl Acad Sci USA 1985; 82:7202-7206.

32. Lee WH, Morton RA, Epstein JI et al. Cytidine methylation of regulatory sequences near the π-class glutathione S-transferase gene accompanies human prostatic carcinogenesis. Proct Natl Acad Sci USA 1994; 91:11733-11737.

33. Brooks JD, Weinstein M, Lin X et al. CG island methylation changes near the GSTP1 gene in prostatic intraepithelial neoplasia. Cancer Epidemiol Biomark Prev 1998; 7:531-536.

34. Goessl C, Krause H, Müller M et al. Fluorescent Methylation-Specific Polymerase Chain Reaction for DNA-based detection of prostate cancer in bodily fluids. Cancer Res 2000; 60:5941-5945.

35. Goessl C, Müller M, Heicapell R et al. Methylation-specific PCR for detection of neoplastic DNA in biopsy washing. J Pathol 2002; 196(3):331-334.

36. Chu DC, Chuang CK, Fu JB et al. The use of real time quantitative polymerase chain reaction to detect hypermethylation of the CpG islands in the promoter region flanking the GSTP1 gene to diagnose prostate cancer. J Urol 2002; 167(4):1854-1858.

37. Hatada I, Hayashizaka Y, Hirotsune S et al. A genomic scanning method for higher organisms using restriction sites as landmarks. Proc Natl Acad Sci USA 1991; 88:9523-9527.

38. Hayashizaki Y, Shibata H, Hirotsune S et al. Identification of an imprinted U2af binding protein related sequence on mouse chromosome 11 using the RLGS method. Nat Genet 1994; 6:33-40.

39. Costello JF, Frühwald MC, Smiraglia DJ et al. Aberrant Cpl-island methylation has nonrandom and tumor-type specific patterns. Nat Genet 2000; 25:132-138.

40. Kuick T, Asakawa J, Neel J et al. High yield of restriction fragment length polymorphisms in two-dimensional separation of human genomic DNA. Genomics 1995; 13:1095-1107.

41. Huang THM, Perry M, Laux D. Methylation profiling of CpG islands in human breast cancer cells. Hum Mol Genet 1999; 8(3):459-470.

42. Gonzalgo ML, Liang G, Spruck III CH et al. Identification and characterization of differentially methylated regions of genomic DNA by methylation-sensitive arbitrarily primed PCR. Cancer Res 1999; 57:594-599.

43. Toyota M, Coty H, Ahuja N et al. Identification of differentially methylated sequences in colorectal cancer by methylated Cpl island amplification. Cancer Res 1999; 59:2307-2312.

44. Adorjan P, Distler J, Lipscher E et al. Tumour class prediction and discovery by microarray-based DNA methylation analysis. Nuc Acids Res 2002; 30(5)e21.

45. Frommer M, Mc Donald LE, Millar DS et al. A genomic sequencing protocol that yields a positive display of 5-methylcytosine residues in individual DNA strands. Proc Natl Acad Sci USA 1992; 89:1827-1831.

46. Mendenhall W, Sincich T. Statistics for Engeneering and the Sciences NJ: Prentice Hall, 1995.

47. Lipshutz RJ, Fodor S PA, Gingeras TR et al. High density synthetic oligonucleotide arrays. Nature Genet 1999; 21(suppl):25-32.

48. Palmisano W, Divine KK, Saccomanno G et al. Predicting Lung Cancer by detecting aberrant promoter methylation in sputum. Cancer Res 2000; 60:5954-5958.

49. Eads CA, Danenberg KD, Kawakami K et al. Methylight: a high-throughput assay to measure DNA methylation. Nuc Acid Res 2000; 28(8):e32.

50. Cotrell SE, Distler J, Goodman NS et al. unpublished data.

CHAPTER 10

Regulation of DNA Methyltransferases in Cancer

Nancy Detich and Moshe Szyf

Abstract

The DNA methyltransferases (DNMTs) are critical proteins involved in establishing proper control of epigenetic information. They are responsible for maintaining the cell's methylation pattern, as well for transcriptional repression through both methylation dependent and independent mechanisms. It is therefore fitting that the cell has evolved a number of layers of regulation to manage the appropriate expression of the DNMTs. While transcriptional control is the major player in regulation of DNMT1 by signaling pathways, post-transcriptional mechanisms appear to be critical for regulation during cell cycle progression and differentiation. In addition, regulatory interactions between DNMT1 and proteins involved in replication and cell cycle progression, as well as between all three DNMTs, have recently been elucidated. This review will discuss cellular processes in which these various mechanisms are involved, and provide suggestions as to how misregulation at these levels might lead to the development of certain pathologies.

Introduction

DNA methylation is carried out by DNA methyltransferase (DNMT) enzymes, which catalyze the transfer of a methyl group from S-adenosyl-methionine (AdoMet) onto the 5' position of the cytosine ring[1] The products of the reaction are S-adenosyl-homocysteine (AdoCys) and methylated DNA. DNMT1 was the first DNA methyltransferase to be cloned from vertebrates,[2] and has been extensively studied for many years. It is believed to be the enzyme responsible for replicating the DNA methylation pattern during cell division due to its preference for hemi-methylated DNA[3-5] and as demonstrated by mouse knockout experiments.[6] The roles of DNMT1 in DNA replication[7-9] and cellular transformation[10] are also well documented.

During development new patterns of methylation are laid out, and must be catalyzed by enzymes that are not guided by the methylation of the parental stand. DNMT3a and DNMT3b are believed to be partly responsible for de novo methylation during embryogenesis, as well as de novo methylation of proviral sequences.[11,12] They have been termed the de novo methyltransferases since they show no preference for hemi- versus non- methylated DNA in vitro.[13] However DNMT3a and 3b cannot explain all the de novo methylation occurring during embryogenesis since *dnmt3a-/-* and *dnmt3b-/-* are not completely devoid of methylation. DNMT3a and 3b are probably involved in methylation of specific satellite sequences. The rest of de novo DNA methylation must be carried out by either DNMT1 or other DNA methyltransferases that have not yet been identified. Although the DNMT2 and DNMT3L proteins do not possess any methyltransferase activity, DNMT3L may be involved in establishing maternal methylation imprints.[14]

DNA Methylation and Cancer Therapy, edited by Moshe Szyf. ©2005 Eurekah.com and Kluwer Academic/Plenum Publishers.

However this simple division of DNA methyltransferase activities into maintenance methylation, which is dictated by the methylation pattern of the template and replicates the DNA methylation pattern during cell division, and de novo methylation, which generates new methylation patterns, might not be accurate. Studies illustrate that the difference between maintenance and de novo DNMTs is blurred in some instances. Recent findings also uncover a functional cooperation between DNMT1 and the DNMT 3 family members,[12,15,16] which may also involve physical interactions, as will be discussed below. Overexpression of DNMT1 can lead to de novo methylation[17,18] and recent studies suggest that DNMT3a and DNMT3b are required for maintenance methylation of repetitive sequences.[12] Thus the maintenance of DNA methylation patterns in living cells must involve other mechanisms in addition to the biochemical property of substrate discrimination, which was identified in vitro. A number of data suggest that chromatin structure plays an important role in targeting DNA methyltransferases and demethylase. Thus, it was proposed that maintenance and de novo methylation have to be understood within the context of chromatin.[19] Chromatin structure rather than the DNA methylation pattern of the template might be responsible for guiding both maintenance and de novo DNA methylation.[19] Active chromatin enhances accessibility of demethylases,[20,21] whereas inactive chromatin, or proteins associated with inactive chromatin such as histone deacetylases (HDACs)[22] and K9-methyltransferases, might either recruit or control DNA methylation[23] (see further discussion in the chapter by Szyf et al, in this book).

In addition to their role as methylating enzymes, DNMTs have also been shown to regulate gene expression through their interaction with other chromatin modulating proteins, such as MeCP2 and HDACs.[24,25] It was therefore proposed that DNMTs silence genes through their protein-protein interactions by recruiting factors that modify and inactivate chromatin. In accordance, DNMT1 was shown to repress the expression of tumor suppressor genes by mechanisms that do not involve DNA methylation of the gene.[19,26]

Since maintaining the integrity of the epigenome is critical for maintaining the proper gene expression profile of cells and the entire organism, it stands to reason that multiple mechanisms have evolved to coordinate the replication of the genetic and epigenetic information.[19,27-29] DNMT1 is present in the replication fork and is associated with the replication fork protein PCNA,[30] replication and DNA methylation occur concurrently,[31] initiation of DNA replication requires the presence of DNMT1,[7] and knockdown of DNMT1 activates a checkpoint that triggers an intra S phase arrest of DNA replication.[9] Another important route of guarding the integrity of the epigenome is by maintaining the correct levels and activity of the DNMTs during different cellular events such as cell cycle progression and differentiation. The following sections will describe various examples of transcriptional, posttranscriptional, and protein interactions involved in regulating the expression and activity of the DNMTs.

Transcriptional Regulation of DNMTs

Several years ago, an analysis of the *DNMT1* gene was performed by Bigey et al. This study demonstrated that *DNMT1* contains four transcription start sites, each regulated by independent, TATA-less, promoter and enhancer elements.[32] The first promoter lies in a CpG island 5' to the first exon, possesses the highest basal activity, and can be repressed by the retinoblastoma protein Rb. In contrast, the three downstream promoters, which are located 5' to the second, third and fourth exons, are within CpG poor regions and have a low basal activity, but can be induced by the transcription factor C-JUN. In addition, three enhancer elements were identified, and were found to be either dependent or induced on the ectopic expression of C-JUN.[32] These enhancers may thus be able to activate transcription from all four promoters. The biological significance of these AP-1 recognition regulatory elements will be discussed in further detail below.

Recently, the promoters of *DNMT 3a* and *3b* have also been characterized,[33] and were found to bear a striking resemblance in structure to that of *DNMT1* (Fig. 1). This study determined that *DNMT3a* is composed of at least four transcription start sites controlled by three different promoters. Two of these promoters contain CpG islands, and all three of them lack

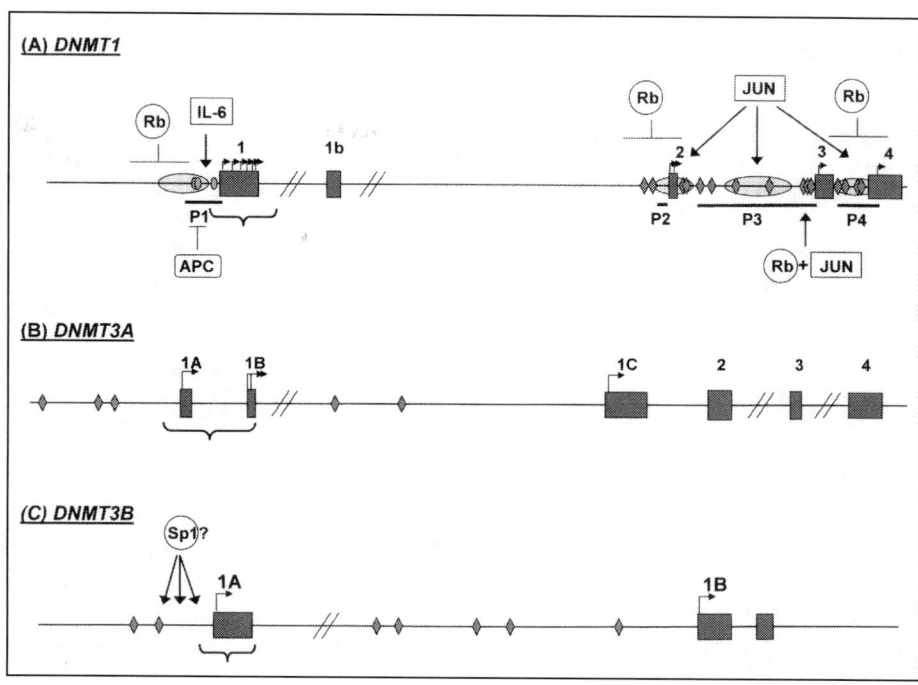

Figure. 1. DNMT promoter structure and regulation. Schematic diagrams of the *DNMT1* (A) *DNMT3A* (B) and *DNMT3B* (C) genes are shown. Boxes (numbered above) designate exons and horizontal arrows indicate transcription initiation sites. DNMT1 promoter boundaries (P1-P4) and enhancers (horizontal ovals) are shown. An open bracket denotes CpG rich regions. Potential binding sites for AP-1 (diamonds) and Fli-1 (vertical ovals) are also indicated. The different regulatory pathways that are postulated to regulate the *DNMT1* and *DNMT3B* genes are noted above and below the map of the gene. A vertical arrow specifies gene activation, while repression is indicated by a blunted line. It must be noted that the upregulation by Rb+Jun was only found to occur with the mouse *dnmt1* in P19 cells.

TATA sequences. The *DNMT3b* gene possesses at least two start sites that are controlled by different TATA-less promoters, one of which also contains a CpG rich region. All the *DNMT3a* and *3b* promoters also contain AP-1 sites, although regulation by these sites was not examined. However, it was shown that many binding sites for the transcription factor Sp1 are located within the *DNMT3b* upstream promoter, and progressive deletion of these sites results in a corresponding decrease in promoter activity.[33] Interestingly, the upstream CpG rich promoter of the mouse *dnmt1* gene was also found to contain an element that is required for *dnmt1* expression and is activated by both Sp1 and Sp3.[34]

The promoter similarities between the three *DNMT*s suggest that similar cellular signals might coordinately regulate the three enzymes through transcription. It is tempting to speculate that the regulatory mechanisms described below, which have only been shown for *DNMT1*, may play a role in the regulation of *DNMT3a* and *3b* as well.

Regulation by the RAS Signaling Pathway

As was mentioned above, the *DNMT1* promoters and enhancers contain multiple AP-1 regulatory elements. These elements can promote transcription upon the binding of dimeric complexes of the FOS and JUN oncoproteins, both of which are activated by the mitogenic RAS signaling pathway. The initial molecular link between DNMT1 and oncogenic signaling was the discovery that the protooncogenic RAS signaling pathway regulates *DNMT1*.[32,35,36] It was first demonstrated that the murine *dnmt1* promoter is induced by JUN or Ha-RAS but not by a dominant negative mutant of JUN in P19 cells.[35] The human *DNMT1* promoter is similarly induced by the RAS signaling pathway.[32] In addition, over expression of *Ha-ras* in P19 cells induced transcription and steady state levels of endogenous *dnmt1* mRNA.[35] It was then shown that the levels of cellular *dnmt1* mRNA could be reduced by downregulating the RAS signaling pathway by ectopic expression of either a human GAP or a dominant negative C-JUN in Y1 cells.[36] This reduction was accompanied by a reversion of the transformed morphology of Y1 cells. If Ha-RAS was then introduced into the GAP transfectants, DNMT1 levels increased and cells regained a transformed phenotype.[36] These data demonstrating that *dnmt1* is downstream to the AP-1 signaling pathway is further supported by a study showing that *dnmt 1* is one of the genes induced by forced expression of *c-fos,* and that inhibition of *dnmt1* by antisense expression reverses cellular transformation induced by *c-fos.*[37] Since it has been shown that overexpression of *DNMT1* can lead to hypermethylation,[38,39] the hypermethylation observed in certain cancer cells with RAS mutations may be a result of increased DNMT1 levels. This data is consistent with fact that tumor suppressor hypermethylation is correlated with activating mutations of RAS in some human colon tumors.[40,41]

DNMT1 regulation by the RAS pathway has also been found to occur in human lymphoid (T) cells. This was initially demonstrated by the finding that mitogenic T-cell stimulation increases *DNMT1* mRNA and enzyme activity.[42] Overexpression of Ha-RAS in T-cells was then found to increase DNMT1 levels, while inhibiting signaling through the RAS-MAPK pathway decreases DNMT1.[43] Interestingly, it has been shown that inhibiting T-cell DNMT1 leads to a lupus-like disease by altering DNA methylation and gene expression,[44,45] and T cells from patients with lupus possess reduced DNMT1 enzyme activity, hypomethylated DNA and modified gene expression.[46] It is therefore consistent with our hypothesis to learn that signaling through the RAS pathway was reduced in these patients.[47] Taken together, these data indicate that misregulation of DNMT1 by the RAS pathway may lead to cellular transformation, lupus-like conditions, and perhaps other as of yet unknown pathogenic states.

APC-TCF Pathway

Another possible link between *DNMT1* and critical cellular control pathways is the APC-β-CATENIN-TCF pathway.[48-50] When β-CATENIN associates with nuclear TCF factors, they form a transcriptional activator. The tumor-suppressor protein APC, which binds to β-CATENIN and causes its destruction, negatively regulates this factor. The *APC* (adenomatus polyposis coli) gene is mutated in many cases of familial colon cancer. In APC-deficient colon carcinoma cells, β-catenin accumulates and is constitutively complexed with TCF, resulting in transcriptional activation of TCF protooncogenic target genes such as *C-MYC.*[51] In addition, *Min* mice bearing a mutation in the mouse homologue of the *APC* gene spontaneously develop colonic adenomatous polyps.[52] However, when *Min* mice are genetically crossed with heterozygous *dnmt1* knockouts, they show a reduction in polyp formation,[52] suggesting that *dnmt1* is a downstream target of APC signaling.

Support for the link between APC and DNMT1 has recently been demonstrated.[53] It was found that ectopic expression of the wild type APC in HT-29 Apc-/- colon carcinoma cells resulted in down regulation of both a *DNMT1* promoter driven reporter construct, as well as the endogenous *DNMT1* mRNA This was further confirmed though the use of a dominant negative mutant of TCF, which was also found to suppress the *DNMT1* mRNA. These results suggest that a mutated APC protein, through its inability to degrade β-CATENIN, leads to a TCF dependent transcriptional upregulation of *DNMT1.* The causal role of DNMT1 in transformation in this system was also demonstrated by the finding that DNMT1 knockdown by antisense treatment resulted in the inhibition of anchorage independent growth of HT-29 cells.[53] Although the *DNMT1* promoter does not contain any consensus TCF binding sites, it is possible that TCF either binds non-consensus sites, or that it functions through an intermediary transcription factor. Further studies are required to determine the exact mechanism involved.

Feedback Regulation

Since DNMT1 is critical for so many cellular events, it is not surprising to learn that cells have developed a feedback mechanism that is dependent on methylation levels.

Several pieces of evidence are in support of this: Rats subjected to methionine deficient, hypomethylation inducing, diets exhibit increased DNMT1 activity in the liver,[54] and competitive inhibition of DNMT1 in T-cells increases *DNMT1* mRNA and activity.[42] In addition, inhibition of DNMT1 by 5-aza-2'-deoxycytidine (5-aza-CdR) in P19 cells also leads to its induction.[55] In this latter study, it was demonstrated that the AP-1 element upstream to the third exon of *dnmt1* is heavily methylated in P19 cells, but becomes demethylated upon 5-aza-CdR treatment. Furthermore, a CpG region upstream the AP-1 element was shown to attract a different set of binding factors depending on its methylation status. It has therefore been proposed that demethylation of this CpG region by inhibition or lack of DNMT1 leads to the formation of different protein-DNA complexes, which in turn allows AP-1 binding, thus increasing *dnmt1* expression.[55] The involvement of AP-1 in feedback regulation of the *DNMT1* promoter by methylation was also shown by another study in T cells, where it was found that treatment with a DNA methylation inhibitor increases transcription regulated by a putative *DNMT1* promoter and that this process requires AP-1 sites.[43]

In summary, the above examples illustrate that transcriptional regulation of *DNMT1* is a response to a critical signaling pathways, as well as to its own methylation state, and involves multiple elements within its promoter (Fig. 1).

Differential Regulation of the DNMTs during Cell Growth

Deregulated expression of *DNMT1* was previously suggested to play a causal role in cellular transformation.[56,57] Several lines of evidence are in accordance with this hypothesis, such as the presence of elevated levels of *DNMT1* mRNA in tumors and cancer cell lines.[58-60] and the fact that ectopic expression of *DNMT1* results in cellular transformation.[38,39] The question of whether *DNMT1* expression is induced in tumor cells, or whether the increase in DNMT1 merely reflects the increase in DNA synthesis activity, has been previously raised.[61] To resolve these contradictory hypotheses, we have previously proposed that the increase in DNMT1 expression is the cause of the increase in DNA synthetic activity.[29]

Two mechanisms have been suggested to explain how DNMT1 influences cellular transformation. It has been proposed that high levels of DNMT1 can lead to ectopic methylation and silencing of tumor suppressor genes.[62] However, a clear correlation between general *DNMT1* overexpression and tumor suppressor hypermethylation has not been established.[63,64] Alternatively, it has been suggested that deregulated expression of *DNMT1* during the cell cycle might be critical for DNMT1's effects on cell growth as discussed above.[19,27-29] In support of the latter, it has been shown that the coordinated cell cycle regulation of DNMT1 is disrupted in colorectal cancer cells in vivo,[65] as well as in estrogen receptor negative breast cancer cells.[66]

DNA methyltransferase activity is regulated with the cell cycle.[67] In 1991, it was demonstrated that murine *dnmt1* mRNA is not present in arrested BALB/c 3T3 cells but is highly induced at the G1-S boundary. Expression remains high during the S-phase, accompanied by an increase in enzyme activity, and is then reduced.[68] Subsequent runoff transcription experiments demonstrated that *dnmt1* is transcribed throughout the cell cycle.[68] This suggested that the levels of *dnmt1* mRNA are regulated with the cell cycle at the posttranscriptional level.

The mechanism involved in this regulation was unraveled many years later, when it was demonstrated that the 3' untranslated region (3'UTR) of the *dnmt1* mRNA plays a role in regulating its levels with the cell cycle, and that deregulation at this level has an effect on cellular transformation by DNMT1.[69] The 3' UTR of the *dnmt1* mRNA can confer a growth dependent mRNA regulation at the posttranscriptional level, and a 54 nucleotide highly conserved element within the 3'UTR is necessary and sufficient to mediate this regulation. Cell free mRNA decay experiments demonstrated that this element increases mRNA turnover rates, and does so to a greater extent in the presence of extracts prepared from arrested cells. A specific

RNA-protein complex is formed within the 3'UTR only in growth-arrested cells, and UV crosslinking analysis revealed a 40 kDa protein (p40), whose binding is dramatically increased in growth arrested cells, and is inversely correlated with *dnmt1* mRNA levels as cells are induced into the cell cycle.[69] A model emerges where the regulation of DNMT1 with the cell cycle involves the degradation of the *dnmt1* mRNA in growth arrested cells, through the interaction of p40 with the *dnmt1* 3'UTR. However, the causal role of p40 has not been determined as of yet.

Most importantly, while previous findings that ectopic expression of human *DNMT1* lacking the 3'UTR can transform NIH-3T3 cells[38] were confirmed, inclusion of the 3'UTR prevented transformation.[69] These results support the alternative hypothesis stated above, that deregulated expression of *DNMT1* with the cell cycle and not the total amount of DNMT1 is important for cellular transformation. This can also explain why some studies showed that *DNMT1* is not overexpressed in tumors when its expression is normalized to other cell cycle associated genes.[61] We propose that ectopic expression of *DNMT1* at the wrong phase of the cell cycle leads to aberrant entry into S phase, which in turn stimulates the expression of other cell cycle associated genes.

In addition to the posttranscriptional mechanism just described, it appears that DNMT 1 may also be regulated at S phase at the translational and posttranslational levels. It was found that upon S phase arrest of MEL 11A2 cells (using an aphidicolin block), both an increase in the synthesis of DNMT1 protein, as well as an increase in protein half-life, takes place.[70] The finding that DNMT1 is controlled during the cell cycle by more than one mechanism further emphasizes the importance of this regulation, so that DNA replication does not proceed in the absence of DNMT1 and ensuring that the replicating DNA is properly methylated.

Our data suggests that deregulated expression of *DNMT1* during the cell cycle could cause cellular transformation. How could ectopic expression of *DNMT1* in arrested cells cause transformation?[69] It is possible that ectopic expression of *DNMT1* in arrested cells results in de novo methylation and silencing of a critical tumor suppressor gene. An alternative hypothesis is that ectopic expression of *DNMT1* causes cell transformation by interfering with cell cycle regulatory circuits through DNMT1 protein-protein interactions. Since DNMT1 forms a complex with Rb and E2F[25] as well as histone deacetylase 1 and 2[22,71] it can inhibit the expression of tumor suppressors by a mechanism that does not involve DNA methylation, as has been previously shown.[26] In addition, since DNMT1 has been shown to bind to the replication protein PCNA at the same site as the cell cycle inhibitor p21, DNMT1 could displace p21 from PCNA during the Go/G1 phase and allow replication to occur.[30] Thus, aberrant expression of *DNMT1* during the Go/G1 phase may override the silencing of tumor suppressors and eliminate normal arrest signals, leading to the uncontrolled growth that is observed in cancer cells (Fig. 2). It is also possible that both mechanisms, abnormal methylation and the elimination of growth arrest signals, occur. This is supported by studies in which inhibition of poly (ADP-ribosyl)ation was found to induce genomic hypermethylation, as well as increase the level of DNMT1 specifically at the G1/S border, accompanied by the premature formation of a DNMT1-PCNA complex.[72]

The growth regulation of DNMTs 3a and 3b has not been studied as extensively. It has been shown that *DNMT3b* mRNA is also regulated with the cell cycle, and its profile is similar as to that of *dnmt1*. The mRNA levels of *DNMT3a*, on the other hand, display less of a down regulation upon cell arrest.[73] Whether the *DNMT3a* and *DNMT3b* mRNAs are controlled transcriptionally, or at a posttranscriptional level as for *dnmt1*, is still unresolved.

Regulation by Viral Infection

It has been shown that the DNA of endogenous and exogenous retroviruses can be highly methylated in the host genome.[74-76] In addition, infected cells can exhibit increased DNA methylation in other regions of the genome as well.[77,78] It is therefore not surprising that the expression of the DNMTs is regulated by certain viruses or viral proteins, both at the transcriptional and posttranscriptional levels.

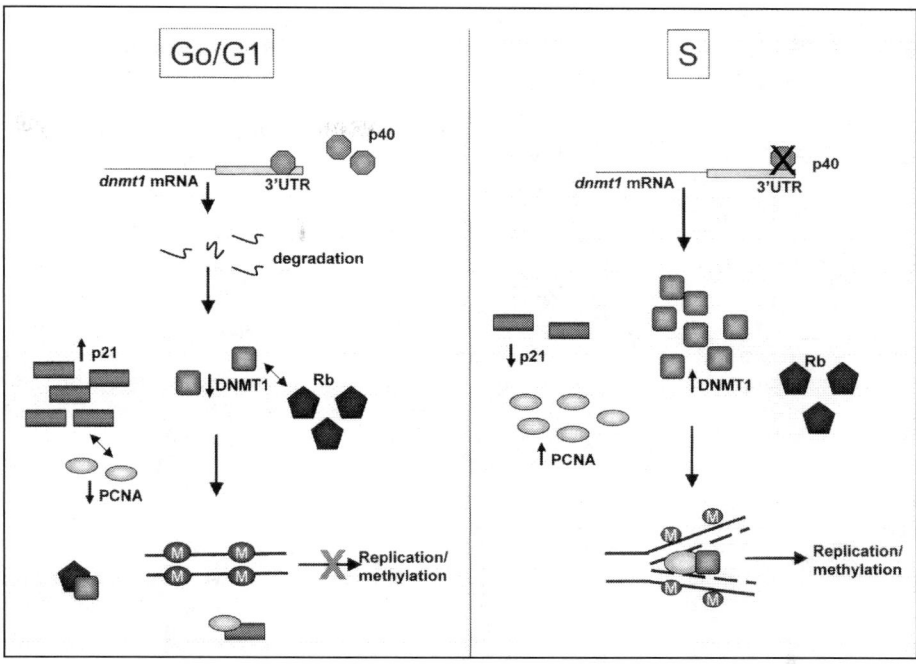

Figure 2. A model for the regulation of DNMT1 with the cell cycle. In Go/G1, p40 binds to the *dnmt1* 3'UTR and destabilizes the mRNA, either directly or by targeting it to mRNA degrading enzymes. This leads to the production of a minimal amount of DNMT1 protein. Both DNMT1 and p21 can bind PCNA, but since p21 levels are elevated in Go/G1, it competes out DNMT1 for PCNA binding. In addition, Rb can bind DNMT1 and prevent it form interacting with any PCNA that is not bound to p21. The end result is a lack of DNMT1-PCNA complexes, and therefore DNA replication and methylation are inhibited. During S phase, the absence of p40 results in the stabilization of the *dnmt1* mRNA, which thus leads to an increase in DNMT1 protein. In this phase, p21 levels are reduced and PCNA levels are increased, so that now DNMT1 competes out p21 for binding to PCNA. High levels of PCNA also compete out Rb for binding to DNMT1. The major complex formed is therefore DNMT1-PCNA, which can then carry out replication of both the DNA and methylation pattern.

Simian Virus 40 (SV40)

T antigen is a protein product of the SV40 virus that can immortalize primary lines,[79] transform immortalized lines.[80] and induce tumors in mice.[81] It can also transform primary cells when expressed together with RAS or its pathway components.[82] The transformation induced by T antigen results, at least in part, from its interaction with the tumor suppressor Rb.[83] Following the observation that two SV40 transformed cell lines displayed increased DNMT1 levels,[30] the regulation of *dnmt1* by SV40 T antigen was examined. This study demonstrated that ectopic expression of SV40 T antigen induces *dnmt1* expression, protein levels, and global DNA methylation.[84] This was found to depend on T antigen's interaction with Rb, since a mutated T antigen that is incapable of binding Rb is also unable to induce *dnmt1*. Furthermore, inhibiting *dnmt1* expression by antisense oligonucleotides[84] could reverse cellular transformation by T antigen. These data demonstrate that a viral oncoprotein can upregulate *dnmt1*, and that this upregulation plays a causal role in T antigen induced transformation. In contrast to the transcriptional upregulation of *dnmt1* by the RAS pathway, T-antigen was found to influence *dnmt1* at the posttranscriptional level, by increasing the stability of the *dnmt1* mRNA.[84] Although the exact mechanism involved in the T-antigen induced mRNA stability was not established, the possibility exists that it involves the same components as found in the cell cycle regulation of DNMT1, i.e., the 3'UTR and p40. One possible model is that Rb, a negative regulator of the cell cycle, acts upstream p40 to downregulate *dnmt1* in Go/ G1, and thus prevents methylation in the absence of replication. Following entrance into S phase, or ectopic expression of T-antigen, Rb is inactivated, p40 levels are decreased, and *dnmt1* mRNA is stabilized. In the case of T-antigen, its constitutive expression would lead to increased DNMT1 throughout the cell cycle, thus promoting cellular transformation. Although this is only speculative, further studies defining the identity of p40 should help to uncover the precise mechanism involved.

Epstein-Barr Virus (EBV)

Recently, the latent membrane protein 1 (LMP1), which is an oncogenic product of EBV, was shown to upregulate *DNMTs* 1, 3a, and 3b.[85] EBV is a human herpes virus that has been implicated in several cancers.[86,87] LMP1 is expressed in approximately 70% of nasopharyngeal carcinomas, and is also capable of transforming cells.[88] In line with its ability to promote transformation, it was found that LMP1 can induce the mRNA and protein levels, as well as activity, of DNMT 1, 3a, and 3b, and that this results in hypermethylation and downregulation of E-cadherin, a gene which is often hypermethylated in cancer.[85] Because LMP1 has been shown to induce AP-1 activity, and given that all three of the *DNMT* promoters contain multiple AP-1 sites, it was proposed that this is a possible mechanism for LMP activation.[85] However, although a steady state increase in the *DNMT* mRNA was demonstrated, it was not determined whether this occurs through increased transcription. As a result, one cannot rule out the participation of a posttranscriptional mechanism in this regulation.

Human Immunodeficiency Virus Type 1 (HIV-1)

Results from another study demonstrate that *DNMT1* regulation is not limited to cancer promoting viruses. The role of the HIV-1 virus in regulating DNMT1 levels was examined, and it was found that this virus is able to upregulate *DNMT1* mRNA and DNMT1 activity, and that this is accompanied by an increase overall DNA methylation. Additionally, de novo methylation of a CpG within the gamma interferon (IFN- γ) promoter and its subsequent downregulation, was also observed [89]. Thus, DNMT1 may be involved in the loss of the type 1 immune response (which involves IFN- γ) observed in AIDS patients.[89,90] This study did not determine the mode, transcriptional or posttranscriptional, by which HIV-1 increases *DNMT1* mRNA. Since this investigation was also carried out before the identification of DNMTs 3a and 3b, it is also not known whether HIV-1 is able to regulate the de novo DNMTs as well.

Cell Differentiation and DNMT1

A number of studies have shown that DNMT1 is upregulated during cellular differentiation.[91-93] *dnmt1* expression is maintained throughout PC12 neuronal differentiation,[94] while *dnmt1* expression is critical for differentiation, since *dnmt1* null embryonic stem cells die when induced to differentiate.[95] On the other hand, DNMT1 is also downregulated in certain differentiation processes.[93,94,96,97] Thus, it appears that the role of DNMT1 in determining cell fate is a complex one, and may depend on cell type. Several modes of DNMT1 regulation during differentiation have been discovered, including transcriptional, post transcriptional, and posttranslational mechanisms. A number of examples where up or downregulation of DNMT1 occur during this process are discussed below.

Upregulation of DNMT1

Recently, it was demonstrated that DNMT1 might play a role in the differentiation of K562 erythroleukemia cells into megakaryocytes. When K562 cells are treated with Interleukin 6 (IL-6), they enter a megakaryocytic differentiation pathway. Coincident with this process, is an upregulation of *DNMT1* mRNA and enzyme activity.[91] This upregulation was found to occur at the transcriptional level, since the *DNMT1* promoter is activated following transfection into K562 cells and treatment with IL-6. In addition, this activation appears to involve several Fli-1 binding sites within the *DNMT1* promoter, since the loss of these sites greatly diminishes *DNMT1* promoter induction. The transcription factor Fli-1, a differentiation mediator, was also induced upon IL-6 treatment. Interestingly, Il-6 treatment has also been shown to lead to activation of JUN.[98] It is therefore possible that the induction of *DNMT1* also involves the AP-1 sites within its promoter, however this issue was not addressed.[91] Furthermore, whether DNMT1 plays a role in the K562 megakaryocytic differentiation process remains unknown.

Another example where DNMT1 is upregulated during differentiation occurs when U937 cells are induced to differentiate to a more monocyte-like phenotype by phorbol ester TPA.[93] While treatment of U937 cells with TPA leads to both increased DNMT1 enzyme activity and *DNMT1* mRNA expression, these changes do not occur in parallel. The induction of activity occurs well before the increase in mRNA, which was found to occur at the posttranscriptional level. The role of protein kinase C (PKC) was studied (since TPA is an activator of PKC), and it was found that the induction of DNMT1 activity following TPA treatment is abolished in the presence of a PKC inhibitor. Since it has been shown that PKC can phosphorylate and activate DNMT1 in vitro,[99] it is possible that TPA regulates DNMT1 activity in this manner, however the authors were not able to demonstrate the phosphorylation of DNMT1 under their conditions.[93] As with the previous example, the function of DNMT1 in this differentiation process, and whether changes in genomic methylation occur, is undefined.

One possible link between differentiation and DNMT1 upregulation is the tumor suppressor Rb. Rb is involved in regulating the G1/S transition, the point at which the decision to differentiate must be made. Rb binds E2F and inhibits E2F dependent transcription in cells that are arrested or in G1, and following Rb's progressive phosphorylation during G1, E2F is released and can activate genes necessary for S phase progression.[100] The inhibition of E2F dependent transcription has been shown to involve an interaction between DNMT1 and Rb.[25] In addition, Rb was found to act synergistically with the protooncogene c-JUN to activate *dnmt1* in differentiating P19 embryocarcinoma cells.[92] This activation of *dnmt1* is mediated through a non-canonical AP-1 recognition signal upstream to the third exon. Whereas c-JUN does not bind this site in the absence of Rb, the presence of both Rb and c-JUN results in formation of an AP-1 complex and strong synergistic activation of *dnmt1*.[92] The cooperative activation of *dnmt1* by Rb and c-JUN might play a role in upregulating expression of *dnmt1* during certain types of differentiation, such as in the IL-6 example described above, where JUN is also induced. This would allow for more DNMT1 to interact with Rb and E2F and lead to an increase in repression of the genes required for cell growth.

Downregulation of DNMT1

There are several well documented examples where DNMT1 is downregulated during the differentiation process.[93,94,96,97] At the start of mouse myoblast differentiation, a decrease in both DNMT1 activity and global methylation is observed.[101] It was then shown that posttranscriptional and posttranslational mechanisms are involved.[97] While the *dnmt1* mRNA half-life is around 5 hours in proliferating cells, this is reduced to 1.5 hours in differentiating cells. In addition, protein turnover measurements indicated that DNMT1 is also more stabilized in growing cells. The mechanisms involved in decreasing both types of stability are unknown, but it is again possible that the mRNA regulation involves some components of the cell cycle mechanism, since differentiation is accompanied by a halt in proliferation. Due to the fact that the global methylation in these cells occurs in the absence of replication,[101] it cannot be directly attributed to the decrease in DNMT1, however the possibility of an indirect role has been suggested.[97]

During the differentiation process of F9 mouse embryonal carcinoma cells, a similar post transcriptional decrease in *dnmt1* mRNA is observed, accompanied by a decrease in DNMT1 activity.[96] However, in contrast to myoblast differentiation, global demethylation in F9 cells is found to occur gradually during differentiation after several rounds of replication.[102] DNA replication in the presence of reduced levels of DNMT1 could directly explain this process.

In contrast to myoblast differentiation, where downregulation of DNMT1 is an early event, a reduction in DNMT1 levels occurs late in PC12 differentiation. PC12 cells are induced to differentiate to sympathetic neuron like cells upon treatment with NGF.[103] When the levels of DNMT1 during the progression of differentiation was studied, it was found that *dnmt1* mRNA, as well as DNMT1 protein and activity, are decreased 4 days post induction.[94] An increase in p21 and a decrease in PCNA occurred in parallel to the decrease in dnmt1. Since neurite outgrowth is already present after 2 days, it implies that downregulation of DNMT1 is not required for this differentiation process, but is rather an adaptation to the post mitotic state of the cell, as has been proposed.[94] It is not known at what level DNMT1 is regulated in this process, nor was it determined if PC12 differentiation involves changes in DNA methylation.

It was mentioned above that the differentiation of U937 cells with TPA is accompanied by an upregulation of DNMT1. However, if the same cells are induced to differentiate by dibutyryl cyclic AMP (dbcAMP), a downregulation of DNMT1 is observed.[93] Four hours after dbcAMP treatment, an almost complete reduction in DNMT1 activity is observed. The decrease in *DNMT1* mRNA, which begins at 2 hours, is complete by 20 hours. These events precede the differentiation associated phenotypic changes, which are only seen after 24 hours, and it is therefore possible that DNMT1 plays a functional role in this process. Similar to TPA, dbcAMP regulates *DNMT1* mRNA at the posttranscriptional level, but in addition, a decrease in transcription is also involved.[93] Although in both cases, the cells differentiate to a monocyte like phenotype, the cell surface and internal markers are to some extent different.[104] Since DNMT1 is involved in modulating gene expression, the opposite directions in which DNMT1 levels change could possibly influence the expression of these different markers.

Taken together, the above data clearly illustrates that changes in DNMT1 during differentiation are complex, and depend on cell type and stimuli. It stands to reason that several mechanisms, (post)transcriptional, posttranslational, and regulation of enzyme activity, have evolved to precisely coordinate DNMT1 levels with a particular cell fate.

Regulation through Protein Interactions

Several recent studies demonstrate that the activity of the DNMTs, as well as targeting to different regions of DNA, can be modulated through the interaction of the different DNMT family members with each other, as well as with other proteins which are not part of the methylation machinery.[14,105-107] Although the DNMTs also interact with several other proteins to regulate gene expression, this section will only discuss interactions involved in regulating DNMT binding and enzymatic activities.

PCNA

One manner of coordinating the duplication of the DNA methylation pattern with DNA replication is by positioning of DNMT1 in the replication fork. DNMT1 would therefore be able to methylate DNA simultaneously with its synthesis, ensuring that the DNA methylation pattern is precisely copied. It has been demonstrated that DNMT1 is part of a multicomponent replication complex[8], and accordingly, nascent DNA is immediately fully methylated following its synthesis.[31] Moreover, inhibition of DNMT1 leads to an inhibition of DNA synthesis activity and halts the progression of the cells through the cell cycle.[7,9]

One factor possibly linking concurrent methylation and replication is the proliferating nuclear cell antigen (PCNA), a protein required for DNA replication. A specific domain within DNMT1 targets it to the replication fork.[108] and has also been shown to bind PCNA.[30] A subsequent study determined the functional significance of this interaction.[105] Iida et al demonstrated that DNMT1 has a higher affinity for DNA when it is first bound by PCNA. In addition, PCNA bound DNA is also methylated more efficiently that the corresponding free DNA. This is in contrast to Chuang et al, who did not observe a change in DNMT1 activity in the presence of PCNA. However, the types of assays used in the two studies were quite different, and may explain this discrepancy. It is of interest to note that DNMT1 binds PCNA at the same position as p21, a tumor suppressor that inhibits DNA replication by forming a complex with PCNA.[30] Since DNMT1 cannot bind PCNA in the presence of p21, whose levels are highest when cells are in Go/G1, inappropriate DNA methylation is prevented during this phase of the cell cycle. The additional finding that DNMT1 has a preference for PCNA-bound DNA, and that this DNA is a better substrate for methylation, explains another mechanism of ensuring that DNA is not methylated in the absence of replication (Fig. 2).

Rb

Previously, the involvement of Rb in the transcriptional and posttranscriptional regulation of DNMT1 was discussed. In addition, another mode of DNMT 1 regulation by Rb has recently been elucidated. Rb is able to physically associate with DNMT1 and can inhibit methyltransferase activity by disrupting the formation of the DNA-DNMT1 complex.[106] This interaction was found to involve part of the N-terminal region of DNMT1 and the B and C pocket regions of Rb. Moreover, overexpression of Rb leads to genomic hypomethylation as well as hypomethylation and activation of a transfected reporter plasmid.[106] Since the region of DNMT1 that binds Rb is the same one that binds PCNA, it suggests another mechanism by which the coordination of replication and methylation is achieved. In Go and G1, Rb binds DNMT1 and precludes PCNA binding, thus preventing methylation when cells are not dividing. This also reinforces the effects of p21 described above (Fig. 2). Since Rb is inactivated in several cancers,[109,110] this finding may partly explain the phenomenon of hypermethylation in cancer.

Interactions between the DNMTs

Several studies have demonstrated that DNMTs 1, 3a, and 3b functionally cooperate to generate methylation patterns. Both DNMT1 together with DNMT 3a or 3b are required for the methylation of a certain class of sequences including LINE-1 elements.[111] Disruption of both DNMT1 and 3b in colorectal cancer cell line results in a > 95% reduction of genomic DNA methylation, concomitant with the re expression of the p16 and growth suppression.[112] In addition, functional cooperation has also been observed between DNMT1 and 3a,[15] and between DNMT 3L and 3a.[16]

Two recent studies may provide an explanation as to how this cooperation is established. Kim et al demonstrated that DNMTs 1, 3a, and 3b physically interact with each other.[107] Through a combination of immunoprecipitation and GST pull down assays, it was shown that DNMT1 is able to bind either DNMTs 3a or 3b or both at the same time. Furthermore, DNMTs 3a and 3b can also interact with each other in the absence of DNMT1. These interactions all occur within the N terminal domains of the proteins. Functional cooperation was also demonstrated, since combinations of the three enzymes show increased methylation rates over the individual enzymes. An especially notable increase was observed when DNMT1 was added to DNMTs 3a + 3b. These data point to a model that can perhaps explain how the methylation pattern is established during development. In pre implantation embryos, when DNMT1 is restricted to the cytosol,[113] DNMTs 3a and 3b are active and can interact to establish the initial wave of de novo methylation. At later stages, these enzymes are joined by DNMT1, which then leads to methylation spreading followed by maintenance of the methylation pattern.[107]

Recent data indicate that protein interactions are not limited to the active DNMT enzymes. Hata et al demonstrated that DNMTs 3a and 3b also physically interact with DNMT3L, and that DNMT3L is required for the establishment of maternal methylation imprints and the proper expression of the corresponding genes.[14] Since DNMT3L does not possess methyltransferase activity, it could not be directly responsible. Moreover, *dnmt3a-/-, dnmt3b+/- knockout* mice are also defective in forming proper maternal imprints. Thus it is possible that DNMT3L might be responsible for targeting the de novo DNMTs to imprinting regions.[14] Since a study by Chedin et al also illustrates that DNMT3L can stimulate de novo methylation by DNMT 3a at both imprinted and non-imprinted sequences, it is possible that DNMT3L may also act as a general activator of DNMT3a.[16]

Table 1. Summary of mechanisms involved in DNMT regulation

	DNMT1	DNMT3a	DNMT3b
Transcriptional Regulation	Regulation through mulitple promoters **Rb; APC**- downregulation **Sp1/Sp3; Rb+jun**- upregulation **ras/fos/jun**	Promoter similar in structure to DNMT1- Possible regulation by Sp1	Same as for DNMT3a
Viral Infection	**SV40 T-antigen**- upregulation (posttranscriptional) **EBV (LMP-1)**- upregulation (mechanism?) **HIV**- upregulation (mechanism?)	**EBV (LMP-1)**- upregulation	**EBV (LMP-1)**- upregulation
Cell Cycle	**Go/G1**- downregulation (posttranscriptional) **S**- upregulation (translational, posttranscriptional)	**Go/G1**- slight downregulation (mechanism?)	**Go/G1**- slight downregulation (mechanism?)
Differentiation	**K562**- Upreulation by IL-6 (transcriptional) **Myoblasts**- downregulation (posttranscriptional, posttranslational) **F9**- downregulation (posttranscriptional) **PC12**- downregulation (mechanism?) **U937 +TPA**- upregulation (posttranscriptional) **+dbcAMP**-downregulation (transcriptional and posttranscriptional)	Unknown	Unknown
Protein Interactions	**PCNA**- upregulation of DNA binding and enzyme activities **Rb**- downregulation of DNMT1- DNA interaction **DNMTs 3a, 3b**- upregulation of activity	**DNMTs 1 , 3b**- upregulation of activity **DNMT 3L**- upregulation of activity, establishment of material imprints	**DNMTs 1 , 3b**- upregulation of activity **DNM3L**- as for DNMT3a

Conclusions

Considering the importance of the DNMTs in maintaining proper epigenetic information, the cells must undoubtedly possess mechanisms to regulate their expression. The existence of multiple layers of regulation, such as transcriptional, posttranscriptional, posttranslational, and protein interactions, is an indication that the proper expression of the DNMT is critical (Table 1). By maintaining the correct levels and activity of the DNMTs during different events such as cell signaling, cell cycle progression and differentiation, the proper gene expression for these events can be established. While transcriptional control is the predominant form of regulation by signaling pathways, posttranscriptional and posttranslational mechanisms are most often involved in regulation with the cell cycle and in differentiation. If DNMT regulation is compromised, it might promote improper gene expression, and thus lead to pathogenic conditions. A key example is the regulation of DNMT1 with the cell cycle, where inappropriate expression at Go/G1, by interfering with its mRNA regulation, leads to cellular transformation. In addition, protein interactions involving the DNMTs, as well as other factors not directly involved in methylation, are beginning to emerge as another important level of regulation. Although most of data concerning DNMT regulation pertains to DNMT1, the similarities between the DNMT1, DNMT3a and 3b promoters and expression profiles during the cell cycle, together with the discovery of DNMT-DNMT interactions and functional cooperation, suggests that other similar mechanisms might exist. Future studies should determine whether other modes of regulation, which until now only pertain to DNMT1, will turn out to be common to the DNMT family.

Acknowledgements

The research from the MS laboratory described in this chapter was supported by the National Institute of Cancer research of Canada and the Canadian Institutes of Health Research. ND is a recipient of the Canadian Institute of Health Research Doctoral Fellowship and the McGill Faculty of Medicine Internal Fellowship.

References

1. Kumar S, Cheng X, Klimasauskas S et al. The DNA (cytosine-5) methyltransferases. Nucleic Acids Res 1994; 22(1):1-10.
2. Bestor TH. Cloning of a mammalian DNA methyltransferase. Gene 1988; 74(1):9-12.
3. Stein R, Gruenbaum Y, Pollack Y et al. Clonal inheritance of the pattern of DNA methylation in mouse cells. Proc Natl Acad Sci USA 1982; 79(1):61-65.
4. Gruenbaum Y, Cedar H, Razin A. Substrate and sequence specificity of a eukaryotic DNA methylase. Nature 1982; 295(5850):620-622.
5. Pradhan S, Bacolla A, Wells RD et al. Recombinant human DNA (cytosine-5) methyltransferase. I. Expression, purification, and comparison of de novo and maintenance methylation. J Biol Chem 1999; 274(46):33002-33010.
6. Li E, Bestor TH, Jaenisch R. Targeted mutation of the DNA methyltransferase gene results in embryonic lethality. Cell 1992; 69(6):915-926.
7. Knox JD, Araujo FD, Bigey P et al. Inhibition of DNA methyltransferase inhibits DNA replication. J Biol Chem 2000; 275(24):17986-17990.
8. Vertino PM, Sekowski JA, Coll JM et al. DNMT1 is a Component of a Multiprotein DNA Replication Complex. Cell Cycle 2002; 1(6):416-423.
9. Milutinovic S, Zhuang Q, Niveleau A et al. Epigenomic stress response: Knock-down of DNA methyltransferase 1 triggers an intra S-phase arrest of DNA replication and induction of stress response genes. J Biol Chem 2003.
10. Szyf M. Towards a pharmacology of DNA methylation. Trends Pharmacol Sci. 2001; 22(7):350-354.
11. Okano M, Bell DW, Haber DA et al. DNA methyltransferases Dnmt3a and Dnmt3b are essential for de novo methylation and mammalian development. Cell 1999; 99(3):247-257.
12. Liang G, Chan M, Tomigahara Y et al. Cooperativity between DNA methyltransferases in the maintenance methylation of repetitive elements. Mol Cell Biol 2002; 22:480-491.
13. Okano M, Xie S, Li E. Cloning and characterization of a family of novel mammalian DNA (cytosine-5) methyltransferases. Nat Genet 1998; 19(3):219-220.
14. Hata K, Okano M, Lei H et al. Dnmt3L cooperates with the Dnmt3 family of de novo DNA methyltransferases to establish maternal imprints in mice. Development 2002; 129(8):1983-1993.
15. Fatemi M, Hermann A, Gowher H et al. Dnmt3a and Dnmt1 functionally cooperate during de novo methylation of DNA. Eur J Biochem 2002; 269(20):4981-4984.
16. Chedin F, Lieber MR, Hsieh CL. The DNA methyltransferase-like protein DNMT3L stimulates de novo methylation by Dnmt3a. Proc Natl Acad Sci USA 2002; 99(26):16916-16921.
17. Vertino PM, Yen RW, Gao J et al. De novo methylation of CpG island sequences in human fibroblasts overexpressing DNA (cytosine-5-)-methyltransferase. Mol Cell Biol 1996; 16(8):4555-4565.

18. Mikovits JA, Young HA, Vertino P et al. Infection with human immunodeficiency virus type 1 upregulates DNA methyltransferase, resulting in de novo methylation of the gamma interferon (IFN-gamma) promoter and subsequent downregulation of IFN-gamma production. Mol Cell Biol 1998;18(9):5166-5177.

19. Szyf M. Targeting DNA methylation in cancer. Aging Res Rev 2003; 56:1-30.

20. Cervoni N, Szyf M. Demethylase activity is directed by histone acetylation. J Biol Chem 2001; 276(44):40778-44087.

21. Cervoni N, Detich N, Seo S et al. The oncoprotein Set/TAF-1beta, an inhibitor of histone acetyltransferase, inhibits active demethylation of DNA, integrating DNA methylation and transcriptional silencing. J Biol Chem 2002; 277(28):25026-25031.

22. Fuks F, Burgers WA, Brehm A et al. DNA methyltransferase Dnmt1 associates with histone deacetylase activity. Nat Genet 2000; 24(1):88-91.

23. Tamaru H, Selker EU. A histone H3 methyltransferase controls DNA methylation in Neurospora crassa. Nature 2001; 414:277-283.

24. Kimura H, Shiota K. Methyl-CpG binding protein, MeCP2, is a target molecule for maintenance DNA methyltransferase, Dnmt1. J Biol Chem 2002.

25. Robertson KD, Ait-Si-Ali S, Yokochi T et al. DNMT1 forms a complex with Rb, E2F1 and HDAC1 and represses transcription from E2F-responsive promoters. Nat Genet 2000; 25(3):338-342.

26. Milutinovic S, Knox JD, Szyf M. DNA methyltransferase inhibition induces the transcription of the tumor suppressor p21(WAF1/CIP1/sdi1). J Biol Chem 2000; 275(9):6353-6359.

27. Szyf M. The role of DNA methyltransferase 1 in growth control. Front Biosci 2001; 6:D599-609.

28. Szyf M. Towards a pharmacology of DNA methylation. Trends Pharmacol Sci 2001; 22(7):350-354.

29. Szyf M, Detich N. Regulation of the DNA methylation machinery and its role in cellular transformation. Prog Nucleic Acid Res Mol Biol 2001;69:47-79.

30. Chuang LS, Ian HI, Koh TW et al. Human DNA-(cytosine-5) methyltransferase-PCNA complex as a target for p21WAF1. Science 1997; 277(5334):1996-2000.

31. Araujo FD, Knox JD, Szyf M et al. Concurrent replication and methylation at mammalian origins of replication [published erratum appears in Mol Cell Biol 1999; 19(6):4546]. Mol Cell Biol 1998; 18(6):3475-3482.

32. Bigey P, Ramchandani S, Theberge J et al. Transcriptional regulation of the human DNA Methyltransferase (dnmt1) gene. Gene 2000; 242(1-2):407-418.

33. Yanagisawa Y, Ito E, Yuasa Y et al. The human DNA methyltransferases DNMT3A and DNMT3B have two types of promoters with different CpG contents. Biochim Biophys Acta 2002; 1577(3):457-465.

34. Kishikawa S, Murata T, Kimura H et al. Regulation of transcription of the Dnmt1 gene by Sp1 and Sp3 zinc finger proteins. Eur J Biochem 2002; 269(12):2961-2970.

35. Rouleau J, MacLeod AR, Szyf M. Regulation of the DNA methyltransferase by the Ras-AP-1 signaling pathway. J Biol Chem 1995; 270(4):1595-1601.

36. MacLeod AR, Rouleau J, Szyf M. Regulation of DNA methylation by the Ras signaling pathway. J Biol Chem 1995; 270(19):11327-11337.

37. Bakin AV, Curran T. Role of DNA 5-methylcytosine transferase in cell transformation by fos. Science 1999; 283(5400):387-390.

38. Wu J, Issa JP, Herman J et al. Expression of an exogenous eukaryotic DNA methyltransferase gene induces transformation of NIH 3T3 cells [see comments]. Proc Natl Acad Sci USA 1993; 90(19):8891-8895.

39. Vertino PM, Yen RW, Gao J, Baylin SB. De novo methylation of CpG island sequences in human fibroblasts overexpressing DNA (cytosine-5-)-methyltransferase. Mol Cell Biol 1996; 16(8):4555-4565.

40. Guan RJ, Fu Y, Holt PR et al. Association of K-ras mutations with p16 methylation in human colon cancer. Gastroenterology 1999; 116(5):1063-1071.

41. Toyota M, Ohe-Toyota M, Ahuja N et al. Distinct genetic profiles in colorectal tumors with or without the CpG island methylator phenotype. Proc Natl Acad Sci USA 2000; 97(2):710-715.

42. Yang J, Deng C, Hemati N et al. Effect of mitogenic stimulation and DNA methylation on human T cell DNA methyltransferase expression and activity. J Immunol 1997; 159(3):1303-1309.

43. Deng C, Yang J, Scott J et al. Role of the ras-MAPK signaling pathway in the DNA methyltransferase response to DNA hypomethylation. Biol Chem 1998; 379(8-9):1113-1120.

44. Richardson B, Powers D, Hooper F et al. Lymphocyte function-associated antigen 1 overexpression and T cell autoreactivity. Arthritis Rheum 1994; 37(9):1363-1372.

45. Cornacchia E, Golbus J, Maybaum J et al. Hydralazine and procainamide inhibit T cell DNA methylation and induce autoreactivity. J Immunol 1988; 140(7):2197-2200.

46. Yung RL, Richardson BC. Role of T cell DNA methylation in lupus syndromes. Lupus 1994; 3(6):487-491.

47. Deng C, Kaplan MJ, Yang J, et al. Decreased Ras-mitogen-activated protein kinase signaling may cause DNA hypomethylation in T lymphocytes from lupus patients. Arthritis Rheum 2001; 44(2):397-407.

48. Clevers H, van de Wetering M. TCF/LEF factor earn their wings. Trends Genet 1997; 13(12):485-489.

49. Korinek V, Barker N, Morin PJ et al. Constitutive transcriptional activation by a beta-catenin-Tcf complex in APC-/- colon carcinoma [see comments]. Science 1997; 275(5307):1784-1787.

50. Rubinfeld B, Robbins P, El-Gamil M et al. Stabilization of beta-catenin by genetic defects in melanoma cell lines [see comments]. Science 1997; 275(5307):1790-1792.

51. He TC, Sparks AB, Rago C, et al. Identification of c-MYC as a target of the APC pathway [see comments]. Science 1998; 281(5382):1509-1512.

52. Laird PW, Jackson-Grusby L, Fazeli A et al. Suppression of intestinal neoplasia by DNA hypomethylation. Cell 1995; 81(2):197-205.

53. Campbell PM, Szyf M. Human DNA methyltransferase gene DNMT1 is regulated by the APC pathway. Carcinogenesis 2003; 24(1):17-24.

54. Christman JK, Sheikhnejad G, Dizik M et al. Reversibility of changes in nucleic acid methylation and gene expression induced in rat liver by severe dietary methyl deficiency. Carcinogenesis 1993; 14(4):551-557.

55. Slack A, Cervoni N, Pinard M et al. Feedback regulation of DNA methyltransferase gene expression by methylation. Eur J Biochem 1999; 264(1):191-199.

56. el-Deiry WS, Nelkin BD, Celano P et al. High expression of the DNA methyltransferase gene characterizes human neoplastic cells and progression stages of colon cancer. Proc Natl Acad Sci USA 1991; 88(8):3470-3474.

57. Szyf M. DNA methylation properties: consequences for pharmacology. Trends Pharmacol Sci 1994; 15(7):233-238.

58. Kautiainen TL, Jones PA. DNA methyltransferase levels in tumorigenic and nontumorigenic cells in culture. J Biol Chem 1986; 261(4):1594-1598.

59. Issa JP, Vertino PM, Wu J et al. Increased cytosine DNA-methyltransferase activity during colon cancer progression. J Natl Cancer Inst 1993; 85(15):1235-1240.

60. Belinsky SA, Nikula KJ, Baylin SB et al. Increased cytosine DNA-methyltransferase activity is target-cell- specific and an early event in lung cancer. Proc Natl Acad Sci USA 1996; 93(9):4045-4050.

61. Lee PJ, Washer LL, Law DJ et al. Limited up-regulation of DNA methyltransferase in human colon cancer reflecting increased cell proliferation. Proc Natl Acad Sci USA 1996; 93(19):10366-10370.

62. Baylin SB, Herman JG, Graff JR et al. Alterations in DNA methylation: a fundamental aspect of neoplasia. Adv Cancer Res 1998; 72:141-196.

63. Jurgens B, Schmitz-Drager BJ, Schulz WA. Hypomethylation of L1 LINE sequences prevailing in human urothelial carcinoma. Cancer Res 1996; 56(24):5698-5703.

64. Eads CA, Danenberg KD, Kawakami K et al. CpG island hypermethylation in human colorectal tumors is not associated with DNA methyltransferase overexpression [published erratum appears in Cancer Res 1999 Nov 15;59(22):5860]. Cancer Res 1999; 59(10):2302-2306.

65. De Marzo AM, Marchi VL, Yang ES et al. Abnormal regulation of DNA methyltransferase expression during colorectal carcinogenesis. Cancer Res 1999; 59(16):3855-3860.

66. Nass SJ, Ferguson AT, El-Ashry D et al. Expression of DNA methyl-transferase (DMT) and the cell cycle in human breast cancer cells. Oncogene 1999; 18(52):7453-7461.

67. Szyf M, Kaplan F, Mann V et al. Cell cycle-dependent regulation of eukaryotic DNA methylase level. J Biol Chem 1985;260(15):8653-8656.

68. Szyf M, Bozovic V, Tanigawa G. Growth regulation of mouse DNA methyltransferase gene expression. J Biol Chem 1991; 266(16):10027-10030.

69. Detich N, Ramchandani S, Szyf M. A conserved 3'-untranslated element mediates growth regulation of dna methyltransferase 1 and inhibits its transforming activity. J Biol Chem 2001; 276(27):24881-24890.

70. Suetake I, Kano Y, Tajima S. Effect of aphidicolin on DNA methyltransferase in the nucleus. Cell Struct Funct 1998; 23(3):137-142.

71. Rountree MR, Bachman KE, Baylin SB. DNMT1 binds HDAC2 and a new co-repressor, DMAP1, to form a complex at replication foci. Nat Genet 2000; 25(3):269-277.

72. Zardo G, Reale A, Passananti C et al. Inhibition of poly(ADP-ribosyl)ation induces DNA hypermethylation: a possible molecular mechanism. FASEB J 2002; 16(10):1319-1321.

73. Robertson KD, Keyomarsi K, Gonzales FA et al. Differential mRNA expression of the human DNA methyltransferases (DNMTs) 1, 3a and 3b during the G(0)/G(1) to S phase transition in normal and tumor cells. Nucleic Acids Res 2000; 28(10):2108-2113.

74. Gunthert U, Schweiger M, Stupp M et al. DNA methylation in adenovirus, adenovirus-transformed cells, and host cells. Proc Natl Acad Sci USA 1976; 73(11):3923-3927.
75. Masucci MG, Contreras-Salazar B, Ragnar E et al. 5-Azacytidine up regulates the expression of Epstein-Barr virus nuclear antigen 2 (EBNA-2) through EBNA-6 and latent membrane protein in the Burkitt's lymphoma line rael. J Virol 1989; 63(7):3135-3141.
76. Youssoufian H, Hammer SM, Hirsch MS et al Methylation of the viral genome in an in vitro model of herpes simplex virus latency. Proc Natl Acad Sci USA 1982; 79(7):2207-2210.
77. de Bustros A, Nelkin BD, Silverman A et al. The short arm of chromosome 11 is a "hot spot" for hypermethylation in human neoplasia. Proc Natl Acad Sci USA 1988; 85(15):5693-5697.
78. Jahner D, Jaenisch R. Retrovirus-induced de novo methylation of flanking host sequences correlates with gene inactivity. Nature 1985; 315(6020):594-597.
79. Tevethia MJ. Immortalization of primary mouse embryo fibroblasts with SV40 virions, viral DNA, and a subgenomic DNA fragment in a quantitative assay. Virology 1984; 137(2):414-421.
80. Aaronson SA, Todaro GJ. SV40 T antigen induction and transformation in human fibroblast cell strains. Virology 1968; 36(2):254-261.
81. Brinster RL, Chen HY, Messing A et al. Transgenic mice harboring SV40 T-antigen genes develop characteristic brain tumors. Cell 1984; 37(2):367-379.
82. Land H, Parada LF, Weinberg RA. Tumorigenic conversion of primary embryo fibroblasts requires at least two cooperating oncogenes. Nature 1983; 304(5927):596-602.
83. DeCaprio JA, Ludlow JW, Figge J et al. SV40 large tumor antigen forms a specific complex with the product of the retinoblastoma susceptibility gene. Cell 1988; 54(2):275-283.
84. Slack A, Cervoni N, Pinard M, Szyf M. DNA methyltransferase is a downstream effector of cellular transformation triggered by simian virus 40 large T antigen. J Biol Chem 1999; 274(15):10105-10112.
85. Tsai CN, Tsai CL, Tse KP et al. The Epstein-Barr virus oncogene product, latent membrane protein 1, induces the downregulation of E-cadherin gene expression via activation of DNA methyltransferases. Proc Natl Acad Sci USA 2002; 99(15):10084-10089.
86. Klein G, Giovanella BC, Lindahl T et al. Direct evidence for the presence of Epstein-Barr virus DNA and nuclear antigen in malignant epithelial cells from patients with poorly differentiated carcinoma of the nasopharynx. Proc Natl Acad Sci USA 1974; 71(12):4737-4741.
87. Shibata D, Weiss LM. Epstein-Barr virus-associated gastric adenocarcinoma. Am J Pathol 1992; 140(4):769-774.
88. Wang D, Liebowitz D, Kieff E. An EBV membrane protein expressed in immortalized lymphocytes transforms established rodent cells. Cell 1985; 43(3Pt 2):831-840.
89. Mikovits JA, Young HA, Vertino P, et al. Infection with human immunodeficiency virus type 1 upregulates DNA methyltransferase, resulting in de novo methylation of the gamma interferon (IFN-gamma) promoter and subsequent downregulation of IFN-gamma production. Mol Cell Biol 1998; 18(9):5166-5177.
90. Shearer GM, Clerici M. Early T-helper cell defects in HIV infection. Aids. Mar 1991; 5(3):245-253.
91. Hodge DR, Xiao W, Clausen PA et al. Interleukin-6 regulation of the human DNA methyltransferase (HDNMT) gene in human erythroleukemia cells. J Biol Chem 2001; 276(43):39508-39511.
92. Slack A, Pinard M, Araujo FD et al. A novel regulatory element in the dnmt1 gene that responds to co-activation by Rb and c-Jun. Gene 2001; 268(1-2):87-96.
93. Soultanas P, Andrews PD, Burton DR et al. Modulation of human DNA methyltransferase activity and mRNA levels in the monoblast cell line U937 induced to differentiate with dibutyryl cyclic AMP and phorbol ester. J Mol Endocrinol 1993; 11(2):191-200.
94. Deng J, Szyf M. Downregulation of DNA (cytosine-5-)methyltransferase is a late event in NGF-induced PC12 cell differentiation. Brain Res Mol Brain Res 1999; 71(1):23-31.
95. Tucker KL, Talbot D, Lee MA, Leonhardt H, Jaenisch R. Complementation of methylation deficiency in embryonic stem cells by a DNA methyltransferase minigene. Proc Natl Acad Sci USA 1996; 93(23):12920-12925.
96. Teubner B, Schulz WA. Regulation of DNA methyltransferase during differentiation of F9 mouse embryonal carcinoma cells. J Cell Physiol 1995; 165(2):284-290.
97. Liu Y, Sun L, Jost JP. In differentiating mouse myoblasts DNA methyltransferase is posttranscriptionally and posttranslationally regulated. Nucleic Acids Res 1996; 24(14):2718-2722.
98. Nakajima K, Kusafuka T, Takeda T et al. Identification of a novel interleukin-6 response element containing an Ets-binding site and a CRE-like site in the junB promoter. Mol Cell Biol 1993; 13(5):3027-3041.
99. DePaoli-Roach A, Roach PJ, Zucker KE et al. Selective phosphorylation of human DNA methyltransferase by protein kinase C. FEBS Lett 1986; 197(1-2):149-153.
100. Ferreira R, Naguibneva I, Pritchard LL et al. The Rb/chromatin connection and epigenetic control: opinion. Oncogene 2001; 20(24):3128-3133.

101. Jost JP, Jost YC. Transient DNA demethylation in differentiating mouse myoblasts correlates with higher activity of 5-methyldeoxycytidine excision repair. J Biol Chem 1994; 269(13):10040-10043.
102. Razin A, Webb C, Szyf M, et al. Variations in DNA methylation during mouse cell differentiation in vivo and in vitro. Proc Natl Acad Sci USA 1984; 81(8):2275-2279.
103. Fujita K, Lazarovici P, Guroff G. Regulation of the differentiation of PC12 pheochromocytoma cells. Environ Health Perspect 1989;80:127-142.
104. Harris P, Ralph P. Human leukemic models of myelomonocytic development: a review of the HL-60 and U937 cell lines. J Leukoc Biol 1985; 37(4):407-422.
105. Iida T, Suetake I, Tajima S et al. PCNA clamp facilitates action of DNA cytosine methyltransferase 1 on hemimethylated DNA. Genes Cells 2002; 7(10):997-1007.
106. Pradhan S, Kim GD. The retinoblastoma gene product interacts with maintenance human DNA (cytosine-5) methyltransferase and modulates its activity. EMBO J 2002; 21(4):779-788.
107. Kim GD, Ni J, Kelesoglu N et al. Co-operation and communication between the human maintenance and de novo DNA (cytosine-5) methyltransferases. EMBO J 2002; 21(15):4183-4195.
108. Leonhardt H, Page AW, Weier HU et al. A targeting sequence directs DNA methyltransferase to sites of DNA replication in mammalian nuclei. Cell 1992; 71(5):865-873.
109. Weinberg RA. Tumor suppressor genes. Science 1991; 254(5035):1138-1146.
110. Ohtani-Fujita N, Dryja TP, Rapaport JM et al. Hypermethylation in the retinoblastoma gene is associated with unilateral, sporadic retinoblastoma. Cancer Genet Cytogenet 1997; 98(1):43-49.
111. Liang G, Chan MF, Tomigahara Y et al. Cooperativity between DNA methyltransferases in the maintenance methylation of repetitive elements. Mol Cell Biol 2002; 22(2):480-491.
112. Rhee I, Bachman KE, Park BH et al. DNMT1 and DNMT3b cooperate to silence genes in human cancer cells. Nature 2002; 416(6880):552-556.
113. Carlson LL, Page AW, Bestor TH. Properties and localization of DNA methyltransferase in preimplantation mouse embryos: implications for genomic imprinting. Genes Dev 1992; 6(12B):2536-2541.

CHAPTER 11

Inhibition of Poly(ADP-Ribosyl)ation Allows DNA Hypermethylation

Anna Reale, Giuseppe Zardo, Maria Malanga, Jordanka Zlatanova and Paola Caiafa

Abstract

This chapter emphasizes that along the chain of events that induce DNA methylation-dependent chromatin condensation, a post-synthetic modification other than histone acetylation, poly(ADP-ribosyl)ation, participates in the establishment and maintenance of methylation-free regions of chromatin. In fact, several lines of in vitro and in vivo evidence have shown that poly(ADP-ribosyl)ation is involved in the control of DNA methylation pattern, protecting genomic DNA from full methylation. More recent studies have provided some clues to the understanding of the molecular mechanism(s) connecting poly(ADP-ribosyl)ation with DNA methylation. We aim here to demonstrate the direct correlation existing between inhibition of poly(ADP-ribose) polymerases and DNA hypermethylation, and to describe some possible mechanisms underlying this molecular link. We will then present our hypothesis that the inhibition of the poly(ADP-ribosyl)ation process in the cell may be responsible for the anomalous hypermethylation of oncosuppressor gene promoters during tumorigenesis and to suggest the possibility that an active poly(ADP-ribosyl)ation process is also involved in maintaining the unmethylated state of CpG islands in normal cells.

Introduction

It is well known that post-synthetic modifications of both DNA and chromatin proteins regulate DNA function, and epigenetic phenomena are involved in the modulation of the replication process and in the control of the appropriate program of gene expression. The post-synthetic modifications are intricately interconnected in a complex network to define specific chromatin conformations that tune chromatin function to the needs of the cell. It is a challenge to understand how epigenetic phenomena take part in the normal functioning of the cell. The recognition that these phenomena also play a role in malignant transformation has led, in the post-genomic era, to a major effort aimed at understanding how epigenetic modifications are involved in inducing the tumorigenic process.

The DNA methylation process, through DNA methylase and demethylase activities, is involved in carcinogenesis in a paradoxical way. It is, in fact, possible to identify in the same tumor sample two contrasting events: a general pattern of DNA hypomethylation and, at this background, a high level of aberrant methylation of the CpG islands of some specific oncosuppressor genes.[1-3] Nothing is known about the mechanism(s) whereby CpG islands—which remain protected from methylation in normal cells—become susceptible to methylation in tumor cells. Changes in the methylation pattern of CpG islands[4,5] and/or the methylation-dependent chromatin condensation/decondensation are two mechanisms through which DNA methylases/demethylases control gene expression.[6-8]

DNA Methylation and Cancer Therapy, edited by Moshe Szyf. ©2005 Eurekah.com and Kluwer Academic/Plenum Publishers.

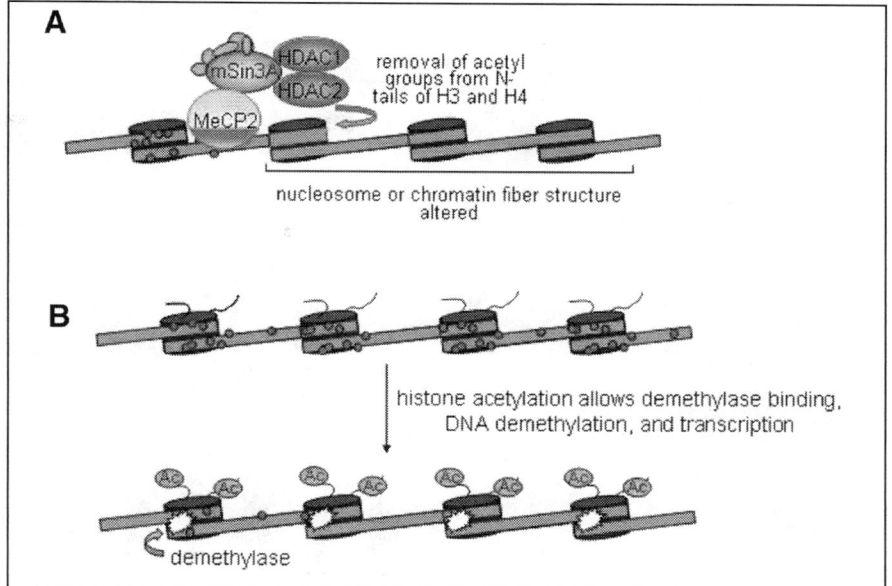

Figure 1. The mechanisms connecting DNA methylation/histone deacetylation to gene silencing and histone acetylation/DNA demethylation to gene reactivation are schematically presented in panels A and B, respectively.

Epigenetics involves several modifications. It is not simple to identify the role of a single modification since these modifications often work together to reach a definite molecular goal; furthermore, it is extremely difficult to establish the time sequence in which these modifications act.

If we consider DNA methylation and histone deacetylation, they could be connected to each other through the general phenomenon of DNA methylation-dependent chromatin condensation which induces gene silencing. The two modifications work together both in gene activation, where acetylation of histones may induce DNA demethylase activity,[9,10] and in gene silencing, where MeCP2—a member of the family of methyl-DNA-binding proteins— recruits histone deacethylase onto chromatin to deacetylate core histones,[11,12] (Fig. 1). It is clear that in the latter case DNA must be previously methylated to allow MeCP2 binding.

There is evidence that poly(ADP-ribosyl)ation controls the normal DNA methylation pattern, including the unmethylated state of CpG islands. Below, we will summarize our data that indicate the involvement of poly(ADP-ribosyl)ation in protecting the unmethylated state of CpG dinucleotides. We will further describe the experimental results that give insight into the possible molecular mechanism(s) involved in inducing DNA hypermethylation both during replication and in nonreplicating chromatin. We will present a unified model that may explain—in this scenario—the mechanisms involved in the aberrant DNA methylation in cancer.

Poly(ADP-Ribosyl)ation

Poly(ADP-ribose) polymerases (PARPs) are enzymes that by introducing ADP-ribose polymers onto chromatin proteins change their charge, and possibly alter their conformation. The reaction, catalyzed by PARPs, uses the respiratory coenzyme NAD^+ as a source of ADP-ribose moieties to synthesize protein-bound polymers of variable size (from 2 to over 200 residues) and structural complexity (linear or branched)[13] (Fig. 2A). The intracellular level of ADP-ribose polymers is under tight control so that increased synthesis is coupled to a higher degradation rate that reduces the half life of the ADP-ribose polymers to less than 1 min.[14]

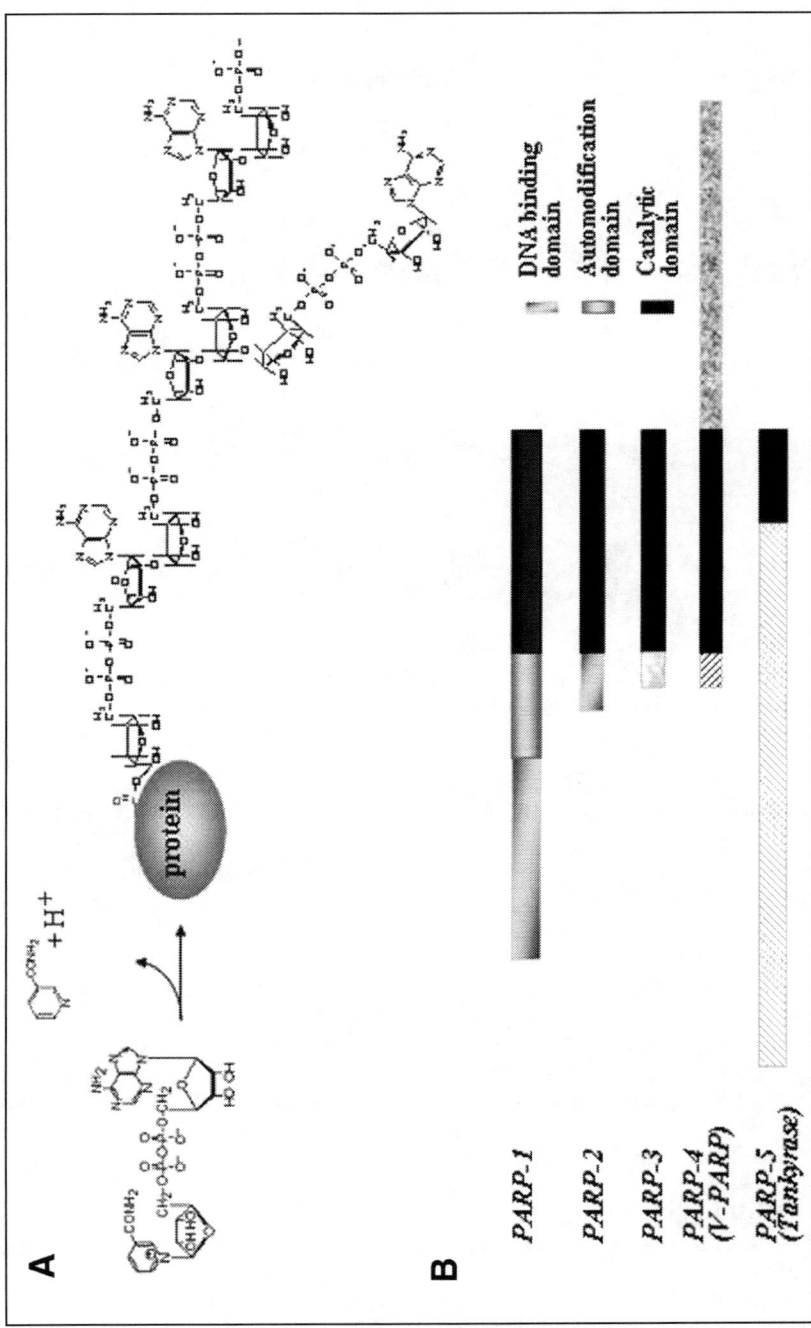

Figure 2. A) Scheme of the enzymatic reaction by which PARPs transfer ADP-ribose from NAD⁺ to a glutamic acid residue on a protein acceptor; subsequently, new units are added, allowing the formation of ADP-ribose polymers on the acceptor protein. B) Schematic structure of members of the poly(ADP-ribose) polymerase family. The homology with the catalytic domain of PARP1 is indicated in black. All discovered PARPs are sensitive to the inhibitors of PARP1.

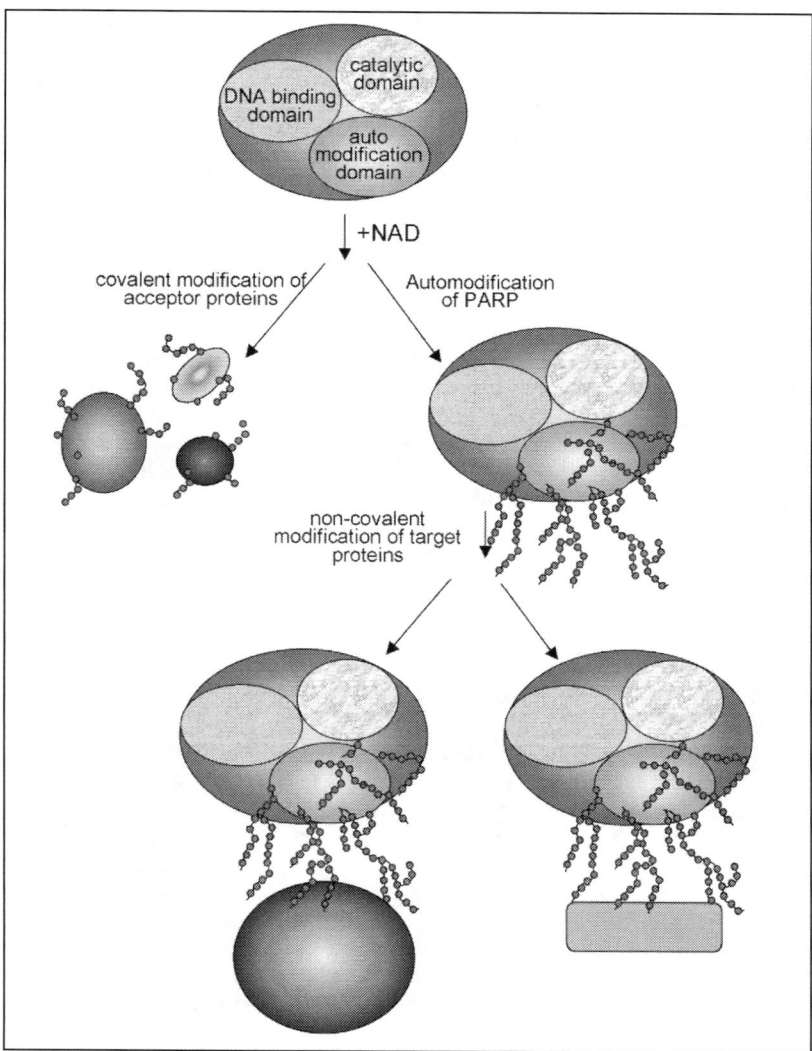

Figure 3. The schematic shows PARP1 involved in modifying chromatin proteins in covalent or noncovalent way. The enzyme is the best acceptor of the covalent modification and forms long and branched ADP-ribose polymers on its central domain. In this form, PARP1 can associate with other proteins that interact noncovalently with the enzyme-bound polymers.

PARP1, the most abundant and best characterized member of the PARP family (Fig. 2B), has a nuclear localization, exhibits high affinity for nicked DNA and has been considered to have regulatory functions in DNA repair and transcription. Upon binding to ss/ds-DNA breaks, PARP1 becomes activated to catalyze the covalent poly(ADP-ribosyl)ation of a number of nuclear proteins, with diverse roles in chromatin structure (e.g. histones, HMGs, nuclear matrix proteins) and DNA metabolism (DNA and RNA polymerases, topoisomerases, transcription factors).[15,16] The process by which PARPs introduce covalently-bound ADP-ribose polymers onto other proteins is known as heteromodification (Fig. 3). PARP1 itself has been found to be the main acceptor of ADP-ribose polymers, both in vitro and in vivo in a reaction termed automodification. In fact, the very first protein that undergoes poly(ADP-ribosyl)ation is PARP1 itself. The enzyme acts as a dimer,[17] and each monomer builds long and branched—up to 200 residues[18]—ADP-ribose polymers on up to 28 automodification sites[19,20] in the central domain of the associated monomer. Once the enzyme is activated, along with automodification, it can start a series of heteromodification reactions that modulate the functions of chromatin proteins.[21-24]

In addition to the automodification domain, a DNA binding domain that contains zinc-finger motifs has been recognized in the N-terminal portion of the molecule and the recognition sites are DNA strand-breaks rather than specific polynucleotide sequences.[25-28] The C-terminal domain contains the catalytic site[29] (see Fig. 2B for the domain structure of PARPs).

Automodification has long been regarded exclusively as a self-regulatory mechanism, allowing PARP1 to cycle on and off DNA; in fact, poly (ADP-ribosyl)ation causes the enzyme to detach from DNA with consequent loss of activity. However, besides binding to DNA, PARP1 interacts with other proteins[13] and it has been shown that such interactions may depend on its automodification state.[30] Moreover, it has been shown that automodified PARP1 may be a potent modulator of chromatin structure, promoting its decondensation.[31] PARP-bound ADP-ribose polymers play an active role in such a process, by being able to interact noncovalently with histones (Fig. 3), thus competing with their binding to DNA; in fact, the same effect could be observed with protein-free polymers. Further studies demonstrated that all the histones are indeed able to bind to ADP-ribose polymers, even in the presence of a large excess of DNA, with histone H1 exhibiting the highest binding affinity.[32-35] Interestingly, ADP-ribose polymers target the histone domains that play key roles in higher order chromatin organization, i.e., the C-terminal part of H1 and the N-terminal tails of core histones.[33]

The potential for ADP-ribose polymers to affect cellular functions through noncovalent interactions is further emphasized by the finding that other proteins, besides histones, are targets for ADP-ribose polymer binding (Table 1). It is interesting to note that recently, a shared sequence motif has been identified in proteins that bind noncovalently to ADP-ribose polymers.[36] That ADP-ribose polymers binding to these proteins may have functional consequences has been suggested by several studies.[37-39]

These observations, associated with in vivo evidence of increased sensitivity to genotoxic stress in the absence of functional PARP1[40] and PARP2,[41] have led to the proposal that ADP-ribose polymers could play a role in DNA damage signalling:[42] by interacting with selected proteins, ADP-ribose polymers may promote the rapid assembly of repair complexes at the damaged DNA sites, as well as modulate the function of downstream effectors of the DNA damage response.

Block of Poly(ADP-Ribosyl)ation Induces in Vivo DNA Hypermethylation

Based on a series of in vivo experiments, our laboratory came to the realization that there exists a negative correlation between poly(ADP-ribosyl)ation and DNA methylation.[43-46] We studied the DNA methylation level of L929 and/or NIH/3T3 mouse fibroblasts that were pretreated with 3-aminobenzamide (3-ABA), a well known inhibitor of PARPs. Four different strategies were used.

First strategy: as an experimental approach, we performed methyl-accepting ability assays on isolated nuclei and/or DNA purified from control L929 mouse fibroblasts or fibroblasts preincubated with 3-ABA.[43] In one variant of the assay, isolated nuclei were incubated in the presence of radioactively-labeled donor of methyl groups, and the level of DNA methylation achieved as a result of the activity of the endogonous DNA methyltransferases was measured. A consistent increase in the DNA methylation level was observed in the drug-treated cells as early as after 24 hours of 3-ABA treatment. The level of methyl groups incorporated into total DNA was found to be 60% higher in the 3-ABA treated cells than in the control cells. In a second variant of the methyl-accepting assay, DNA was isolated from the two cell populations and sequently methylated in vitro by an exogenous enzyme. The ability of the DNA isolated from

Table 1. Proteins that bind non-covalently to ADP-ribose polymers

Target Protein	Interaction Sites
H1	C-terminus
H2A	N-terminus (aa 11-36)
H2B	N-terminus (aa 23-47)
H3	N-terminus (aa 51-72)
H4	N-terminus (aa 16-40)
XP-A	C-terminus (aa 215-237)
MSH-6	N-terminus (aa 295-317)
DNA ligase III	N-terminus (aa 12-34)
DNA polymerase ε	N-terminus (aa 691-709)
DNA-PKcs	core region (aa 2728-2752)
Ku70	core region (aa 243-264)
XRCC-1	BRCT domain (aa 379-400)
NF-kB (p52 subunit)	Rel homology domain (aa 179-199)
p21	PCNA binding domain (aa 140-163)
p53	DNA binding domain (aa 153-178; aa 231-253)
	oligomerization domain (aa 326-348)
iNOS	calmodulin binding domain (aa 505-525)
CAD (Caspase Activated DNase)	core region (aa 148-169)
MARCKS/MRP	effector domain (aa 151-175)
Telomerase (TERT)	C-terminus (aa 962-983)
Caspase 7	n.d.
Nuclear matrix proteins	n.d.
20S Proteosome	n.d.

the drug-treated cell to be methylated in vitro was severely reduced, presumably as a consequence of its already increased in vivo methylation level.

Second strategy: we examined the possibility that poly(ADP-ribosyl)ation was directly involved in maintaining the unmethylated state of CpG islands.[44] To that end, the methylation status of the DNA from both control and 3-ABA-treated cells was assessed by using either methylation-dependent restriction enzymes on purified genomic DNA or a sequence-dependent restriction enzyme on an aliquot of same DNA, previously modified by the bisulphite reaction. The first method was introduced by Bird[47,48] to reveal clusters of unmethylated CpG dinucleotides in CpG islands: the exclusive presence of closely spaced unmethylated CpGs in the islands gives, upon HpaII digestion of genomic DNA, the so-called "HpaII tiny fragments", easily recognizable by gel electrophoresis. These fragments were present when the DNA was purified from control cells, but were greatly decreased when the DNA was purified from cells preincubated with 3-ABA.

For the second set of experiments, fragment 1482-1773 of the CpG island from the promoter region of the mouse *Htf9* gene was amplified by PCR after bisulphite reaction,[49] that converts cytosine to uracil, but does not affect 5-methylcytosine. Thus, the bisulphite reaction retains the memory of the original methylation pattern even after PCR amplification. During PCR amplification of the chemically-modified DNA fragment, uracil is amplified as thymine, while 5mC residues are amplified as cytosines: thus, a sequence-dependent restriction enzyme could be used in lieu of methylation-dependent restriction enzymes. Using BstUI which recognizes and cuts CGCG sequences, alterations in the methylation pattern would be observed only if both cytosines were methylated in the sequence (the methylation of only one cytosine or the absence of 5mC would give rise to sequences nonrecognizable by BstUI).

Following digestion of the PCR-amplified fragment with BstUI, it was possible to observe an anomalous methylation pattern when the Htf9 promoter region was purified from fibroblasts treated with 3-ABA. In fact, southern blot analysis of the digestion products showed the presence of a 55 bp fragment only in the 3-ABA-treated sample, identifying the aberrantly methylated CGCG sequence as the sequence in position 1526.

These data confirm the hypothesis that, at least for the *Htf9* promoter region, an active poly(ADP-ribosyl)ation process protects the unmethylated state of the CpG island.

Third strategy: The role played by poly(ADP-ribosyl)ation in protecting genomic DNA from full methylation was studied on cytological preparations from control and 3-ABA-treated L929 and NIH/3T3 cells. The cells were indirectly immunolabelled with anti-5-methylcytosine (anti-5mC) monoclonal antibodies,[45] and microscope analysis was performed on a cell-by-cell basis. Images of individual interphase nuclei were recorded by a CCD camera and quantitatively analysed with the help of a computer. Cells preincubated with 3-ABA consistently showed an increased number of often enlarged heterochromatic regions; the level of anti-5mC antibody binding to these regions was also elevated.

Fourth strategy: This approach was used to verify whether the inhibition of poly(ADP-ribosyl)ation would introduce an anomalous methylation pattern on transfected DNA.[46] Plasmid pVHCk containing the SV40 early promoter linked to the bacterial chloramphenicol acyltransferase (CAT) gene was used in our experiments since this prokaryotic vector, containing a high density of CpG pairs, is CpG island-like. The methylation pattern of the transfected plasmid was directly analyzed by sequencing a fragment of its DNA according to Frommer's method.[49] The results clearly demonstrated that when the plasmid was transfected into cells with drug-inhibited poly(ADP-ribosyl)ation, an anomalous methylation pattern characterized its DNA: nearly all cytosines, and not only those present in CpG dinucleotides, were now methylated.

Experiments were also carried out to test the transcription status of the originally unmethylated plasmid as a function of time following transfection into control or 3-ABA-treated cells. Literature reports have shown that in vitro methylation of a plasmid significantly reduces the transcription of the associated reporter gene after transfection, and that this inhibition is dependent on the extent of methylation and on the time after transfection. The time dependence of the transcriptional inhibition was shown to reflect the time needed to assemble chromatin on the plasmid DNA,[50,51] and to spread the inactive chromatin structure from a focus of methylation.[52]

The data showed that the expression of the CAT reporter gene as measured by the CAT activity at 24 or 48 hours after transfection was decreased by about 30% in the 3-ABA-treated cells relative to the untreated controls.

The results of these four different experimental strategies taken together allowed us to propose the first method to induce DNA hypermethylation in vivo: treatment of cells in culture with 3-aminobenzamide.[53]

Atomic Force Microscopy (AFM) Studies of the Effect of DNA Methylation on Chromatin Fiber Structure

In an effort to understand better the structural changes imposed on the chromatin fiber by elevated levels of DNA methylation, we decided to make use of the imaging capabilities of AFM to compare chromatin fibers of varying degrees of methylation. To that end, we used both in vivo and in vitro approaches.

In the in vivo approach, we used our method to induce DNA hypermethylation by treatment of cultured cells with 3-ABA. Chromatin fibers were isolated from nuclei of control and treated cells, were dialysed against low ionic strength buffers, and were imaged in air on mica following glutaradehyde fixation.[54] Bringing the isolated fibers to low salt was necessary to allow resolution of individual nucleosomal particles, and thus better assessment of the fiber structural parameters; glutaraldehyde fixation was needed to preserve the extended fiber conformation during the deposition and washing steps preceding imaging. A difference between control fibers and fibers isolated from drug-treated cells was immediately apparent, the treated fibers being much more condensed, with individual nucleosomes situated much closer to each other, often overlapping. Measurements of parameters such as center-to-center internucleosome distances, fiber heights, or number of nucleosomes per unit fiber length confirmed the visual impression: the treatment of the cells with 3-ABA resulted in an almost two-fold compaction of the chromatin fiber.

These results, although very clear-cut, had to be confirmed by an independent approach since the effects observed could be due to the direct inhibition of the poly(ADP-ribose) polymerase activity, rather than to the associated DNA methylation. Thus, we substantiated the in vivo results with in vitro experiments in which we reconstituted defined nucleosomal arrays on either unmethylated or methylated DNA sequences. We made use of the well-characterized 208-12 system,[55] in which nucleosomes were assembled by salt dialysis on a tandemly repeated sequence. Each unit in the tandem positions one histone octamer, thus resulting in the formation of a rather regular nucleosomal array. Arrays reconstituted on unmethylated 208-12 or on the same sequence methylated in vitro with SssI, were imaged and analysed as outlined above. In one set of experiments, the reconstitution was done with histone octamers only; in another set, linker histones were added too. Rather to our surprise, when the fibrers contained no linker histone bound, they were indifferent to the methylation status of the underlying DNA. Only when the linker histone was present, did the methylated chromatin fiber look more compact. The quantitative analysis again confirmed the visual impression. Gratifyingly, the degree of compaction of the linker histone-containing methylated chromatin fiber was exactly the same as the compaction of the chromatin fiber extracted from cell treated with 3-ABA. Thus, based on the in vivo and in vitro results, we concluded that DNA methylation induces chromatin compaction only when assisted by linker histone.

How Is Poly(ADP-Ribosyl)ation Involved in Protecting DNA Methylation Pattern

As far as the correlation between DNA methylation and gene expression is concerned, the CpG islands, that range from 0.5 to 2 kbp in size, are usually found in the 5' promoter regions of housekeeping genes, overlapping the genes to various extents.[56] There is evidence that the transcription of genes associated with CpG islands is inhibited when these regions are methylated.[57] The fact that CpG dinucleotides are present in an unmethylated state in CpG islands is very intriguing since the CpG density is six-ten times higher than in bulk DNA. The mechanisms involved in protecting the unmethylated status of CpG islands in genomic DNA remain far from understood. Experiments suggest the existence of sequence motifs which, by binding transcription or other protein factors, protect against de novo methylation.[58,59] The presence of some cis-acting "boundaries of methylation", capable of preventing the methylation of flanking DNA sequences, has also been suggested.[58,60-66] The simple plausible explanation that trans-acting protein factors associated with CpG islands prevent access of methylases to those DNA regions, has been difficult to demonstrate so far, although recently some trans-acting factors have been identified or suggested to play a general or a specific role in maintaining the DNA methylation pattern.[67-69]

On the basis of our combined data, we propose that a nonprotein factor is involved in this trans-acting role: long and branched molecules of ADP-ribose polymers. To explain how poly(ADP-ribosyl)ation is involved in maintaining DNA methylation pattern, two hypotheses can be put forward.

The first hypothesis considers that the modification is not directly involved in modifying some trans-acting protein factor capable of binding DNA, but rather induces a deregulation of Dnmt1 expression. As the expression of Dnmt1 is normally cell-cycle dependent,[70,71] its overexpression in an anomalous cell-cycle phase could represent one of the molecular events involved in DNA hypermethylation (Fig. 4). The second hypothesis considers that a modified

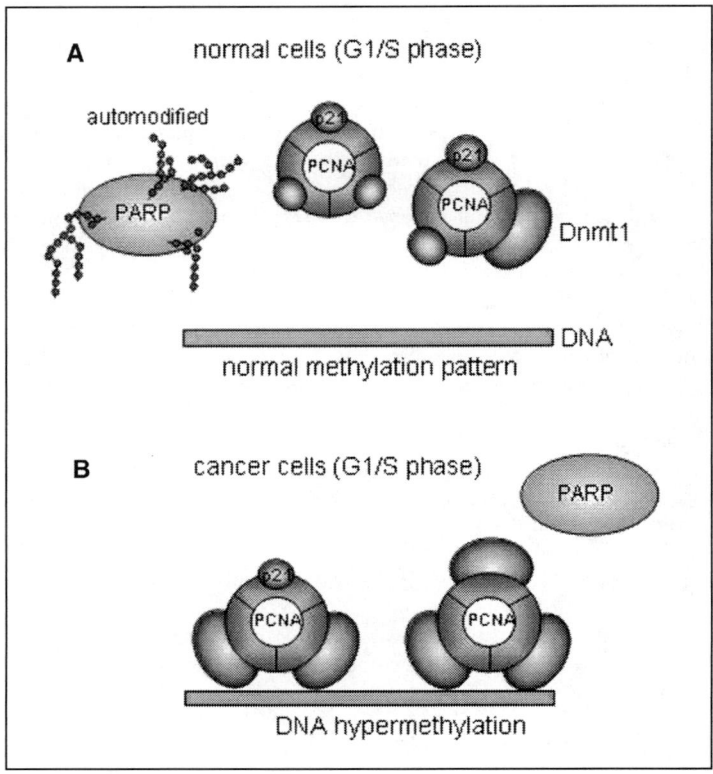

Figure 4. A) In normal cells, automodified PARP plays a role in controlling the expression of Dnmt1 in the appropriate cell-cycle phase, and thus the level of Dnmt1-PCNA active complex is low in G1/S phase. B) Inhibition of PARPs induces overexpression of Dnmt1 in G1/S phase of the cell-cycle, causing anomalously high level of the active Dnmt1-PCNA complex in this phase. This mechanism may be involved in inducing the hypermethylation of CpG islands in oncosuppressor genes during malignant transformation.

protein is a trans-acting factor able to bind directly to DNA, to inhibit the access of Dnmt1 to DNA (Fig. 5).

The first hypothesis suggests that poly(ADP-ribosyl)ation may be involved, for example, in modulating the binding between the proliferating cellular nuclear antigen (PCNA) and DNA methyltransferase.[72] Our data show that inhibition of poly(ADP-ribose) polymerase(s) at different cell-cycle phases increases both the mRNA and the protein levels of the major maintenance DNA methyltransferase (Dnmt1).[73] This increase in Dnmt1 results in more PCNA-Dnmt1 complex formation, which is expected to facilitate the maintenance as well as the de novo DNA methylation processes during the G1/S phase, perhaps leading to abnormal hypermethylation of early replicating DNA sequences. The observation of higher levels of PCNA-Dnmt1 complexes in cells treated with inhibitors of PARPs may also provide a possible explanation as to how anomalous hypermethylation of CpG islands occurs during neoplasia: a precocious formation of the Dnmt1-PCNA complex at the G1/S border may modify the unmethylated state of the CpG islands in the promoter regions of housekeeping genes[73] that are present in early replicating DNA.[74,75] Since there are important oncosuppressor genes among the housekeeping genes, our results could provide evidence supporting the model proposed by Baylin[67] to explain hypermethylation of oncosuppressor genes in cancer cells. According to this model—in which PCNA has an all-important role—in normal cells the CpG islands are

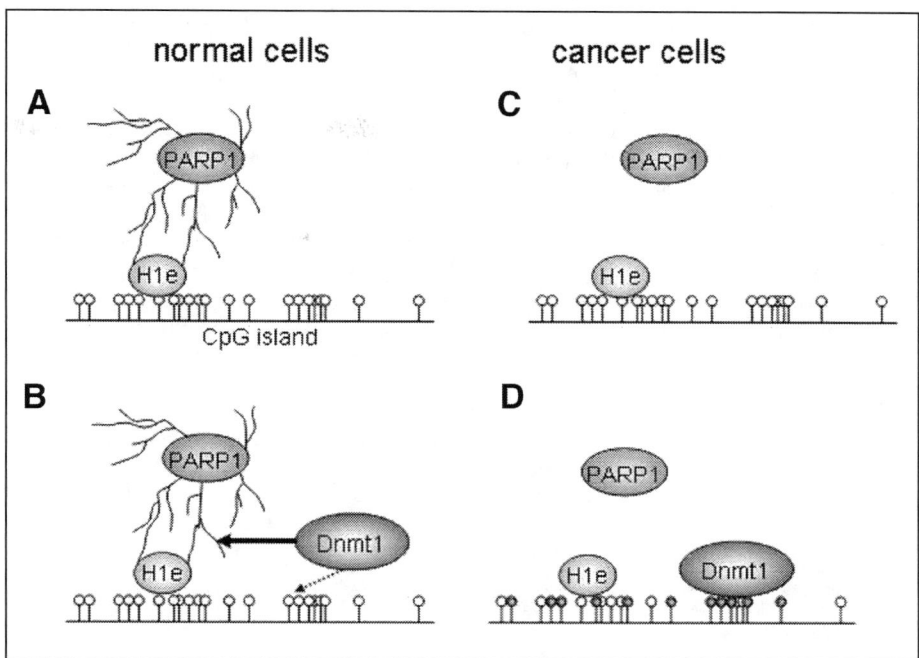

Figure 5. A,B) Possible role of H1e, noncovalently modified by PARP1, as a marker of CpG islands in normal cells, where poly(ADP-ribosyl)ated H1e protects the islands from methylation. C,D) In the absence of poly(ADP-ribose) polymers Dnmt1 can introduce new methyl groups into the DNA. In cancer cells, inhibition of PARPs may be responsible for the anomalous hypermethylation of the CpG islands associated with the oncosuppressor genes in chromatin.

protected from methylation in early S-phase by the higher amount of p21 present relative to Dnmt1. The increase in the level of Dnmt1 in tumor cells could favour the premature formation of the Dnmt1-PCNA complex in G1/S phase allowing the anomalous methylation of oncosuppressor gene promoters, thus suppressing their functions. Thus, malignant transformation could occur even if Dnmt1 is not overexpressed, but is expressed in the "wrong" phase of the cell-cycle (a phase in which it is not normally expressed) (Fig. 4).

Another possibility is that the affinity between PCNA, p21 and Dnmt1 may be modulated by the poly(ADP-ribosyl)ation process, as in vivo both unmodified and modified isoforms of PCNA and p21 have been reported.[76,36] The DNA hypermethylation induced by blocking poly(ADP-ribosyl)ation could be explained by assuming that the unmodified form of PCNA is the only one capable of binding DNA methyltransferase1 and of transferring the enzyme onto DNA. The absence of poly(ADP-ribosyl)ation could also be involved in the competition between DNA methyltransferase1 and p21, for the same domain on PCNA.[73] Dnmt1 could have the upper hand in the absence of active PARPs. Thus, inhibition of PARPs could help the association between PCNA and Dnmt1.

The second hypothesis considers the nonallelic somatic variant H1e of histone H1, in its poly(ADP-ribosyl)ated isoform, as a nuclear trans-acting factor involved in maintaining the unmethylated state of CpG islands. A possible mechanism considers that in normal cells H1e, in its poly(ADP-ribosyl)ated isoform, is capable of recognizing and binding to CpG islands, thus protecting them from DNA methyltransferase action (Fig. 5A,B). By contrast, in cells in which the poly(ADP-ribosyl)ation process is inhibited this protection from methylation cannot occur, since the poly(ADP-ribose)-free protein is unable to inhibit the DNA methyltransferase and, therefore, the enzyme can now carry out its reaction (Fig. 5C,D). In suggesting this model, we have considered that: (a) H1 histone, through its genic variant H1e, is capable of inhibiting in vitro DNA methylation;[77] (b) the inhibitory effect of H1 histone on in vitro DNA methylation seems to be due to its poly(ADP-ribosyl)ated isoform;[43] (c) H1e is the only H1 histone variant able to bind CpG island-like DNA sequences;[77] (d) H1e variant, in its poly(ADP-ribosyl)ated isoform, could be present in decondensed chromatin;[77] (e) the inhibition of poly(ADP-ribose) polymerases introduces anomalous methylation pattern in CpG islands,[44] and (f) both total H1 histone and each H1 histone variant are able to bind long and branched ADP-ribose polymers in a noncovalent way.[77]

Thus, in this model it is assumed that ADP-ribose polymers are present in CpG islands bound to histone H1e; H1e is both a marker for CpG islands and a target for the noncovalent binding to long and branched ADP-ribose polymers (these may be free or still attached to the automodified PARP, see above). We have already presented evidence that Dnmt1 is another member of the family of poly(ADP-ribose) binding proteins, and that the affinity of Dnmt1 for ADP-ribose polymers is higher than for DNA (Reale et al, manuscript submitted for publication). Thus, Dnmt1 will be attracted by and hosted on ADP-ribose polymers, thus being prevented from introducing new methyl groups onto DNA (Fig. 5).

In conclusion, while several studies have indicated a direct involvement of PARP1 in transcription, with either positive or negative effects,[78-80] our model assigns to PARP1 an indirect impact on gene expression, through a mechanism whereby ADP-ribose polymers exclude methylation from CpG islands.

Acknowledgements

This work was supported by Ministero Italiano dell'Università (Areneo FIRB, COFIN), by Ministero della Sanità, by Istituto Pasteur Fondazione Cenci Bolognetti and by AIRC (Associazione Italiana Ricerca sul Cancro).

References

1. Robertson KD, Jones PA. DNA methylation: Past, present and future directions. Carcinogenisis 2000; 21:461-467.
2. Baylin SB, Herman JG. DNA methylation in tumorigenesis. Epigenetics joins genetics. Trends Genet 2000; 16:168-174.
3. Costello JF, Plass C. Methylation matters. J Med Genet 2001; 38:285-303.
4. Bird AP, Wolffe AP. Methylation-induced repression-belts, braces and chromatin. Cell 1999; 99:451-454.
5. Bird A. DNA methylation patterns and epigenetic memory. Genes Dev 2002; 16:6-21
6. Kass SU, Pruss D, Wolffe AP. How does methylation repress trnscription? Trends Genet 1997; 12: 444-449.
7. Razin A, Shemer R Epigenetic control of gene expression. Results Probl Cell Differ 1999; 25:189-204.
8. Szyf M. The role of DNA methyltransferase 1 in growth control. Front Biosci 2001; 6:599-609.
9. Cervoni N, Szyf M. Demethylase activity is directed by histone acetylation. J Biol Chem 2001; 276:40778-40787.
10. Cervoni N, Detich N, Seo S-B et al. The oncoprotein Set/TAF-1b, an inhibitor of histone acetyltransferase, inhibits active demethylation of DNA, integrating DNA methylation and transcriptional silencing. J Biol Chem 2002; 277:25026-25031.
11. Nan X, Ng H-H, Johnson CA et al. Transcriptional repression by the methyl-CpG binding protein MeCP2 involves a histone deacetylase complex. Nature 1998; 393:386-389.
12. Jones PL, Veenstra GJC, Wade PA et al. Methylated DNA and MeCP2 recruit histone deacetylase to repress transcription. Nat Genet 1998; 19:187-191.
13. de Murcia G, Shall S, eds. From DNA damage and stress signaling to cell death Poly ADP-ribosylation reactions. Oxford University Press, 2000.
14. Alvarez-Gonzalez R, Althaus FR. Poly (ADP-ribose) catabolism in mammalian cells exposed to DNA-damaging agents. Mutation Res 1989; 218:67-74.
15. Althaus FR, Richter C. ADP-Ribosylation of proteins: Enzymology and Biological Significance. Berlin: Springer-Verlag, 1987.

16. D'Amours D, Desnoyers S, D'Silva I et al. Poly(ADP-ribosyl)ation reactions in the regulation of nuclear functions. Biochem J 1999; 342:242-268.
17. Mendoza-Alvarez H, Alvarez-Gonzalez R. Poly (ADP-ribose) polymerase is a catalytic dimer and the automodification reaction is intermolecular. J Biol Chem 1993; 268:22575-22580.
18. Alvarez-Gonzalez R, Jacobson MK. Characterization of polymers of adenosine diphosphate ribose generated in vitro and in vivo. Biochemistry 1987; 26:3218-3224.
19. Kawaichi M, Ueda K, Hayaishi O. Multiple poly(ADP-ribosyl)ation of rat liver poly(ADP-ribose) synthetase. Mode of modification and properties of automodified synthetase. J Biol Chem 1981; 256:9483-9489.
20. Desmarais Y, Menard L, Lagueux J et al. Enzymological properties of poly(ADP-ribose)polymerase : characterization of automodification sites and NADase activity. Biochim Biophys Acta 1991; 1078:179-186.
21. Ferro AM, Higgins NP, Olivera BM. Poly (ADP-ribosylation) of a DNA topoisomerase. J Biol Chem 1983; 258:6000-6003.
22. Yoshihara K, Itaya A, Tanaka Y et al. Inhibition of DNA polymerase a, DNA polymerase b, terminal deoxynucleotidyl transferase, and DNA ligase II by poly(ADP-ribosyl)ation reaction in vitro. Biochem Biophys Res Commun 1985; 28:1 61-67.
23. Boulikas T. DNA str, breaks alter histone ADP-ribosylation. Proc Natl Acad Sci USA 1989; 86:3499-3503.
24. Scovassi AI, Mariani C, Negroni M et al. ADP-ribosylation of nonhistone proteins in HeLa cells: Modification of topoisomerase II. Exp Cell Res 1993; 206:177-181.
25. Ménissier-de Murcia J, Molinete M, Gradwohl G et al. Zinc-binding domain of poly(ADP-ribose) polymerase participates in the recognition of single strand breaks on DNA. J Mol Biol 1989; 210:229-233.
26. Gradwohl G, de Murcia JM, Molinete M et al. The second zinc-finger domain of poly(ADP-ribose) polymerase determines specificity for single-stranded breaks in DNA. Proc Natl Acad Sci USA 1990; 87:2990-2994.
27. Ikejma M, Noguchi S, Yameshita R et al. The zinc-fingers of human poly(ADP-ribose) polymerase are differentially required for the recognition of DNA breaks and nicks and the consequent enzyme activation. J Biol Chem 1990; 265:21907-21913.
28. de Murcia G, Ménissier-de Murcia J. Poly(ADP-ribose) polymerase: a molecular nick sensor. Trends Biochem Sci 1994; 19:172-176.
29. de Murcia G, Jacobson M, Shall S. Regulation by ADP-ribosylation. Trends Cell Biol 1995; 5:78-81.
30. Griesenbeck J, Oei SL, Mayer-Kuckuk P et al. Protein-protein interaction of the human poly(ADP-ribosyl)transferase depends on the functional state of the enzyme. Biochemistry 1997; 36:7297-7304.
31. Realini C, Althaus FR. Histone shuttling by poly (ADP-ribosylation). J Biol Chem 1992; 267:18858-188621.
32. Panzeter PL, Realini CA, Althaus FR. Noncovalent interactions of poly(adenosine diphosphate ribose) with histones. Biochemistry 1992; 31:1379-1385.
33. Panzeter PL, Zweifel B, Malanga M et al. Targeting of histone tails by poly(ADP-ribose). J Biol Chem 1993; 268:17662-17664.
34. Malanga M, Atorino L, Tramontano F et al. Poly(ADP-ribose) binding properties of histone H1 variants. Biochim Biophys Acta 1998; 1399:154-160.
35. Reale A, Malanga M, Zardo G et al. In vitro induction of H1-H1 histone cross-linking by adenosine diphosphate-ribose polymers. Biochemistry 2000; 39:10413-10418.
36. Pleschke JM, Kleczkowska HE, Strom M et al. Poly(ADP-ribose) binds to specific domains in DNA damage checkpoint proteins. J Biol Chem 2000; 275:40974-40980.
37. Mayer-Kuckuk P, Ullrich O, Ziegler M et al. Functional interaction of poly(ADP-ribose) with the 20S proteasome in vitro. Biochem Biophys Res Commun 1999; 259:576-581.
38. Germain M, Affair EB, D'Amours D et al. Cleavage of automodified poly(ADP-ribose) polymerase during apoptosis. J Biol Chem 1999; 274:28379-28384.
39. Malanga M, Pleschke JM, Kleczkowska HE et al. Poly(ADP-ribose) binds to specific domains of p53 and alters its DNA binding functions. J Biol Chem 1998; 273:11839-11843.
40. Shall S, de Murcia G. Poly(ADP-ribose) polymerase-1: what have we learned from the deficient mouse model? Mutation Res 2000; 460:1-15.
41. Schreiber V, Amé JC, Dollé P et al. Poly(ADP-ribose) polymerase-2 (PARP-2) is required for efficient base excision DNA repair in association with PARP-1 and XRCC1. J Biol Chem 2002; 277:23028-23036.
42. Althaus FR, Kleczkowska HE, Malanga M et al. Poly ADP-ribosylation: A DNA break signal mechanism. Mol Cell Biochem 1999; 193:5-11.

43. Zardo G, D'Erme M, Reale A et al. Does poly(ADP-ribosyl)ation regulate the DNA methylation pattern? Biochemistry 1997; 36:7937-7943.
44. Zardo G, Caiafa P The unmethylated state of CpG islands in mouse fibroblasts depends on the poly(ADP-ribosyl)ation process J Biol Chem 1998; 273:16517-16520.
45. de Capoa A, Febbo Giovannelli FR, Niveleau A et al. Reduced levels of poly(ADP-ribosyl)ation result in chromatin compaction and hypermethylation as shown by cell-by-cell computer assisted quantitative analysis. FASEB J 1999; 13:89-93.
46. Zardo G, Marenzi S, Perilli M et al. I. nhibition of poly(ADP-ribosyl)ation introduces an anomalous methylation pattern in transfected foreign DNA. FASEB J 1999; 13:1518-1522.
47. Bird AP. CpG islands as gene markers in the vertebrate nucleus. Trends Genet 1987; 3:342-347.
48. Bird AP. CpG-rich islands and the function of DNA methylation. Nature 1986; 321:209-213.
49. Frommer M, McDonald lE, Millar DS et al. A genomic sequencing protocol that yields a positive display of 5-methylcytosine residues in individual DNA strands. Proc Natl Acad Sci USA 1992; 89:1827-1831.
50. Buschhausen G, Wittig B, Graessmann M et al. Chromatin structure is required to block transcription of the methylated herpes simplex virus thymidine kinase gene. Proc Natl Acad Sci USA 1987; 84:1177-1181.
51. Kass SU, Landsberger N, Wolffe AP. DNA methylation directs a time-dependent repression of transcription initiation. Curr Biol 1997; 7 157-165.
52. Kass SU, Goddard JP, Adams RLP. Inactive chromatin spreads from a focus of methylation. Mol Cell Biol 1993; 13:7372-7379.
53. Zardo G, Marenzi S, Caiafa P. Correlation between DNA methylation and poly(ADP-ribosylation) process. Gene Therapy Mol Biol 1998; 1:661-679.
54. Karymov MA, Tomschik M, Leuba SH et al. DNA methylation-dependent chromatin fiber compaction in vivo and in vitro: Requirement for linker histone. FASEB J 2001; 15:2631-2641.
55. Simpson RT, Thoma F, Brubaker JM. Chromatin reconstituted from tandemly repeated cloned DNA fragments and core histones; a model for study of higher order structure. Cell 1985; 42:799-808.
56. Antequera F, Bird A. CpG islands as genomic footprints of promoters that are associated with replication origins. Curr Biol 1999; 9:661-667.
57. Keshet I, Ysraeli J, Cedar H. Effect of regional DNA methylation on gene expression. Proc Natl Acad Sci USA 1985; 82:2560-2564.
58. Szyf M, Tanigawa G, McCarthy PL Jr. A DNA signal from the Thy-1 gene defines de novo methylation patterns in embryonic stem cells. Mol Cell Biol 1990; 10:4396-4400.
59. Tollefsbol TO, Hutchinson III CA. Control of methylation spreading in synthetic DNA sequences by the murine DNA methyltransferase. J Mol Biol 1997; 269:494-504.
60. Szyf M. DNA methylation patterns: an additional level of information? Biochem Cell Biol 1991; 69:764-767.
61. Mummaneni P, Bishop PL, Turker MS A cis-acting element accounts for a conserved methylation pattern upstream of the mouse adenine phosphoribosyltransferase gene. J Biol Chem 1993; 268:552-558.
62. Brandeis M, Frank D, Keshet I et al. Sp1 elements protect a CpG island from de novo methylation. Nature 1994; 371:435-438.
63. Hasse A, Schulz WA. Enhancement of reporter gene de novo methylation by DNA fragments from the alpha-fetoprotein control region. J Biol Chem 1994; 269:1821-1826.
64. MacLeod D, Charlton J, Mullins J et al. Sp1 sites in the mouse aprt gene promoter are required to prevent methylation of the CpG island. Genes Dev 1994; 8:2282-2292.
65. Magewu AN, Jones PA. Ubiquitous and tenacesous methylation of the CpG sites in codon 248 of the p53 gene may explain its frequent appearance as a mutational hot spot in human cancer. Mol Cell Biol 1994; 14:4225-4232.
66. Mummaneni P, Walker KA, Bishop PL et al. Epigenetic gene inactivation induced by a cis-acting methylation center. J Biol Chem 1995; 270:788-792.
67. Baylin SB. Tying it all together: epigenetics, genetics, cell cycle and cancer. Science 1997; 277:1948-1949.
68. Pradhan S, Kim GD. The retinoblastoma gene product interacts with maintenance human DNA (cytosine-5) methyltransferase and modulates its activity. EMBO J 2002; 21:1-10.
69. Di Croce L, Raker VA, Corsaro M et al. Methyltransferase recruitment and DNA hypermethylation of target promoters by an oncogenic transcription factor. Science 2002; 295:1079-1082.
70. Szyf M, Bozivic V, Tanigawa G. Growth regulation of mouse DNA methyltransferase gene expression. J Biol Chem 1991; 266:10027-10030.

71 Robertson KD, Keyomarsi K, Gonzales FA et al. Differential mRNA expression of the human DNA methyltransferase (DNMTs) 1, 3a and 3b during the G0/G1 to S phase transition in normal and tumor cells. Nucleic Acids Res 2000; 28:2108-2113.

72. Chuang LS, Ian HI, Koh TW et al. Human DNA-(cytosine-5) methyltransferase-PCNA complex as a target for p21 WAF1. Science 1997; 26:1996-2000.

73. Zardo G, Reale A, Passananti C et al. Inhibition of poly(ADP-ribosyl)ation induces DNA hypermethylation: A possible molecular mechanism. FASEB J 2002; 21:1319-1321.

74. Goldman MA, Holmquist GP, Gray MC et al. Replication timing of genes and middle repetitive sequences. Science 1984; 224:686-692.

75. Selig S, Okumura K, Ward DC et al. Delineation of DNA replication time zones by fluorescence in situ hybridization. EMBO J 1992; 11:1217-1225.

76. Simbulan-Rosenthal CM, Rosenthal DS, Boulares AH et al. Regulation of the expression or recruitment of components of the DNA synthesome by poly(ADP-ribose) polymerase. Biochemistry 1998; 37:9363-9370.

77. Zardo G, Marenzi S, Caiafa P. H1 histone as a trans-acting factor involved in protecting genomic DNA from full methylation. Biol Chem 1998; 353:647-654.

78. Meisterernst M, Stelzer G, Roeder RG. Poly(ADP-ribose) polymerase enhances activator-dependent transcription in vitro. Proc Natl Acad Sci USA 1997; 94:2261-2265.

79. Oei SL, Griesenbeck J, Schweiger M et al. Regulation of RNA polymerase II-dependent transcription by poly(ADP-ribosyl)ation of transcription factors. J Biol Chem 1998; 273:31644-31647.

80. Oei SL, Griesenbeck J, Ziegler M et al. A novel function of poly(ADP-ribosyl)ation: silencing of RNA polymerase II-dependent transcription. Biochemistry 1998; 37:1465-1469.

CHAPTER 12

The Role of Active Demethylation in Cancer and Its Therapeutic Potential

Moshe Szyf, Paul M. Campbell, Nancy Detich, Jing Ni Ou, Stefan Hamm and Veronica Bovenzi

Abstract

Regional hypermethylation and global hypomethylation coexist in cancer cells. Understanding the mechanisms responsible for global hypomethylation and regional hypermethylation in cancer is required for the proper design of therapeutic strategies targeting the DNA methylation machinery. This chapter discusses different models explaining this paradox. Global hypomethylation is proposed to be associated with activation by demethylation of metastasis-associated genes. Thus, anticancer therapy directed at DNA methyltransferase might have the untoward effect of promoting metastasis. Inhibition of demethylase activity on the other hand could potentially inhibit metastasis. It is therefore important to identify and characterize the enzymes responsible for global hypomethylation in cancer.

Introduction: The Paradox of DNA Methylation Patterns in Tumors

The main question that bewildered us when we try to come to grips with the therapeutic implications of DNA methylation in cancer is the coexistence of regional hypermethylation and global hypomethylation in almost all tumors as discussed in previous chapters in this book. The general occurrence of DNA methylation aberrations in cancer is a powerful indication that DNA methylation is a pivotal player in cancer, however the paradoxical pattern of changes defies a simple mechanistic implication. As a consequence, defining the correct therapeutic response to DNA methylation aberrations in cancer is puzzling.

Several groups, as was reviewed in previous chapters in this book, are considering inhibitors of DNA methyltransferase 1 (DNMT1) as candidate anticancer drugs. Although DNMT1 might be involved in transformation by methylation independent mechanisms,[1-4] the main therapeutic goal of currently developed inhibitors is to induce demethylation and reexpression of tumor suppressor genes. Preclinical data supports the hypothesis that DNMT inhibitors have antitumorigenic effects.[5-7] However, since global hypomethylation is also prevalent in cancer, we must address the following question: are regional hypermethylation and global hypomethylation both important for transformed cells? If both processes are important, then any therapeutic interference that inhibits DNA methylation with the aim of inducing methylation silenced tumor suppressor genes will promote the transformation process by causing global hypomethylation. To truly deal with this critical issue we need to understand the mechanisms responsible for the coexistence of regional hypermethylation and global hypomethylation in cancer and identify the important components of these processes for cellular transformation. Although we do not have as of yet a clear understanding of all the mechanisms involved, some recent data might help us establish a working hypothesis. In this chapter I will review the

DNA Methylation and Cancer Therapy, edited by Moshe Szyf. ©2005 Eurekah.com and Kluwer Academic/Plenum Publishers.

pertinent data on the possible role of DNA hypomethylation in cancer that might guide us in formulating a working hypothesis. I will also discuss the therapeutic implications of available data and alternative working hypotheses.

Global Hypomethylation in Cancer

Once the tight association of DNA methylation and silencing of gene expression was established, it stood to reason that a possible connection between DNA methylation and cancer would be sought, as it had been accepted for the last three decades that aberrant gene expression programming is involved in cancer. Since the first focus of cancer molecular biology was on oncogenes, which are aberrantly activated in cancer, it is clear why the first experiments that tested the relation between cancer and DNA methylation focused on hypomethylation of oncogenes in cancer.[8,9] It was postulated that aberrant DNA hypomethylation was an additional mechanism of ectopic activation of protooncogenes. Aberrant hypomethylation is an attractive mechanism for ectopic oncogene activation since it might occur at a higher frequency than physical changes in genes caused by rearrangements and mutations, and it is potentially reversible. The reversibility of the process might provide therapeutic and pharmacological opportunities. These first studies unveiled that hypomethylation in cancer is not limited to oncogenes but, what was proven to be by later studies, hypomethylation is global and is a general hallmark of cancer cells.[10]

Global Hypomethylation in Cancer; Single Copy and Multiple Copy Sequences Are Hypomethylated in Multiple Tumor Types

During the last two decades, the state of methylation of both general DNA and specific gene sequences in multiple tumor types and stages of malignancy were studied as discussed in previous chapters. In general, sparsely distributed CG sequences in DNA from tumors have reduced methylation levels in comparison with their normal tissue counterparts. This conclusion has been supported by numerous studies in a variety of cancers from different cellular origins such as Moris hepatoma,[11] metastatic prostate cancer,[12] breast cancer,[13,14] colorectal cancer,[15] cervical dysplasia,[16] ovarian epithelial tumors,[17] metastatic variants of human melanoma cell lines,[18] Wilms tumor,[19] hepatocellular carcinoma[20] and premalignant stages of gastric carcinoma.[21]

It was also observed that hypomethylation of repeat sequences such as Line 1 retroviral elements occurs at a higher frequency in urothelial carcinoma than in paired normal tissues[22] and it was suggested that there is some selectivity in methylation of certain classes of Line 1 elements in malignant tissues.[23] Another class of repetitive sequences that are hypomethylated in tumors such as breast adenocarcinoma,[24] ovarian carcinoma, and Wilms tumor[25] are satellite 2 repeat sequences found in the pericentromeric region of chromosome 1 and 16. In some studies a correlation was established between the state of malignancy and the extent of hypomethylation of these sequences.[17]

In addition to global hypomethylation of repetitive sequences, a number of studies have shown that single copy genes are hypomethylated in tumors relative to paired tissue. This group of genes includes certain oncogenes; *Ras* in colonic adenocarcinoma and small cell lung carcinoma,[8] *c-Myc* and epidermal growth factor receptor in hepatocellular carcinoma[26] and in bladder cancer,[27] and *Ornithine Decarboxylase* and *Erb-A1* oncogene in chronic lymphatic leukemia.[28]

Another interesting group of genes that are hypomethylated in certain cancers encode proteins that stimulate tumor cell motility and invasion such as the metastasis associated Ca^{++} binding protein MTS1/S100A4 in colonic adenenocarcinoma and pancreatic ductal adenocarcinoma[29,30] and the protease urokinase-type plasminogen activator *uPA* in metastatic breast cancer.[31] An important class of genes that might have critical therapeutic applications is composed of genes encoding proteins that confer multidrug resistance such as *MDR1,* whose hypomethylation was correlated with increased drug resistance in acute myeloid leukemia.[32] Hypomethylation and increased drug resistance might compromise chemotherapy regimens.

Mechanisms Responsible for Hypomethylation in Cancer

To be able to address the potential role of global hypomethylation in tumorigenesis, we ought to understand how hypomethylation comes about and how it coexists with regional hypermethylation. Two basic rules emerge from the vast literature that analyzed global hypomethylation in cancer. First, hypomethylation is global, wide regions of the genomes are hypomethylated. Second, hypomethylation is heterogeneous, the extent of demethylation and its sequence specificity differ even among tumors from the same type. This profile of hypomethylation is consistent with a general defect in the DNA methylation machinery. A general change in the DNA methylation machinery is consistent with a global but stochastic loss of methylation, which is observed in tumors.

The commonly accepted model of inheritance of DNA methylation pattern is based on two fundamental principles. The first principle is that during cell division the maintenance DNA methyltransferase DNMT1 is guided exclusively by the methylation pattern of the parental strand. This guarantees faithful inheritance of the methylation pattern.[33] Second, the DNA methylation reaction is enzymatically irreversible. The only feasible enzymatic reaction is methylation, which is catalyzed by DNA methyltransferases. According to this model hypomethylation could come about only by replication in the absence of DNA methyltransferase.[33] If this model is correct, then global hypomethylation in cancer might be a consequence of a general reduction in DNA methyltransferase activity. However, one persistent observation in cancer is increased DNA methyltransferase activity[34] as discussed in other chapters in this book. It is possible however that although DNMT1 levels are elevated in developed tumors, its activity is inhibited at an early stage in transformation causing demethylation. Once DNA methylation is lost, DNMT1 might be unable to correct this defect by de novo methylation. However, the progressive loss of methylation seen during progression of some cancers is inconsistent with such a model.[14,16,20]

An alternative explanation is that global hypomethylation in tumors is caused by a reduction in the intracellular concentration of the DNA methyl donor S-adenosylmethionine (AdoMet) or an increase in the concentration of the product and inhibitor of the DNA methylation reaction S-adenosylhomocysteine (AdoHcy).[35] Tetrahydrofolate is required for AdoMet synthesis and it has been shown that folate deficiency occurs in squamous cell lung carcinoma cells.[36] Nevertheless, a reduction in AdoMet was well documented only in animal models that were fed a methyl-deficient diet[37] or chemically induced rat liver tumors.[38] Thus, whereas it is feasible that low methyl diet can precipitate reduction in AdoMet concentrations and global hypomethylation, it is not clear yet that this is a mechanism generally involved in tumorigenesis.

In summary, the classical model of maintenance DNA methylation implicating exclusively DNA methyltransferase in enzymatic DNA methylation, does not explain the global and heterogeneous hypomethylation observed in tumors.

The Possible Role of a Demethylase; DNA Methylation Is a Reversible Reaction

A simple explanation for global hypomethylation in tumors could theoretically be a general increase in demethylase activity.[39,40] However, there has been a general reluctance to accept that it is possible to remove methyl groups from DNA by an enzymatic reaction because of the predicted stability of the carbon-carbon bond between the methyl moiety and the cytosine ring. However, it was obvious two decades ago that demethylation was involved in the shaping of DNA methylation patterns during development. It was originally proposed that demethylation came about by a passive mechanism, by masking methylatable sites from the DNA methyltransferase during replication.[33] In support of this hypothesis it has recently been shown that binding of a transcription factor which has high affinity to its recognition element on a stable episome can result in demethylation, possibly by masking of the site from DNA methyltransferase.[41-43]

However, a passive demethylation mechanism could not explain the global demethylation observed during development[44-46] and differentiation[47] since it is hard to believe that specific binding proteins protect vast sections of the genome from DNA methylation. Moreover, there is data demonstrating that demethylation of specific genes[48-51] as well as global hypomethylation during differentiation[52] occurs in the absence of DNA replication. Rapid global hypomethylation has been demonstrated in the paternal genome well before initiation of the first round of DNA replication.[53]

To be able to explain replication independent demethylation without having to presume an enzymatic removal of methyl groups from DNA, it has been proposed that active demethylation comes about by a repair process. Two repair processes were proposed. First, a glycosylase activity cleaves the bond between the 5-methylcytosine base and the deoxyribose moiety in DNA. The abasic site is then repaired by resident repair activity in the absence of DNA methyltransferase resulting in replacement of a 5-methylcytosine with an unmethylated cytosine.[54-57] A second mechanism proposed that the methylated nucleotide was removed by nucleotide excision and was then replaced by an unmethylated cytosine.[58] These models do not differ in principle from the passive demethylation model. Both models suggest that demethylation comes about by DNA synthesis in the absence of DNA methyltransferase. The main difference resides in the mode of DNA synthesis, the first model considers DNA synthesis during cell division while in the other two models unmethylated bases are incorporated by DNA repair.

In support of the glycosylase model, two mismatch repair glycosylases were shown to have methyl CG DNA glycosylase (5-MCDG) activity that results in demethylation in vitro, the cloned G/T mismatch repair enzyme[59] and the methylated binding protein MBD4.[60] Ectopic expression of the 5-MCDG glycosylase in human embryonal kidney cells results in the specific demethylation of a stably integrated ecodysone-retinoic acid responsive enhancer-promoter linked to a beta-galactosidase reporter gene,[61] suggesting that this glycosylase could cause demethylation in vivo. It was also shown that this glycosylase participates in global demethylation during differentiation of C2C12 cells since transfection of cells with an antisense oligonucleotide to 5-methylcytosine DNA glycosylase (G/T mismatch DNA glycosylase) decreases both the activity of the enzyme and genome-wide demethylation.[61]

One interesting property of the 5-MCDG glycosylase is that it demethylates specifically hemimethylated mCGs. The ^{m}CG/CG hemimethylated site resembles the TG/CG mismatch. Both T and ^{m}C are pyrimidines, which are methylated at their 5' position. The critical problem is to define the principal activity of this enzyme in living cells. One possibility is that the bona fide substrate of the enzyme is a G/T mismatch and its in vivo function mismatch repair. In this case the ^{m}CG/CG repair activity is an artifact that is a consequence of high expression in transfection experiments or the high concentrations present in vitro. Alternatively, the main function of this enzyme is demethylation of hemimethylated DNA. It is also possible that the enzyme performs both tasks in the cell. In any case, global demethylation and demethylation during differentiation and tumorigenesis must also involve demethylation of both strands of DNA, which must be therefore catalyzed by a different enzyme. A possible role for the 5-MCDG glycosylase might be correction of aberrant methylated sites introduced during replication and added to one strand of DNA. Methyl groups found only on one strand of the DNA might signal that they are not authentically inherited and should be removed.

One main problem with repair mediated passive demethylation mechanisms is explaining global hypomethylation, especially the global hypomethylation that takes place at the very early stages of embryogenesis as discussed above.[62] If demethylation occurs by a repair mechanism, global hypomethylation would involve global DNA damage. This might significantly harm the integrity of the genome and it stands to reason that such a mechanism would not be utilized early in development.

We therefore pursued a third possibility that an enzyme transforms methylated cytosines to cytosines by removing the methyl moiety catalyzing a bona fide reversal of the DNA methylation reaction. We hypothesized that such an enzyme should be found in abundance in tumor cells because of the consistent global hypomethylation observed in tumors. We then partially purified a demethylase activity from nonsmall cell human lung carcinoma.[63] The partially purified demethylase converts 5-methyl cytosine in the dinucleotide CG to cytosine and releases the methyl moiety in the form of methanol.[63] We showed that the demethylase is a processive enzyme, which can explain how it could potentially demethylate vast regions of the genome to bring about global hypomethylation.[64] Demethylase activity was significantly higher in tumor cells than in normal cells (unpublished data) and we have previously shown that ectopic Ras expression in the embryonal teratocarcinoma cell line P19 induces demethylase activity.[39] Thus, it is possible that increased demethylase activity in tumor cells plays a causal role in global hypomethylation in cancer. We also cloned a methylated DNA binding protein MBD2b from human cervical carcinoma cell line HeLa and showed that it can cause demethylation of methylated CG in DNA.[65] The assignment of a demethylation activity to MBD2, which was cloned and identified as a member of the methylated DNA binding protein family by Hendrich et al,[66] was disputed by several groups, who could not repeat the initial observations of Bhattacharya et al.[65] The more accepted functional role of MBD2 is the suppression of methylated genes, similar to MeCP2. MBD2 is associated with the NuRD[67] chromatin remodeling and gene repression complex and it has been suggested to be the methylated DNA binding component of the MeCP1 complex.[68,69]

It is still unclear what the reasons are for the discrepancy between these results. It is possible that MBD2 acts differently on different substrates, or that an unknown factor that is required for the demethylation reaction copurifies with the enzyme under some purification protocols but not others. It is also possible that MBD2 copurifies with an inhibitor under some conditions but not others. MBD2 is found in multiprotein complexes[68] and might have different activities in different multiprotein contexts. We are obviously actively testing all these possibilities. Nevertheless, recent data from our laboratory indicates that ectopic expression of MBD2 can cause demethylation of cotransfected ectopically methylated DNA[70] and that this activity is promoter dependent.[71] The promoter dependence of MBD2 might explain some of the discrepancies between the results of different laboratories.

It is still unclear whether MBD2 directly suppresses gene expression in living cells. Some of the information that is published is inconsistent with MBD2 repressing methylated DNA. MBD2 was found to bind to unmethylated CG islands but not to methylated CG islands in living cells using the ChIP chromatin immunoprecipitation assay.[72] MBD2 was found bound to methylated CG islands only when proteins are crosslinked to DNA in isolated nuclei but not when proteins are crosslinked to DNA in whole cells. It is feasible that the process of nuclei isolation disrupts the interactions between proteins and DNA seen in living cells.[72] MeCP2 on the other hand was shown to interact with methylated DNA in living cells by numerous publications.[72-74] In *Drosophila*, association of dMBD2/3 with DNA coincides with activation of the embryonic genome and it associates with the activated Y chromosome during activation of the spermatocyte genome.[75] This data is consistent with a role for MBD2 in gene activation, not gene suppression, although it is possible as suggested by the authors of this publication, that MBD2 suppresses specific genes in active domains of the chromosome. While the biological role of MBD2 in either demethylation or gene suppression is still unclear, its critical role in cellular transformation is beginning to emerge as discussed below.

It is clear however that MBD2 is redundant as far as fetal development, viability and fertility are concerned since an *mbd2-/-* knock out mouse is viable and fertile.[76] The *mbd2-/-* knock out mouse does not exhibit either gross changes in gene expression or DNA methylation. This must imply that other demethylases are present in vertebrates. Nevertheless, data from our laboratory suggests that MBD2/demethylase is critical for tumorigenesis.[77] The fact that MBD2/demethylase is critical for tumorigenesis but not for the viability of the healthy animal raises interesting therapeutic opportunities that will be discussed below.

In summary, there is evidence that a bona fide demethylase is present in vertebrates and is possibly overexpressed in tumors. Higher activity of a general demethylase might be

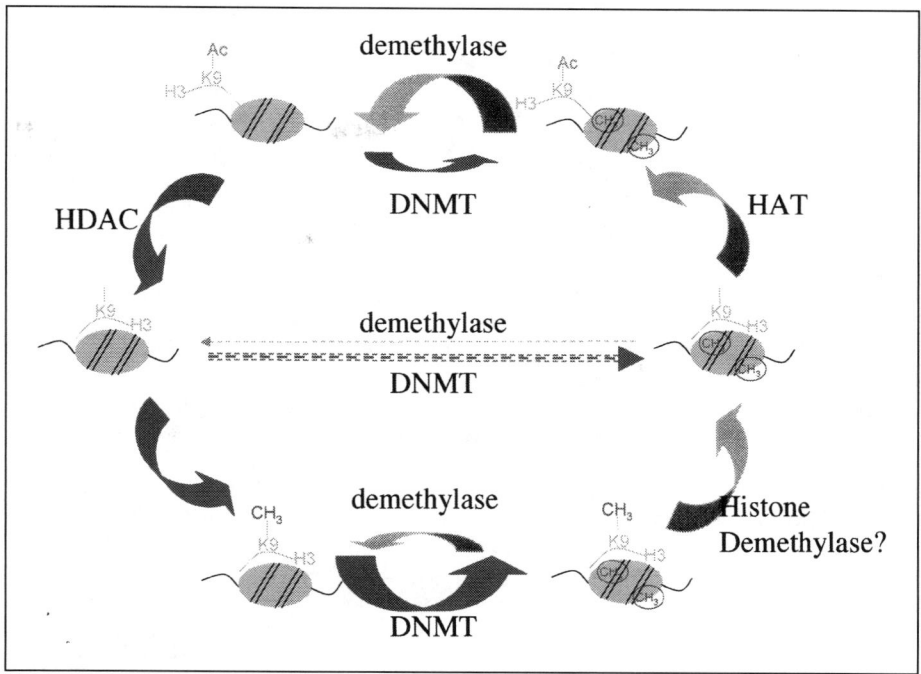

Figure 1. The dynamic cycle of chromatin and DNA methylation states. The state of DNA methylation is determined by a balance of DNA methylation, catalyzed by DNA methyltransferases (DNMT) and demethylation, catalyzed by demethylase. The direction of the reaction is determined by chromatin structure. DNA associated with a nucleosome (yellow oval) that bears an acetylated (Ac) Histone 3 (H30 tail (yellow line) at K9 is a preferred substrate for demethylase and is poorly methylated by DNMT. Methyl CpGs are indicated by an encircled CH_3. Interaction of histone deacetylase (HDAC), which is recruited by corepressors, with the nucleosome results in histone deacetylation and a tight interaction between the tail and DNA. This DNA is inaccessible to demethylase but is a preferred substrate for DNMT, which is recruited by HDACs. Deacetylated histone tails could undergo further modification by histone methyltransferases SUV39. Histone methylation further stabilizes the inactive state. DNA associated with nonacetylated K9-methylated histones is inaccessible to demethylase but is a preferred substrate for DNA methylation. Once the DNA is methylated, the inactive state of the gene is stabilized. Putative histone demethylases recruited by putative transcriptional activators could remove the methylation from H3 K9, which creates a preferred substrate for histone acetyltransferases (HATs). Acetylation creates a preferred accessible substrate for demethylase, resulting in demethylation and stable activation of the gene.

responsible for the global and heterogeneous hypomethylation seen in cancer. The possibility of a demethylase enzyme obliges us to amend our perception of the DNA methylation reaction from a unidirectional reaction involving one enzyme to a reversible reaction involving two sets of enzymes, DNA methyltransferases and demethylases.[2,78,79] The DNA methylation pattern is proposed to represent the steady state balance of methylation and demethylation (Fig. 1). An increase in the activity or inhibition of either enzyme should alter the balance of the reaction and the state of DNA methylation. It is unclear whether demethylases are involved in the regular homeostasis of DNA methylation in normal cells, or whether they are recruited only under certain circumstances such as cellular differentiation or transformation. The special role of MBD2/demethylase in the process is also unclear. An important question is what determines the direction of the methylation reaction, its specificity and its correlation with gene expression.

Resolving the Paradox of the Coexistence of Regional Hypermethylation and Global Hypomethylation in Cancer; Role of Chromatin Structure

As discussed in the preceding chapters both regional hypermethylation and global hypomethylation coexist in cancer. How could regional hypermethylation be possible in the presence of high levels of demethylase activity?

A first clue as to a possible solution to this paradox could be derived from the fundamental differences between hypermethylation and hypomethylation in cancer. The fact that the hypermethylation is regional and not global implies that these changes are not caused by a general defect in the DNA methylation machinery, but rather by a regional change in the properties of these genes as substrates for DNA methylation. Alternatively, regional changes are consistent with a change in activity of a DNA methyltransferase that recognizes specific regions in the genome, and which is upregulated in cancer. There is no evidence for such a mechanism. However, there is emerging evidence that CpG islands undergo significant changes in chromatin structure and that these changes might precede changes in DNA methylation. For example, an analysis of the *E-CADHERIN* CpG island which is silenced in multiple tumor cell lines reveals that the methylation pattern is extremely heterogeneous in different cell lines in which the *E-CADHERIN* gene is silenced.[72] In one of the cell lines, the gene is silenced in spite of the fact that the promoter is completely unmethylated. However, in all cases where the gene is silenced, the histones associated with the promoter are deacetylated.[72] This detailed analysis provides us with a picture of different stages in the silencing of *E-CADHERIN* in cancer. Since histone acetylation is the change observed in all cases, it most probably precedes hypermethylation, which seems to be a response to the inactive state of chromatin.

DNMT1 and DNMT3b were shown to associate with both HDAC1 and HDAC2[80,81] and we propose that gene silencing by factors that recruit HDAC1 or HDAC2 to CpG islands eventually leads to recruitment of DNMTs to the silenced genes and their eventual methylation. Factors that silence CpG islands by altering chromatin structure are proposed to be the primary cause for silencing and hypermethylation, and their expression is hypothesized to alter in cancer cells (Fig. 1). An example of an oncogenic factor leukemia-promoting PML-RAR fusion protein, which specifically recruits DNMT1 and HDAC1 to target promoters has been recently described.[82] One important area of research in the future should be uncovering additional factors that suppress expression of tumor suppressor genes in cancer.

In contrast to the hypothesis proposed here, experiments with the histone deacetylase inhibitor TSA and the DNA methylation inhibitor 5-aza-CdR have led to the suggestion that DNA methylation is the primary cause of silencing of methylated tumor suppressor genes. The experiments show that while 5-aza-CdR activates methylated tumor suppressor genes, TSA is unable to activate them on its own.[83,84] TSA nevertheless does have a synergistic effect with 5-aza-CdR.[83,84] I would like to suggest that this data does not necessarily imply that chromatin modification is secondary to DNA methylation. It is possible that TSA treatment is insufficient to fully activate the chromatin structure and therefore chromatin structure might still be the primary reason of tumor suppressor gene inactivation.

It is now clear that chromatin structure involves multiple silencing modifications in addition to histone acetylation, such as histone methylation, which might not be relieved by TSA treatment exclusively (Fig. 2). Recent detailed mapping of the silenced and methylated tumor suppressor p16 5' region by ChIP analysis revealed methylation of K9 in the H3-histone tail. H3-K9 methylation is associated with gene silencing and is believed to preclude H3-acetylation.[74,85] 5-aza-CdR treatment results in rapid reversal of histone H3-K9 methylation and demethylation and activation of the p16 gene.[86,87] TSA does not inhibit histone methylation and is therefore incapable of activating H3-K9 methylated chromatin.

An important related issue is to understand how 5-aza-CdR, a DNA methylation inhibitor reverses histone methylation. There are two possible explanations. First, inhibition of

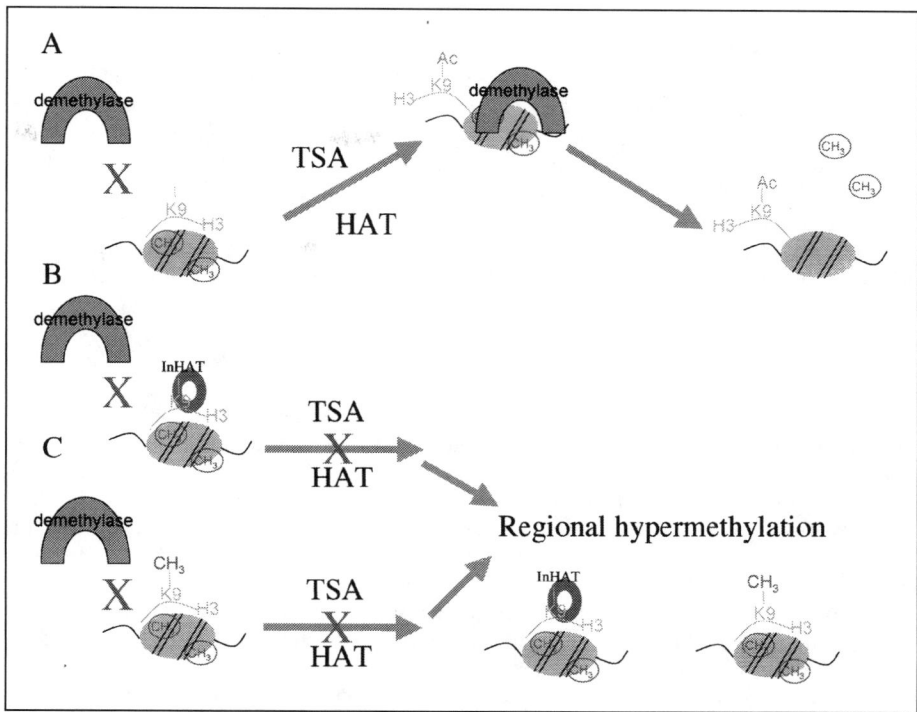

Figure 2. Coexistence of regional hypermethylation and global demethylation, a model. Nonacetylated histone tails inhibits the access of demethylase to DNA. The histone deacetylase inhibitor TSA induces histone acetylation and accessibility to demethylase (A) resulting in DNA demethylation as indicated by the encircled (CH_3) removed from DNA. Gene B however escapes the global increase in acetylation since its histone tails are associated with inhibitors of histone acetylation, InHATs, which mask K9 from histone acetyl transferases (HATs). This explains why under condition of excess global demethylation certain genes remain hypermethylated, and why methylated tumor suppressors are not induced and demethylated by TSA treatment. An additional mechanism that might mask specific genes (C) from global demethylation and TSA treatment is histone H3-K9 methylation. H3-K9 methylation renders the gene inaccessible to demethylase.

SUV39 histone methyltransferases (which catalyze H3-K9 methylation) might be a side effect of 5-aza-CdR. Second, histone methylation might require the presence of MeCP2, which was recently shown to associate with histone methyltransferase.[88] Inhibition of DNA methylation by 5-aza-CdR results in removal of MeCP2 and the histone methyltransferase associated with it.

In addition to histone methylation, some genes, as will be described below, might be protected from histone acetylation even in the presence of TSA by InHATs, a complex of proteins that bind histone tails and prevent their acetylation[89] (Fig. 2). In summary, the lack of response to TSA does not indicate that a gene is not silenced by chromatin modification.

The model proposed here that local alterations in chromatin structure are responsible for regional hypermethylation, still leaves the unanswered question of what protects these CpG islands from global hypomethylation? This question relates to the general question of what delimits the activity of demethylase? What determines the balance of the DNA methylation reaction towards either methylation or demethylation? Since different genes exhibit different methylation patterns, what are the signals that determine the steady state of the DNA methylation reaction?

To address this question we ought to consider the basic rules that define DNA methylation patterns in vertebrates. One of the most consistent properties of DNA methylation is the tight correlation between methylated DNA and an inactive chromatin configuration and between hypomethylated DNA and an active chromatin configuration.[90] We have therefore previously hypothesized that the steady state methylation patterns of a gene is determined by the chromatin structure. Active chromatin structure targets demethylase activity while inactive chromatin structure targets DNMT activity[2,78] (Fig. 1). The possible mechanisms for recruitment of DNMTs to inactive chromatin were discussed above.

My laboratory focused on the possibility that active chromatin targets demethylase. We proposed that the hypomethylated state of active genes is a consequence of their active chromatin structure, which attracts demethylase activity. To test the hypothesis we had to identify a system where active demethylation could be studied in isolation from passive demethylation caused by inhibition of DNA methylation during replication, and de novo methylation. Numerous studies followed the state of methylation of endogenous genes in vivo and in cell culture, but it was impossible to study active demethylation in isolation from DNA replication and DNA methylation. Thus, whereas in vivo studies contributed to the characterization of the steady state methylation pattern of genes, they did not unravel the mechanisms that defined these DNA methylation patterns. To study mechanisms responsible for the steady state methylation pattern of a sequence, we need to introduce the sequence into a cell at a defined state of methylation, and then determine the mechanisms, which lead to different steady state patterns of methylation.

According to the accepted model described above, DNA methylation is a unidirectional reaction determined by the state of methylation of the parental strand.[33] Therefore, if a methylated DNA is introduced into a cell, it should remain methylated irrespective of its potential state of activity. However, if DNA methylation is a reversible reaction and is determined by chromatin structure then methylated DNA introduced into a cell will acquire, with time, a state of methylation that reflects its state of chromatin activation in the cell.

We found that transformed human embryonal kidney HEK293 cells are ideal for studying the mechanisms defining active demethylation in vertebrate cells.[70] These cells are readily transfected by exogenous DNA, and the exogenous transiently transfected DNA does not replicate and does not undergo de novo methylation in these cells under these conditions. DNA is methylated in vitro before transfection and is introduced into the cell by calcium phosphate precipitation method. The state of methylation of the transfected DNA is determined three days after transfection. To separate differences in methylation caused by variation in the primary sequence from changes caused by the state of activity of the gene, we studied the state of methylation of an identical reporter gene placed under different promoters. Thus, we compare the state of methylation of identical sequences. We first introduced methylated reporter DNA that was associated with either an active or inactive promoter and we found that the reporter sequence was demethylated only when it was associated with an active promoter. This data demonstrates that the DNA methylation is dynamic in vertebrate cells, and that it reflects the state of activity of genes. This data is inconsistent with the hypothesis that DNA methylation patterns are defined by the DNA methylation of the parental strand. Thus, the DNA methylation pattern is not fixed and its steady state reflects the state of gene activity (Fig. 1). We then determined whether the active demethylation of the reporter gene is dependent either on the sequence of the promoter or the state of modification of chromatin.

To address this question, we modified the state of histone acetylation pharmacologically using the general histone deacetylase inhibitor TSA and found that histone acetylation is the primary determinant of the state of methylation irrespective of the promoter used. Histone acetylation triggers demethylation even in the absence of a promoter.[70] To finally demonstrate a direct relation between histone acetylation and demethylation we used a chromatin immunoprecipitation assay to show that the transfected DNA, which is associated with acetylated histones is demethylated, whereas DNA associated with deacetylated histones is fully methylated.

One simple mechanism explaining the relation between state of acetylation of histones and demethylation is that the deacetylated histone tails, which form tight association with DNA, inhibit the accessibility of demethylase to methylated DNA (Fig. 2).

These experiments provide a simple mechanism explaining the maintenance of the unmethylated state of active genes in vertebrate cells. Active genes are associated with acetylated histones in all instances and are therefore an excellent substrate for the demethylase, which maintains them unmethylated. A shift in the state towards deacetylation of a gene will tilt the steady state of the reaction towards methylation. However, since the process of de novo methylation is inefficient, DNA methylation of a silenced genes lags behind its silencing and chromatin inactivation as discussed above. This mechanism also can explain the demethylation of genes during development. Binding of transacting factors to cis elements recruits histone acetyltransferases to the gene, which then targets the gene for demethylation. For example, the immunoglobulin kappa gene is specifically demethylated during development in a process requiring the transcription factor NFκB as well as the intronic kappa enhancer and the matrix attachment region.[91] Similarly, it was recently shown in maize that the transcription factor TpnA, which binds cis acting sequences in Suppressor-mutator (Spm) transposon also promotes its demethylation.[92]

The fact that demethylation by demethylase is dependent on regional and gene specific acetylation states can explain how certain genes can remain hypermethylated in cancer even in the presence of high levels of demethylase activity. Regional hypermethylation is caused by regional changes in chromatin structure that inhibit the accessibility of the gene even to highly abundant demethylase (Fig. 2).

Support for the hypothesis that regional changes in chromatin structure can protect from active demethylation comes from determining the effect of proteins that inhibit acetyltransferase activity (InHATs) might have on demethylation. InHATs bind lysine 9 on histone tails and protect them from acetylation even in the presence of the pharmacological inhibitor of histone deacetylase TSA.[89] Ectopic expression of InHAT proteins protects exogenously introduced methylated DNA from demethylation even in the presence of high concentrations of TSA.[93] An increase in InHAT binding to certain CG islands can explain how regional hypermethylation can persist in the presence of high levels of demethylase. Interestingly, one of the proteins comprising the InHAT complex is the oncoprotein Set/TAF-1β, which we show to be elevated in tumors.[93]

This line of experiments might also explain why TSA does not cause global induction of gene expression and global demethylation and why TSA does not induce expression of tumor suppressor genes as a single agent.[84] We propose that InHATs as well as other histone modification such as H3-K9 methylation protect certain CG sequences from acetylation, and as a consequence from demethylation (Fig. 2). These regional sequences are masked from demethylase even when it is abundant. Demethylation of such sequences would be possible only if these additional histone modifications are removed.

In summary, including demethylase in the DNA methylation equilibrium can help us explain the global hypomethylation observed in tumors. New data showing that demethylase acts only on DNA that is found in an active chromatin structure can explain how regional hypermethylation persists even in the presence of high levels of demethylase. If one assumes that the DNA methylation pattern is dynamic and reversible, the paradox of hypo- and hypermethylation in cancer could be resolved.

Possible Role of Global Hypomethylation in Cancer

As discussed above and in other chapters in this book, global hypomethylation is a consistent property of cancer cells and it occurs independently of regional hypermethylation.[19] The critical question with respect to the therapeutic implications of this phenomenon is whether it plays a causal role in tumorigenesis or whether it is a byproduct of the transformation process. If global hypomethylation is critical for transformation, it implies that the use

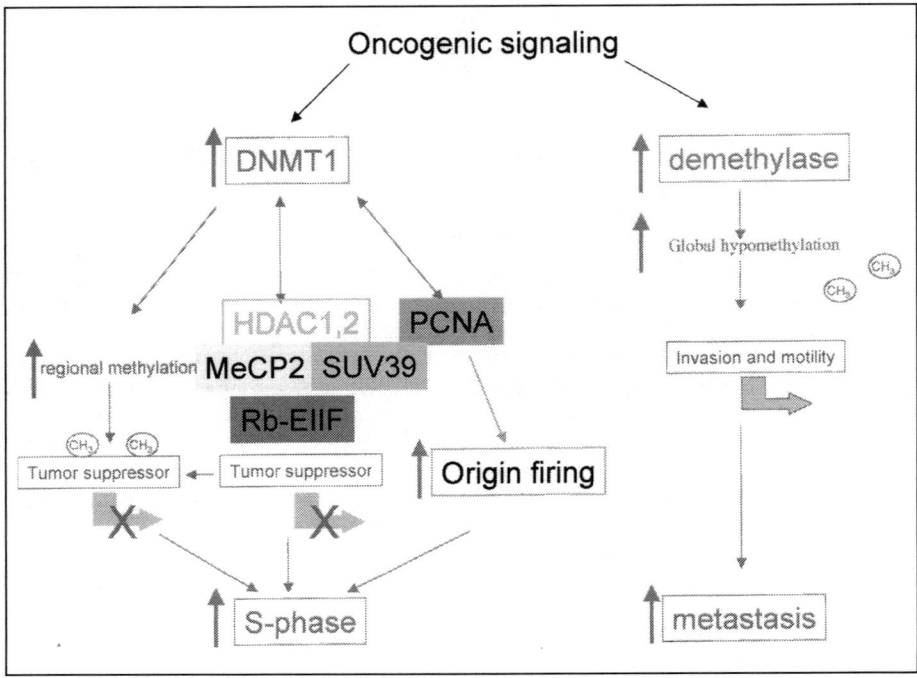

Figure 3. The different roles of DNMT and demethylase in tumorigenesis. Oncogenic signals induce both DNMT and demethylase activities. Increase in DNMT activity stimulates oncogenesis by three mechanisms. First, DNMT interaction with the replication factor PCNA stimulates initiation of DNA replication and overriding cell arrest signals. Second, DNMT forms protein-protein interactions with chromatin modifiers and transcriptional repressors resulting in silencing of tumor suppressors. Third, DNMT is required for maintenance of the ectopic methylation of tumor suppressors leading to their silencing. The combined action of DNMT results in stimulating entry into the S phase of the cell cycle. Induction of demethylase results in demethylation and activation of a class of genes involved in tumor invasion and migration. Tumor suppressors are masked from this effect because of regional chromatin modification as explained in Figure 1. The combined action of demethylase and DNMT leads to fully developed tumorigenesis, uncontrolled growth and metastasis.

of DNA methylation inhibitors in anticancer therapy might have serious risks on one hand, and that demethylase inhibitors might have an important therapeutic potential. If global hypomethylation plays a causal role in cancer, it is important to define this role as well as the role of regional hypermethylation of tumor suppressor genes. It stands to reason that these processes target two different but mutually critical steps in tumorigenesis (Fig. 3).

Three lines of evidence support the hypothesis that hypomethylation plays a causal role in cancer. The first line of evidence is correlative and was discussed above. The second line of evidence comes from epidemiological and experimental data correlating cancer with methyl deficient diets. Methyl deficient diets were shown by numerous studies to cause hypomethylation and promote liver cancer in rats.[37,94-97] and liver DNA was shown to be hypomethylated during the early stages of chemical carcinogenesis.[98] On the other hand hepatocarcinogenesis and hypomethylation induced in diethylnitrosamine-initiated rats by "resistant hepatocyte" (RH) protocol is inhibited by exogenous administration of the methyl donor S-adenosyl-L-methionine (AdoMet).[99]

AdoMet is the methyl donor of numerous methylation reactions and it is possible that AdoMet exerts its chemoprotective effects through other biological methylations. However, an experiment showing that the chemoprotective effect of AdoMet is reversed when it is co administered with the DNA methylation inhibitor 5-aza-CdR supports the idea that DNA methylation is critical for the chemoprotective action of AdoMet.[100] Epidemiological data have suggested a correlation between low methyl-group and folate dietary intake and the risk for colorectal adenomas and cancer.[35,97] Similarly, chronic alcohol consumption, which can diminish cellular AdoMet levels, causes genomic hypomethylation and was implicated as an etiologic agent in colorectal carcinogenesis.[101,102] Third, early experiments have shown that 5-aza-CdR treatment can enhance the metastatic potential of tumor cell lines.[103-105] These data indicate that hypomethylation of DNA might cause, in addition to its anticancer effects, induction of genes that promote the invasive capacity of tumors as well as promote formation of new tumors. Although none of these data demonstrate directly that the mechanism of carcinogenesis is through hypomethylation of DNA, and it is possible that both hypomethylating diets and 5-aza-CdR promote carcinogenesis by a mechanism independent of DNA methylation, the convergence of animal, epidemiological, and 5-aza-CdR results strongly supports this hypothesis.

Mechanisms Whereby Hypomethylation Enhances Tumorigenesis

If hypomethylation plays a role in tumorigenesis, what is the mechanism through which it enhances tumorigenesis? The mechanism involved has obvious therapeutic implications for both DNA methylation inhibitors and potential demethylase inhibitors.

Three different mechanisms have been proposed in literature to be induced by hypomethylation. First, oncoprotein gene hypomethylation, which is observed in normal tumors as well as chemically and methyl deficient induced tumors, is proposed to promote aberrant oncogene activity similar to activating mutations. There is evidence for hypomethylation of *c-myc* and *Ha-ras* oncogenes in human tumor samples from colonic adenocarcinoma and small cell lung carcinoma relative to adjacent normal tissue,[8] of *c-myc* and EGF receptor in hepatocellular carcinoma[26,106] in bladder carcinomas[27] and the *erb-A1* gene was found to be hypomethylated in chronic lymphatic leukemia.[28] Similarly, hypomethylation of oncogenes was shown to occur in methyl-deficient induced hepatic cancers in rats[94] and in chemically induced liver cancer in rodents.[98,107-109] However, although oncogene hypomethylation has been observed in tumors and during stages of carcinogenesis, there is no direct evidence that it plays a role in expression of these genes.

A second possible role is that global hypomethylation promotes genomic instability. This hypothesis is supported by genetic and pharmacological evidence. Firstly, DNA hypomethylation has been associated with abnormal chromosomal structures in cells from patients with ICF (Immunodeficiency, Centromeric instability and Facial abnormalities) syndrome as discussed in other chapters in this book. In this syndrome which is most probably caused by mutations in the DNA methyltransferase DNMT3b,[110,111] methylation of satellite 2 in the pericentromeric region of chromosome 1 and 16 is defective.[112] Thus, there is direct genetic evidence showing that a defect in DNA methylation causes chromosomal instability in humans. Second, there is pharmacological evidence for a causal relation between hypomethylation of pericentromeric regions and chromosomal instability since pre B cells treated with the demethylating agent 5-aza-CdR exhibited pericentromeric rearrangements of chromosome 1 at a very high frequency.[113] Third, in addition to this pharmacological evidence, it was reported that *dnmt1-/-* knockout murine embryonic stem cells exhibited genomic instability.[114]

Interestingly, similar to ICF syndrome, many cancers also exhibit defective methylation of satellite 2 and other repetitive sequences as well as pericentromeric chromosomal rearrangements. This similarity between both conditions is consistent with the hypothesis that global hypomethylation in cancer leads to chromosomal instability.[115] For example, almost half of 25 examined breast adenocarcinomas exhibited hypomethylation in satellite 2 DNA which is normally highly methylated.[24]

A third possible mechanism is that hypomethylation induces metastasis by activation of genes required for cell motility, invasion and metastasis. This hypothesis is based on earlier observations that treatment of cancer cells with the DNA demethylating agent 5-aza-CdR leads to increased metastasis as discussed above.[105] Thus, we propose that increased hypomethylation is required for the stages of tumorigenesis involving metastasis, invasion and migration (Fig. 3). This hypothesis is supported by the correlation between the degree of hypomethylation of tumors and their invasive potential observed in some but not all studies. In prostate cancer, hypomethylation of the genome correlates with the metastatic capacity of the tumor.[12] In cervical cancer the degree of hypomethylation increases with the grade of cervical neoplasia.[16] Soares et al, observed a trend for DNA from breast carcinoma with positive axillary nodes to be more hypomethylated than those without nodal involvement and a statistically significant correlation was found between global hypomethylation and disease stage.[14] Similarly, Shen et al, found a significant correlation between global hypomethylation and infiltration and metastatic capacity.[20] Liteplo and Kerbel compared the global methylation level of melanoma cell MelWo and its metastatic variants and found that metastatic variants were hypomethylated relative to the nonmetastatic variants.[18] Although these correlative studies on their own do not demonstrate that global hypomethylation is causal in metastasis, they strongly support a role for global hypomethylation in invasion and metastasis in conjunction with the fact that 5-aza-CdR, a DNA methylation inhibitor, promotes metastasis. However, to fully establish the role of hypomethylation in metastasis we need to demonstrate a feasible mechanism whereby hypomethylation results in increased metastasis. A possible mechanism is that global hypomethylation also results in the hypomethylation and activation of genes required for either cell motility, adhesion molecules or proteases required for invasion. These genes are methylated in nonmetastatic cells and the cell is transformed to a metastatic variant by hypomethylation and activation of these genes (Fig. 3).

A number of examples of genes involved in cell motility and invasion support this hypothesis. The metastasis associated protein Mts1/S110A4, which encodes a calcium binding protein that is involved in cell motility is overexpressed in metastatic variants relative to nonmetastatic variants of tumors.[116] Ectopic expression of this protein in nonmetastatic mouse mammary adenocarcinoma cell line increases in vitro motility but not invasiveness.[117] However transfection of this gene into the non metastatic human breast cancer cell line MCF-7 increased invasiveness in vitro and in vivo.[118] In accordance with the hypothesis proposed here it was shown that *Mts1* expression in human colon carcinoma correlates with its state of methylation[29] and that overexpression of Mts1 in pancreatic ductal carcinoma correlates with hypomethylation and is associated with poor differentiation.[30]

DNA methylation regulates other proteases similar to *Mts1*, which are potentially involved in tumor invasion. Gelatinase B is overexpressed upon induction with interleukin 1 in a metastatic melanoma cell line but not in a nonmetastatic cell line derived from the same patient. 5-aza-CdR treatment of both cell lines results in constitutive activation of Gelatinase B.[119]

Direct evidence that DNA hypomethylation activates a protease involved in tumor invasion has recently been published. MCF7 is a non metastatic human breast cancer line whereas MDA-MB-231 is a highly invasive and metastatic line. The protease Urokinase-Type Plasminogen Activator (uPA), which is required for tumor invasion, is not expressed in MCF-7 and the uPA gene is hypermethylated, whereas it is expressed in MDA-MB-231 and it is hypomethylated. Upon treatment with the DNA methylation inhibitor 5-aza-CdR, the gene is demethylated, uPA is expressed, and cells are transformed to be invasive and metastatic. This study also showed that global demethylase activity is elevated in MDA-MB-231 cells relative to MCF7 cells.[31]

We propose that changes in DNA methylation required for tumorigenesis involve two separate steps. The first step involves an increase in DNMT1 and possibly DNMT3b activity, including perhaps a specific alternative splicing product of DNMT3b.[120] This step leads to bypassing of tumor growth signals, and silencing and hypermethylation of CG islands of growth

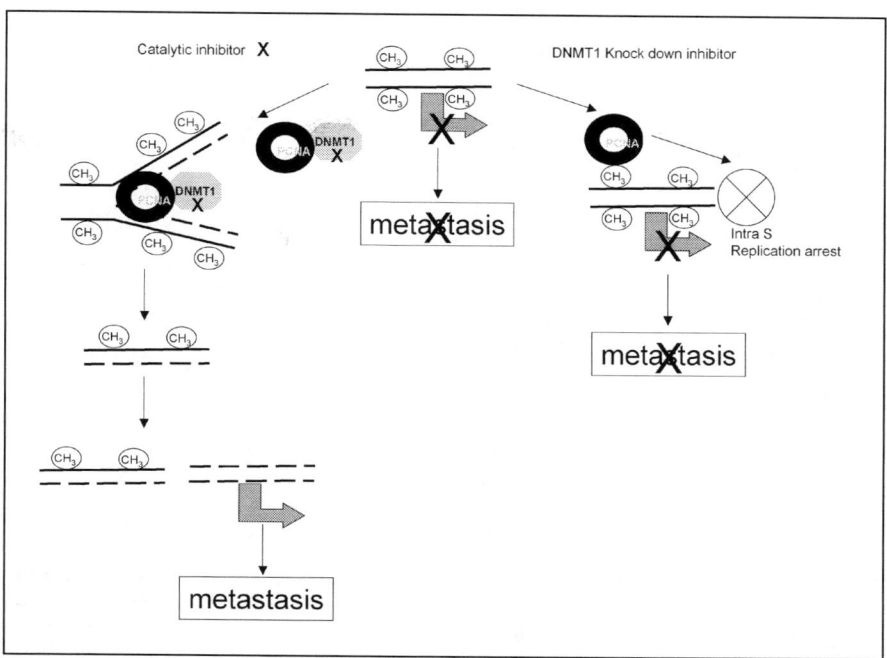

Figure 4. Two putative mechanisms of action of DNMT1 inhibitors. The first class of inhibitors interferes with the catalytic activity of the enzyme but does not disrupt its protein-protein interactions. The replication fork forms in the presence of DNMT1 and replication of the new strand (dashed red line) proceeds in the absence of DNMT1 activity resulting in passive demethylation of the nascent strand. Further rounds of replication in the presence of the inhibitor result in the formation of DNA, which is unmethylated on both strands of DNA. Some of the genes that are demethylated might promote metastasis. The second class of inhibitors either knocks down DNMT1 levels such as antisense oligonucleotides or siRNA or knock down important protein-protein interactions of DNMT1 such as its interaction with PCNA. This results in intra S phase arrest of initiation of DNA replication and therefore little or no demethylation and no induction of pro-metastatic genes.

regulatory genes. We have previously proposed that the immediate effects of DNMT1 on growth control are independent of DNA methylation, and involve protein-protein interactions of DNMT1 and possibly other DNMTs. The second step, once growth control has been breached, is invasion and metastasis. Hypomethylation of growth invasive genes is required for activation of genes involved in this process (Fig. 3).

The Therapeutic Implications of Global Hypomethylation

As discussed in earlier chapters in this book, DNA methyltransferase (DNMT) inhibitors have shown promise as anticancer agents.[2,40,121] Two classes of inhibitors are used: DNMT1 antisense molecules, which knockdown the protein levels,[6,7] and inhibitors of DNA methyltransferase activity such as 5-aza-CdR.[122,123] We have previously proposed that knockdown of DNMT1 protein also causes inhibition of DNA replication by a mechanism, which is independent of demethylation and is dependent on the protein-protein interactions of DNMT1.[1,4] Inhibitors that knock down DNMT1 levels should not cause extensive global hypomethylation since they inhibit replication and therefore the synthesis of unmethylated DNA[4] (Fig. 4). However, catalytic inhibitors of DNA methyltransferase activity will cause global hypomethylation since they do not limit the levels of DNMT1 protein. This hypothesis has been recently supported by comparing the profile of genes induced by 5-aza-CdR and

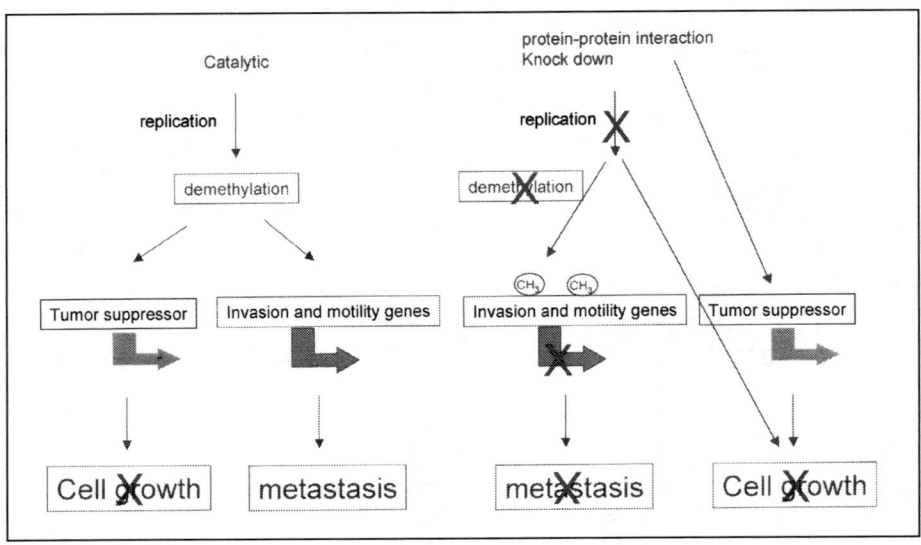

Figure 5. The predicted effects of catalytic inhibitors of DNMT1 and knock down inhibitors of DNMT1 on tumor growth and metastasis. Catalytic inhibitors cause demethylation of tumor suppressor genes, which would result in inhibition of tumor growth as well as induction of pro-metastatic genes. Knock down inhibitors will trigger an arrest of DNA replication as well as activation of certain tumor suppressors by a methylation independent pathway. The combination of these effects results in inhibition of tumor growth. Induction of pro-metastatic genes should not occur since inhibition of DNA replication limits the possibility of passive demethylation.

MG88 *(DNMT1* antisense oligonucleotide) treatment of A549 cells using 12K human gene arrays. Whereas MG88 induces an array of stress response genes that explain its antimitogenic effect, 5-aza-CdR induces a class of methylated testis/cancer specific antigens such as melanoma associated antigen[125] (Fig. 4) consistent with its demethylating activity.

The discussion in the previous section raises the specter of induction of metastasis upon treatment with DNA demethylating drugs. The examples brought above clearly indicate that DNA methyltransferase inhibitors induce metastatic genes such as uPA,[31] Gelatinase B[119] and Mts1.[124] An excellent illustration of the combined blessing and curse in DNA methyltransferase inhibitors is the simultaneous induction of TIMP-1, a protease inhibitor that inhibits metastasis and Gelatinase B[119] a protease that promotes cell invasion, by the demethylating agent 5-aza-CdR. I therefore previously proposed that the main focus of DNMT inhibitor development should be on agents that knockdown DNMT1 but do not cause global hypomethylation.[2] Agents that inhibit specific protein-protein interactions of DNMT1 or those that block its synthesis will accomplish this goal (Fig. 5). While inhibition of DNA methylation by DNA methyltransferase inhibitors might block growth by inducing tumor suppressor genes, it might bring about a clinically serious and unwanted outcome of increased metastasis (Fig. 5). If DNA demethylation can cause metastasis, DNA methylation inhibitors should be used with extreme care in anticancer or other forms of therapy. All DNMT1 inhibitors should be screened for their potential pro-metastatic activity.

Since hypermethylation in cancer occurs independent of global hypomethylation, inhibitors of demethylase activity are potential anticancer agents (Fig. 6). However, to be able to develop such agents, one has to first characterize the demethylase involved in global hypomethylation in tumors. Another question that will have important therapeutic consequencs is whether global hypomethylation affects cell growth or invasion and metastasis exclusively. If

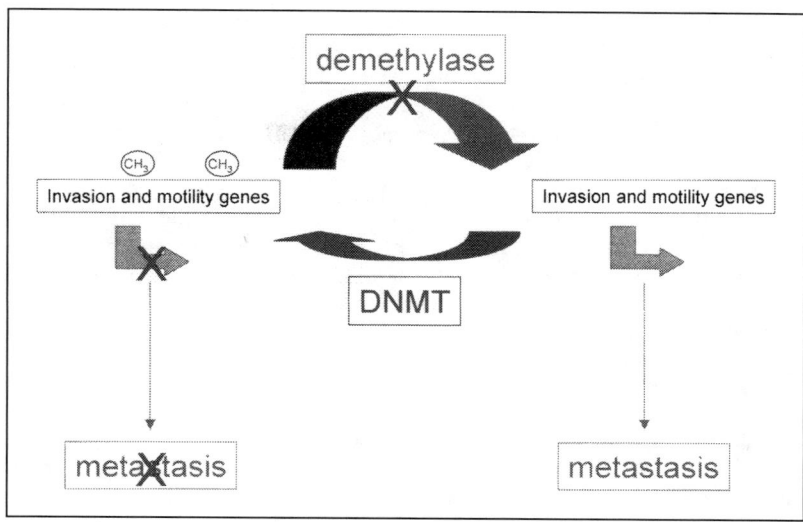

Figure 6. Proposed antitumorigenic effect of demethylase inhibitors, a model. Genes required for invasion, migration and metastasis are maintained in a methylated (indicated by the encircled CH₃) state in nonmetastatic cell since they are preferred substrates for DNMTs. The genes are inactive and metastasis is inhibited. During tumorigenesis, induction of demethylase activity results in demethylation of this class of genes resulting in activation of their transcription and metastasis. Inhibition of demethylase by specific inhibitors will redirect the DNA methylation equilibrium towards methylation and silencing of the pro-metastatic genes resulting in inhibition of metastasis.

global hypomethylation is critical only for tumor invasion and metastasis but not for cell growth, then inhibitors of demethylase would not suffer from the limiting toxicity of other antimitotic drugs. If this is the case, then DNMT1 inhibitors and demethylase inhibitors would act on two fundamentally different regulatory circuits, cell growth and division for DNMT1 inhibitors (Fig. 5) and cell motility and invasion for demethylase inhibitors (Fig. 6). This notion, if true, expands the potential therapeutic applications for a combination of DNMT1 and demethylase inhibitors, as well as guides us in defining the specific use of these inhibitors for distinct therapeutic requirements.

MBD2/Demethylase As an Anticancer Target

We do not know yet which demethylases are critical for maintaining the hypomethylated state of tumor invasion and metastasis genes. However, since the only cloned protein with demethylase activity in our hands is MBD2/demethylase, we tested the hypothesis that MBD2/demethylase plays a critical role in cancer. We addressed two questions: a. Does MBD2/demethylase play a causal role in cancer? b. Is MBD2/demethylase required for cell growth or is it exclusive for the unique requirement of tumorigenic growth such as anchorage independence, invasion and metastasis? To knock down MBD2/demethylase, we expressed the cDNA in the antisense orientation. Antisense expression resulted in knockdown of MBD2/demethylase expression and in a dramatic reduction in the ability of a wide range of human cancer cells to grow in an anchorage independent manner. Interestingly, knockdown of MBD2/demethylase had no effect on the capacity of the cells to grow in an anchorage dependent manner and there was no effect on cell cycle kinetics.[77] This is consistent with the hypothesis that MBD2/demethylase controls genes that are required for tumorigenic growth but not regular growth. Thus, MBD2/demethylase might be an ideal anticancer target since its inhibition would not result in arrest of normal growth, which is the cardinal problem of anticancer therapies.

We then tested whether inhibition of MBD2/demethylase might have an effect on tumor growth in vivo. We used an adenoviral vector to deliver MBD2/demethylase into human nonsmall cell carcinoma line A549 that was passaged as a xenoplast tumor in nude mice. Intratumoral administration of MBD2/demethylase antisense vector resulted in reduction of tumor growth in vivo, supporting the hypothesis that MBD2/demethylase is an anticancer target.[77]

However, to determine whether systemic inhibition of MBD2/demethylase would result in an anticancer effect, we screened and designed second-generation antisense oligonucleotide inhibitors of MBD2/demethylase mRNA. These compounds are also candidate therapeutic agents. The antisense oligonucleotides inhibited MBD2/demethylase mRNA and protein in a dose responsive manner at 100-200 nM. MBD2/demethylase antisense oligonucleotides inhibited anchorage independent growth of human lung and colorectal cancer cells but did not inhibit their growth under anchorage dependent conditions. The oligonucleotides did not have an effect on the cell cycle kinetics of normal skin fibroblasts, supporting the hypothesis that MBD2/demethylase is not required for cell cycle progression. This is consistent with genetic experiments showing that MBD2-/- mice are viable and fertile.[76] MBD2/demethylase antisense oligonucleotides administered in vivo by tail vein injection into nude mice bearing either human lung or colorectal cancer completely inhibited tumor growth at a concentration of 8 mg/kg.[126] The antisense sequence, which has only one mismatch with the murine MBD2/demethylase mRNA, had no toxic side effects on hematopoietic parameters and animal weight which is consistent with the cellular data. The combination of a strong antitumorigenic effect with lack of antimitogenic effects makes this an extremely attractive and unique target. Moreover, this study might provide us with an insight into the different roles of the DNA methylation machinery proteins in tumorigenesis. Future therapeutics targeting the DNA methylation machinery should take advantage of the differential and intertwined roles of its components in cancer.

Our previous studies suggest that DNMT1 has a predominant effect on the cell cycle whereas MBD2 controls tumorigenesis but not the cell cycle. What are the specific steps in tumorigenic growth that are controlled by MBD2/demethylase? It is tempting to speculate that MBD2/demethylase controls genes involved in tumor invasion and metastasis. We have indicated before that multiple sets of data point towards the association of global hypomethylation with tumor invasion and metastasis and that demethylating agents induce metastasis. The possibility that MBD2/demethylase controls genes involved in metastasis is consistent with its demethylase activity. However, at this stage there is no direct evidence to support this hypothesis.

As the roles of demethylase and demethylation in the activation of metastasis-associated genes will become clear in the coming years, the therapeutic potential of demethylase(s) will hopefully be utilized. If MBD2/demethylase is indeed responsible for the tumorigenic potential of cancer cells as exhibited by their anchorage independence and invasive capacity, but not cell cycle regulation, then agents that inhibit this demethylase should specifically inhibit tumors and not other mitotic cells, which is the cardinal drawback of most anticancer agent. Separating tumorigenesis from cell cycle is critical for designing drugs that are highly specific to cancer cells, and for avoiding the common toxicities, which are the consequence of the antimitotic effects of many anticancer agents.

Summary

Global hypomethylation of DNA is a hallmark of cancer. The significance of this phenomenon has not received sufficient attention since hypermethylation of tumor suppressor genes was the focus of most studies. The discovery of demethylase activity in tumors, changes our basic understanding of DNA methylation, from a unidirectional reaction to an equilibrium of methylation and demethylation, and also points towards a new understanding of the functional role of hypomethylation in cancer. Global hypomethylation is proposed to be a consequence of increased demethylase activity in cancer cells. This increased demethylase activity is hypothesized to result in hypomethylation of a panel of genes involved in anchorage independence, and tumor invasion that enable cancer cells to grow in normally restrictive environments. Identifying the demethylase activities responsible for global hypomethylation in cancer might provide us with unique antitumorigenic and antimetastatic targets. Data obtained following knock down of MBD2/demethylase, the first demethylase candidate studied thus far, is consistent with this hypothesis. Antisense inhibitors of MBD2/demethylase achieved highly promising anti tumor effects in mice. I believe that unraveling the demethylases involved in cancer, understanding their mechanism of action and identifying specific inhibitors will lead towards a new paradigm of anticancer therapeutics in the coming years.

Acknowledgements

The research from my lab discussed in this chapter is supported by the National Cancer Institute of Canada and the Canadian institute of Health research.

References

1. Szyf M, Knox DJ, Milutinovic S et al. How does DNA methyltransferase cause oncogenic transformation? Ann N Y Acad Sci 2000; 910:156-74 discussion 75-7.
2. Szyf M. Towards a pharmacology of DNA methylation. Trends Pharmacol Sci 2001; 22:350-4.
3. Milutinovic S, Knox JD, Szyf M. DNA methyltransferase inhibition induces the transcription of the tumor suppressor p21(WAF1/CIP1/sdi1). J Biol Chem 2000; 275:6353-9.
4. Knox JD, Araujo FD, Bigey P et al. Inhibition of DNA methyltransferase inhibits DNA replication. J Biol Chem 2000; 275:17986-90.
5. MacLeod AR, Szyf M. Expression of antisense to DNA methyltransferase mRNA induces DNA demethylation and inhibits tumorigenesis. J Biol Chem 1995; 270:8037-43.
6. Fournel M, Sapieha P, Beaulieu N et al. Down-regulation of human DNA-(cytosine-5) methyltransferase induces cell cycle regulators p16(ink4A) and p21(WAF/Cip1) by distinct mechanisms. J Biol Chem 1999; 274:24250-6.
7. Ramchandani S, MacLeod AR, Pinard M et al. Inhibition of tumorigenesis by a cytosine-DNA, methyltransferase, antisense oligodeoxynucleotide. Proc Natl Acad Sci USA 1997; 94:684-9.
8. Feinberg AP, Vogelstein B. Hypomethylation of ras oncogenes in primary human cancers. Biochem Biophys Res Commun 1983; 111:47-54.
9. Feinberg AP, Vogelstein B. Hypomethylation distinguishes genes of some human cancers from their normal counterparts. Nature 1983; 301:89-92.
10. Ehrlich M. DNA methylation in cancer: Too much, but also too little. Oncogene Aug 12; 21:5400-13.
11. Lu LJ, Randerath E, Randerath K. DNA hypomethylation in Morris hepatomas. Cancer Lett 1983; 19:231-9.
12. Bedford MT, van HPD. Hypomethylation of DNA in pathological conditions of the human prostate. Cancer Res 1987; 47:5274-6.
13. Bernardino J, Roux C, Almeida A et al. DNA hypomethylation in breast cancer: An independent parameter of tumor progression? Cancer Genet Cytogenet 1997; 97:83-9.
14. Soares J, Pinto AE, Cunha CV et al. Global DNA hypomethylation in breast carcinoma: Correlation with prognostic factors and tumor progression. Cancer 1999; 85:112-8.
15. Feinberg AP, Gehrke CW, Kuo KC et al. Reduced genomic 5-methylcytosine content in human colonic neoplasia. Cancer Res 1988; 48:1159-61.
16. Kim YI, Giuliano A, Hatch KD et al. Global DNA hypomethylation increases progressively in cervical dysplasia and carcinoma. Cancer 1994; 74:893-9.
17. Qu G, Dubeau L, Narayan A et al. Satellite DNA hypomethylation vs. overall genomic hypomethylation in ovarian epithelial tumors of different malignant potential. Mutat Res 1999; 423:91-101.
18. Liteplo RG, Kerbel RS. Reduced levels of DNA 5-methylcytosine in metastatic variants of the human melanoma cell line MeWo. Cancer Res 1987; 47:2264-7.
19. Ehrlich M, Jiang G, Fiala E et al. Hypomethylation and hypermethylation of DNA in Wilms tumors. Oncogene 2002; 21:6694-702.
20. Shen L, Fang J, Qiu D et al. Correlation between DNA methylation and pathological changes in human hepatocellular carcinoma. Hepatogastroenterology 1998; 45:1753-9.

21. Cravo M, Pinto R, Fidalgo P et al. Global DNA hypomethylation occurs in the early stages of intestinal type gastric carcinoma. Gut 1996; 39:434-8.

22. Jurgens B, Schmitz-Drager BJ, Schulz WA. Hypomethylation of L1 LINE sequences prevailing in human urothelial carcinoma. Cancer Res 1996; 56:5698-703.

23. Alves G, Tatro A, Fanning T. Differential methylation of human LINE-1 retrotransposons in malignant cells. Gene 1996; 176:39-44.

24. Narayan A, Ji W, Zhang XY et al. Hypomethylation of pericentromeric DNA in breast adenocarcinomas. Int J Cancer 1998; 77:833-8.

25. Qu GZ, Grundy PE, Narayan A et al. Frequent hypomethylation in Wilms tumors of pericentromeric DNA in chromosomes 1 and 16. Cancer Genet Cytogenet 1999; 109:34-9.

26. Kaneko Y, Shibuya M, Nakayama T et al. Hypomethylation of c-myc and epidermal growth factor receptor genes in human hepatocellular carcinoma and fetal liver. Jpn J Cancer Res 1985; 76:1136-40.

27. Del SL., Maestri I, Piva R et al. Differential hypomethylation of the c-myc protooncogene in bladder cancers at different stages and grades. J Urol 1989; 142:146-9.

28. Lipsanen V, Leinonen P, Alhonen L et al. Hypomethylation of ornithine decarboxylase gene and erb-A1 oncogene in human chronic lymphatic leukemia. Blood 1988; 72:2042-4.

29. Nakamura N, Takenaga K. Hypomethylation of the metastasis-associated S100A4 gene correlates with gene activation in human colon adenocarcinoma cell lines. Clin Exp Metastasis 1998; 16:471-9.

30. Rosty C, Ueki T, Argani P et al. Overexpression of S100A4 in pancreatic ductal adenocarcinomas is associated with poor differentiation and DNA hypomethylation. Am J Pathol 2002; 160:45-50.

31. Guo Y, Pakneshan P, Gladu J et al. Regulation of DNA methylation in human breast cancer. Effect on the urokinase-type plasminogen activator gene production and tumor invasion. J Biol Chem 2002 Nov 1; 277:41571-9.

32. Nakayama M, Wada M, Harada T et al. Hypomethylation status of CpG sites at the promoter region and overexpression of the human MDR1 gene in acute myeloid leukemias. Blood 1998; 92:4296-307.

33. Razin A, Riggs AD. DNA methylation and gene function. Science 1980; 210:604-10.

34. Issa JP, Vertino PM, Wu J et al. Increased cytosine DNA-methyltransferase activity during colon cancer progression. J Natl Cancer Inst 1993; 85:1235-40.

35. Potter J. Methyl supply, methyl metabolizing enzymes and colorectal neoplasia. J Nutr 2002; 132:2410S-2S.

36. Piyathilake CJ, Johanning GL, Macaluso M et al. Localized folate and vitamin B-12 deficiency in squamous cell lung cancer is associated with global DNA hypomethylation. Nutr Cancer 2000; 37:99-107.

37. Wilson MJ, Shivapurkar N, Poirier LA. Hypomethylation of hepatic nuclear DNA in rats fed with a carcinogenic methyl-deficient diet. Biochem J 1984; 218:987-90.

38. Simile MM, Pascale R, De Miglio MR et al. Correlation between S-adenosyl-L-methionine content and production of c- myc, c-Ha-ras, and c-Ki-ras mRNA transcripts in the early stages of rat liver carcinogenesis. Cancer Lett 1994; 79:9-16.

39. Szyf M, Theberge J, Bozovic V. Ras induces a general DNA demethylation activity in mouse embryonal P19 cells. J Biol Chem 1995; 270:12690-6.

40. Szyf M. DNA methylation properties: Consequences for pharmacology. Trends Pharmacol Sci 1994; 15:233-8.

41. Hsieh CL. Evidence that protein binding specifies sites of DNA demethylation. Mol Cell Biol 1999; 19:46-56.

42. Lin IG, Tomzynski TJ, Ou Q et al. Modulation of DNA binding protein affinity directly affects target site demethylation. Mol Cell Biol 2000; 20:2343-9.

43. Han L, Lin IG, Hsieh CL. Protein binding protects sites on stable episomes and in the chromosome from de novo methylation. Mol Cell Biol 2001; 21:3416-24.

44. Kafri T, Gao X, Razin A. Mechanistic aspects of genome-wide demethylation in the preimplantation mouse embryo. Proc Natl Acad Sci USA 1993; 90:10558-62.

45. Razin A, Shemer R. DNA methylation in early development. Hum Mol Genet 1995; 4:1751-5.

46. Kafri T, Ariel M, Brandeis M et al. Developmental pattern of gene-specific DNA methylation in the mouse embryo and germ line. Genes Dev 1992; 6:705-14.

47. Razin A, Webb C, Szyf M et al. Variations in DNA methylation during mouse cell differentiation in vivo and in vitro. Proc Natl Acad Sci USA 1984; 81:2275-9.

48. Wilks A, Seldran M, Jost JP. An estrogen-dependent demethylation at the 5' end of the chicken vitellogenin gene is independent of DNA synthesis. Nucleic Acids Res 1984; 12:1163-77.

49. Wilks AF, Cozens PJ, Mattaj IW et al. Estrogen induces a demethylation at the 5' end region of the chicken vitellogenin gene. Proc Natl Acad Sci USA 1982; 79:4252-5.

50. Paroush Z, Keshet I, Yisraeli J et al. Dynamics of demethylation and activation of the alpha-actin gene in myoblasts. Cell 1990; 63:1229-37.
51. Yisraeli J, Adelstein RS, Melloul D et al. Muscle-specific activation of a methylated chimeric actin gene. Cell 1986; 46:409-16.
52. Szyf M, Eliasson L, Mann V et al. Cellular and viral DNA hypomethylation associated with induction of Epstein-Barr virus lytic cycle. Proc Natl Acad Sci USA 1985; 82:8090-4.
53. Oswald J, Engemann S, Lane N et al. Active demethylation of the paternal genome in the mouse zygote. Curr Biol 2000; 10:475-48.
54. Razin A, Feldmesser E, Kafri T et al. Cell specific DNA methylation patterns; formation and a nucleosome locking model for their function. Prog Clin Biol Res 1985; 198:239-53.
55. Razin A, Szyf M, Kafri T et al. Replacement of 5-methylcytosine by cytosine: A possible mechanism for transient DNA demethylation during differentiation. Proc Natl Acad Sci USA 1986; 83:2827-31.
56. Jost JP, Jost YC. Mechanism of active DNA demethylation during embryonic development and cellular differentiation in vertebrates. Gene 1995; 157:265-6.
57. Jost JP, Siegmann M, Sun L et al. Mechanisms of DNA demethylation in chicken embryos. Purification and properties of a 5-methylcytosine-DNA glycosylase. J Biol Chem 1995; 270:9734-9.
58. Weiss A, Keshet I, Razin A et al. DNA demethylation in vitro: Involvement of RNA [published erratum appears in Cell 1998 Nov 13;95(4):following 573]. Cell 1996; 86:709-18.
59. Zhu B, Zheng Y, Hess D et al. 5-methylcytosine-DNA glycosylase activity is present in a cloned G/T mismatch DNA glycosylase associated with the chicken embryo DNA demethylation complex. Proc Natl Acad Sci USA 2000 May 9; 97:5135-9.
60. Zhu B, Zheng Y, Angliker H et al. 5-Methylcytosine DNA glycosylase activity is also present in the human MBD4 (G/T mismatch glycosylase) and in a related avian sequence. Nucleic Acids Res 2000 Nov 1; 28:4157-65.
61. Jost J, Oakeley E, Zhu B et al. 5-Methylcytosine DNA glycosylase participates in the genome-wide loss of DNA methylation occurring during mouse myoblast differentiation. Nucleic Acids Res 2001 Nov 1; 29:4452-61.
62. Oswald J, Engemann S, Lane N et al. Active demethylation of the paternal genome in the mouse zygote. Curr Biol 2000; 10:475-8.
63. Ramchandani S, Bhattacharya SK, Cervoni N et al. DNA methylation is a reversible biological signal [see comments]. Proc Natl Acad Sci USA 1999; 96:6107-12.
64. Cervoni N, Bhattacharya S, Szyf M. DNA demethylase is a processive enzyme. J Biol Chem 1999; 274:8363-6.
65. Bhattacharya SK, Ramchandani S, Cervoni N et al. A mammalian protein with specific demethylase activity for mCpG DNA [see comments]. Nature 1999; 397:579-83.
66. Hendrich B, Bird A. Identification and characterization of a family of mammalian methyl-CpG binding proteins. Mol Cell Biol 1998; 18:6538-47.
67. Zhang Y, Ng HH, Erdjument-Bromage H et al. Analysis of the NuRD subunits reveals a histone deacetylase core complex and a connection with DNA methylation. Genes Dev 1999; 13:1924-35.
68. Ng HH, Zhang Y, Hendrich B et al. MBD2 is a transcriptional repressor belonging to the MeCP1 histone deacetylase complex [see comments]. Nat Genet 1999; 23:58-61.
69. Boeke J, Ammerpohl O, Kegel S et al. The minimal repression domain of MBD2b overlaps with the Methyl-CpG binding domain and binds directly to Sin3A. J Biol Chem 2000.
70. Cervoni N, Szyf M. Demethylase activity is directed by histone acetylation. J Biol Chem 2001; 276:40778-4087.
71. Detich N, Theberge J, Szyf M. Promoter-specific Activation and Demethylation by MBD2/Demethylase. J Biol Chem 2002 Sep 27; 277:35791-4.
72. Koizume S, Tachibana K, Sekiya T et al. Heterogeneity in the modification and involvement of chromatin components of the CpG island of the silenced human CDH1 gene in cancer cells. Nucleic Acids Res 2002; 30:4770-80.
73. Ghoshal K, Datta J, Majumder S et al. Inhibitors of histone deacetylase and DNA methyltransferase synergistically activate the methylated metallothionein I promoter by activating the transcription factor MTF-1 and forming an open chromatin structure. Mol Cell Biol 2002; 22:8302-19.
74. Fournier C, Goto Y, Ballestar E et al. Allele-specific histone lysine methylation marks regulatory regions at imprinted mouse genes. Embo J 2002; 21:6560-70.
75. Marhold J, Zbylut M, Lankenau D et al. Stage-specific chromosomal association of Drosophila dMBD2/3 during genome activation. Chromosoma 2002 Mar; 111:13-21.
76. Hendrich B, Guy J, Ramsahoye B et al. Closely related proteins MBD2 and MBD3 play distinctive but interacting roles in mouse development. Genes Dev 2001; 15:710-23.

77. Slack A, Bovenzi V, Bigey P et al. Antisense MBD2 gene therapy inhibits tumorigenesis. J Gene Med 2002; 4:381–9.
78. Szyf M, Detich N. Regulation of the DNA methylation machinery and its role in cellular transformation. Prog Nucleic Acid Res Mol Biol 2001; 69:47-79.
79. Szyf M. The role of DNA methyltransferase 1 in growth control. Front Biosci 2001; 6:D599-609.
80. Fuks F, Burgers WA, Brehm A et al. DNA methyltransferase Dnmt1 associates with histone deacetylase activity. Nat Genet 2000; 24:88-91.
81. Robertson KD, Ait-Si-Ali S, Yokochi T et al. DNMT1 forms a complex with Rb, E2F1 and HDAC1 and represses transcription from E2F-responsive promoters. Nat Genet 2000; 25:338-42.
82. Di CL, Raker V, Corsaro M et al. Methyltransferase recruitment and DNA hypermethylation of target promoters by an oncogenic transcription factor. Science 2002 Feb 8; 295:1079-82.
83. Ghoshal K, Datta J, Majumder S et al. Inhibitors of histone deacetylase and DNA methyltransferase synergistically activate the methylated metallothionein I promoter by activating the transcription factor MTF-1 and forming an open chromatin structure. Mol Cell Biol 2002 Dec; 22:8302-19.
84. Cameron EE, Bachman KE, Myohanen S et al. Synergy of demethylation and histone deacetylase inhibition in the re expression of genes silenced in cancer. Nat Genet 1999; 21:103-7.
85. Kouzarides T. Histone methylation in transcriptional control. Curr Opin Genet Dev 2002; 12:198-209.
86. Nguyen CT, Weisenberger DJ, Velicescu M et al. Histone H3-lysine 9 methylation is associated with aberrant gene silencing in cancer cells and is rapidly reversed by 5-aza-2'-deoxycytidine. Cancer Res 2002; 62:6456-61.
87. Kondo Y, Shen L, Issa JP. Critical role of histone methylation in tumor suppressor gene silencing in colorectal cancer. Mol Cell Biol 2003; 23:206-15.
88. Fuks F, Hurd PJ, Wolf D et al. The methyl-CpG-binding protein MeCP2 links DNA methylation to histone methylation. J Biol Chem 2002.
89. Seo SB, McNamara P, Heo S et al. Regulation of histone acetylation and transcription by INHAT, a human cellular complex containing the set oncoprotein. Cell 2001; 104:119-30.
90. Razin A, Cedar H. Distribution of 5-methylcytosine in chromatin. Proc Natl Acad Sci USA 1977; 74:2725-8.
91. Kirillov A, Kistler B, Mostoslavsky R et al. A role for nuclear NF-kappaB in B-cell-specific demethylation of the Igkappa locus. Nat Genet 1996; 13:435-41.
92. Cui H, Fedoroff N. Inducible DNA demethylation mediated by the maize suppressor-mutator transposon-encoded TnpA protein. Plant Cell 2002; 14:2883-99.
93. Cervoni N, Detich N, Seo S et al. The oncoprotein Set/TAF-1beta, an inhibitor of histone acetyltransferase, inhibits active demethylation of DNA, integrating DNA methylation and transcriptional silencing. J Biol Chem 2002 Jul 12; 277:25026-31.
94. Bhave MR, Wilson MJ, Poirier LA. c-H-ras and c-K-ras gene hypomethylation in the livers and hepatomas of rats fed methyl-deficient, amino acid-defined diets. Carcinogenesis 1988; 9:343-8.
95. Wainfan E, Dizik M, Stender M et al. Rapid appearance of hypomethylated DNA in livers of rats fed cancer- promoting, methyl-deficient diets. Cancer Res 1989; 49:4094-7.
96. Dizik M, Christman JK, Wainfan E. Alterations in expression and methylation of specific genes in livers of rats fed a cancer promoting methyl-deficient diet. Carcinogenesis 1991; 12:1307-12.
97. Poirier LA. The effects of diet, genetics and chemicals on toxicity and aberrant DNA methylation: An introduction. J Nutr 2002; 132:2336S-9S.
98. Rao PM, Antony A, Rajalakshmi S et al. Studies on hypomethylation of liver DNA during early stages of chemical carcinogenesis in rat liver. Carcinogenesis 1989; 10:933-7.
99. Pascale R, Simile M, De MM et al. Chemoprevention of hepatocarcinogenesis. S-adenosyl-L-methionine. Alcohol 2002; 27:193.
100. Pascale R, Simile MM, Ruggiu ME et al. Reversal by 5-azacytidine of the S-adenosyl-L-methionine-induced inhibition of the development of putative preneoplastic foci in rat liver carcinogenesis. Cancer Lett 1991; 56:259-65.
101. Giovannucci E, Rimm EB, Ascherio A et al. Alcohol, low-methionine—low-folate diets, and risk of colon cancer in men. J Natl Cancer Inst 1995; 87:265-73.
102. Giovannucci E, Stampfer MJ, Colditz GA et al. Folate, methionine, and alcohol intake and risk of colorectal adenoma. J Natl Cancer Inst 1993; 85:875-84.
103. Habets GG, van der Kammen RA, Scholtes EH et al. Induction of invasive and metastatic potential in mouse T-lymphoma cells (BW5147) by treatment with 5-azacytidine. Clin Exp Metastasis 1990; 8:567-77.
104. Alvarez E, Elliott BE, Houghton AN et al. Heritable high frequency modulation of antigen expression in neoplastic cells exposed to 5-aza-2'-deoxycytidine or hydroxyurea: Analysis and implications. Cancer Res 1988; 48:2440-5.

105. Takenaga K. Modification of the metastatic potential of tumor cells by drugs. Cancer Metastasis Rev 1986; 5:67-75.
106. Nambu S, Inoue K, Saski H. Site-specific hypomethylation of the c-myc oncogene in human hepatocellular carcinoma. Jpn J Cancer Res 1987; 78:695-704.
107. Munzel P, Bock KW. Hypomethylation of c-myc proto-oncogene of N-nitrosomorpholine—induced rat liver nodules and of H4IIE cells. Arch Toxicol Suppl 1989; 13:211-3.
108. Vorce RL, Goodman JI. Altered methylation of ras oncogenes in benzidine-induced B6C3F1 mouse liver tumors. Toxicol Appl Pharmacol 1989; 100:398-410.
109. Vorce RL, Goodman JI. Hypomethylation of ras oncogenes in chemically induced and spontaneous B6C3F1 mouse liver tumors. J Toxicol Environ Health 1991; 34:367-84.
110. Hansen RS, Wijmenga C, Luo P et al. The DNMT3B DNA methyltransferase gene is mutated in the ICF immunodeficiency syndrome. Proc Natl Acad Sci USA 1999; 96:14412-7.
111. Okano M, Bell DW, Haber DA et al. DNA methyltransferases Dnmt3a and Dnmt3b are essential for de novo methylation and mammalian development. Cell 1999; 99:247-57.
112. Schuffenhauer S, Bartsch O, Stumm M et al. DNA, FISH and complementation studies in ICF syndrome: DNA hypomethylation of repetitive and single copy loci and evidence for a trans acting factor. Hum Genet 1995; 96:562-71.
113. Ji W, Hernandez R, Zhang XY et al. DNA demethylation and pericentromeric rearrangements of chromosome 1. Mutat Res 1997; 379:33-41.
114. Chen RZ, Pettersson U, Beard C et al. DNA hypomethylation leads to elevated mutation rates. Nature 1998; 395:89-93.
115. Ehrlich M. DNA hypomethylation, cancer, the immunodeficiency, centromeric region instability, facial anomalies syndrome and chromosomal rearrangements. J Nutr 2002 Aug; 132:2424S-9S.
116. Grigorian MS, Tulchinsky EM, Zain S et al. The mts1 gene and control of tumor metastasis. Gene 1993; 135:229-38.
117. Ford HL, Salim MM, Chakravarty R et al. Expression of Mts1, a metastasis-associated gene, increases motility but not invasion of a nonmetastatic mouse mammary adenocarcinoma cell line. Oncogene 1995; 11:2067-75.
118. Grigorian M, Ambartsumian N, Lykkesfeldt AE et al. Effect of mts1 (S100A4) expression on the progression of human breast cancer cells. Int J Cancer 1996; 67:831-41.
119. MacDougall JR, Bani MR, Lin Y et al. 'Proteolytic switching': Opposite patterns of regulation of gelatinase B and its inhibitor TIMP-1 during human melanoma progression and consequences of gelatinase B overexpression. Br J Cancer 1999; 80:504-12.
120. Beaulieu N, Morin S, Chute I et al. An essential role for DNA methyltransferase DNMT3B in cancer cell survival. J Biol Chem 2002; 277:28176-81.
121. Szyf M. The DNA Methylation Machinery as a Therapeutic target. Current Drug Targets 2000; 1:101-18.
122. Jones PA. Altering gene expression with 5-azacytidine. Cell 1985; 40:485-6.
123. Wu JC, Santi DV. On the mechanism and inhibition of DNA cytosine methyltransferases. Prog Clin Biol Res 1985; 198:119-29.
124. Tulchinsky E, Grigorian M, Tkatch T et al. Transcriptional regulation of the mts1 gene in human lymphoma cells: The role of DNA-methylation. Biochim Biophys Acta 1995; 1261:243-8.
125. Milutinovic S, Zhuang Q, Niveleau A et al. Epigenomic stress response. Knockdown of DNA methyltransferase 1 triggers an intra-S-phase arrest of DNA replication and induction of stress response genes. J Biol Chem 2003; 278(17):14985-95.
126. Campbell PM, Bovenzi V, Szyf M. Methylated DNA-binding protein 2 antisense inhibitors suppress tumourigenesis of human cancer cell lines in vitro and in vivo. Carcinogenesis 2004; (4):499-507.

CHAPTER 13

Purine Analogues and Their Role in Methylation and Cancer Chemotherapy

Katherine L. Seley and Sylvester L. Mosley

Despite promising leads in the search for new chemotherapeutic agents, there remains an urgent need to develop more effective and less toxic drugs. Nucleosides and their corresponding nucleobases are the fundamental building blocks of many biological systems[1-3] and as a result, have been extensively investigated due to their inherent structural resemblance to the naturally occurring nucleosides and nucleobases.[1-5] Due to the intertwined relationship between purine and pyrimidine nucleotide metabolism, cell proliferation and tumor cell differentiation, inhibition of key enzymes in nucleotide metabolism and DNA synthesis can be used as a chemotherapeutic approach to treating cancer.[6]

The progression of cancer is the result of a metabolic imbalance in the regulation of cell proliferation and cell differentiation.[4,7,8] Cell differentiation involves no change in the integrity of a cell's genetic information, but rather, involves a change in the way a cell expresses or uses that information.[9] Differentiation is epigenetic; i.e., cells are endowed with the ability to develop into several different cell types following activation of certain genes.[9] Therefore, if tumorigenesis results from epigenetic signals, then it follows that tumor growth can potentially be halted if a medicinal agent can be designed to activate the genes responsible for cell differentiation.[4,7,9] There are three subgroups of antimetabolites of purine and pyrimidine nucleosides that induce tumor cell differentiation;

   i.   those that interfere with the de novo synthesis of the nucleic acid precursors such as the antifolates and inhibitors of rate limiting enzymes such as inosine monophosphate dehydrogenase (IMPDH);

   ii.   inhibitors of DNA synthesis such as ara-C and PMEA, and

   iii.   nucleosides that disrupt methylation patterns such as DNA MeTase and SAHase inhibitors.[4] It is this latter class of compounds that this chapter will mainly focus on.

DNA Methylation and SAHase Inhibition

Methylation is a particularly important aspect of DNA metabolism. Specific patterns of DNA methylation are essential for recognition, gene expression and replication, and controlling methyltransferase activity is one determinant of these DNA methylation patterns.[10] Hypomethylation and hypermethylation of DNA both result in significant cellular consequences. It has been suggested that DNA methylation patterns are strongly dependent upon interplay between the level of DNA MeTase activity and site-specific signals.[10] It has also been shown that proper methylation of the 5'-cap of viral mRNA is necessary for stability of the mRNA, as well as the growth and/or replication of many viruses.[11] Therefore inhibition of these methylations leads to faulty transcription and translation.

As a consequence, disruption of DNA methylation becomes an attractive target for cancer chemotherapy.[7,10,12] This can be accomplished in several ways, including inhibition of DNA methyltransferase (DNA MeTase) and/or S-adenosylhomocysteine hydrolase (SAHase), both

DNA Methylation and Cancer Therapy, edited by Moshe Szyf. ©2005 Eurekah.com and Kluwer Academic/Plenum Publishers.

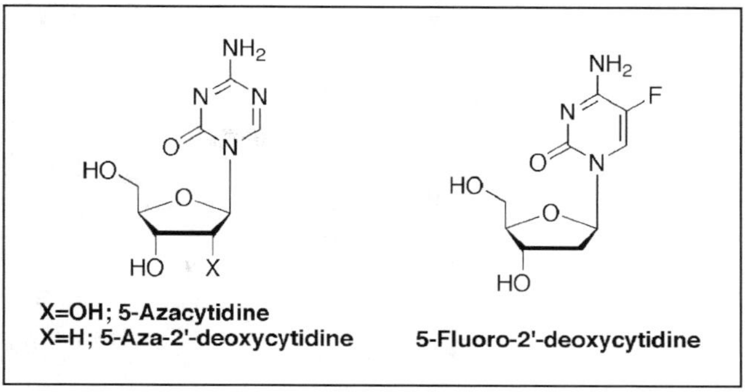

X=OH; 5-Azacytidine
X=H; 5-Aza-2'-deoxycytidine   5-Fluoro-2'-deoxycytidine

Figure 1. Modified pyrimidine DNA methyltransferase inhibitors.

established cellular targets for antiviral, antiparasitic and anticancer agents.[6,13,14] The principal DNA MeTase inhibitors that are currently used are 5-azacytidine, 5-azadeoxycytidine, and 5-fluorocytidine (Fig. 1), all modified pyrimidine nucleoside analogues that resist methylation following incorporation into DNA.[15] Unfortunately, these analogues exhibit many undesirable side effects, including toxicity,[16] therefore the pursuit for new analogues that work by other mechanisms should continue.

Related to this observation, the byproduct of all S-adenosylmethionine (SAM)-promoted methyltransferase reactions is S-adenosylhomocysteine (SAH), therefore SAH is a potent competitive inhibitor of all methylation reactions dependent upon SAM as the methyl donor, including DNA MeTase.[17,18] SAM is the most reactive methyl donor thus SAM dependent methylations are considered to be the most important biologically.[17,18] SAHase is the only known enzyme able to remove SAH in mammals, pointing to the significance of this enzyme. As depicted in Figure 2, SAHase cleaves SAH into its two cellular components, adenosine (Ado) and homocysteine (Hcy), and requires the assistance of an enzyme-bound cofactor, NADH.[19] Inhibition of SAHase by nucleoside inhibitors involves depletion of the NADH cofactor, which causes an intracellular accumulation of SAH, thereby elevating the SAH/SAM ratio. This imbalance in the SAH/SAM ratio results in cessation of SAM-dependent methylations, which leads to improperly methylated DNA and as a result, reduction in cell proliferation.

Mammalian SAHase is a homotetramer of ~48,000 M_r subunits, each of which contains one mole of NAD$^+$.[20] The crystal structures for both the human[20] and rat[21] SAHases were only

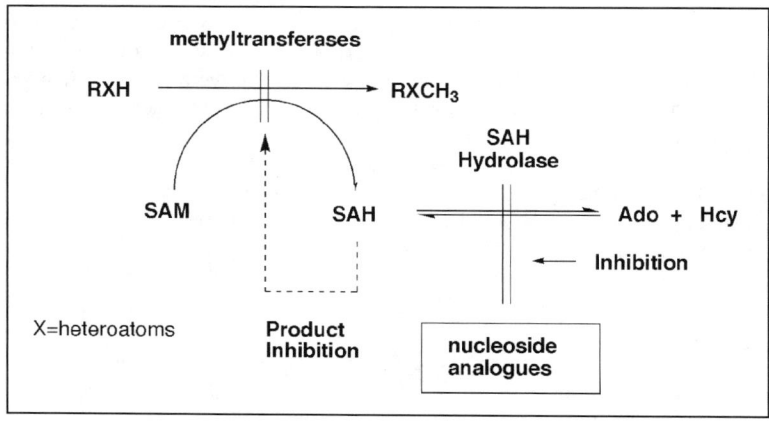

Figure 2. SAM-dependent methyltransferase and SAH metabolism.

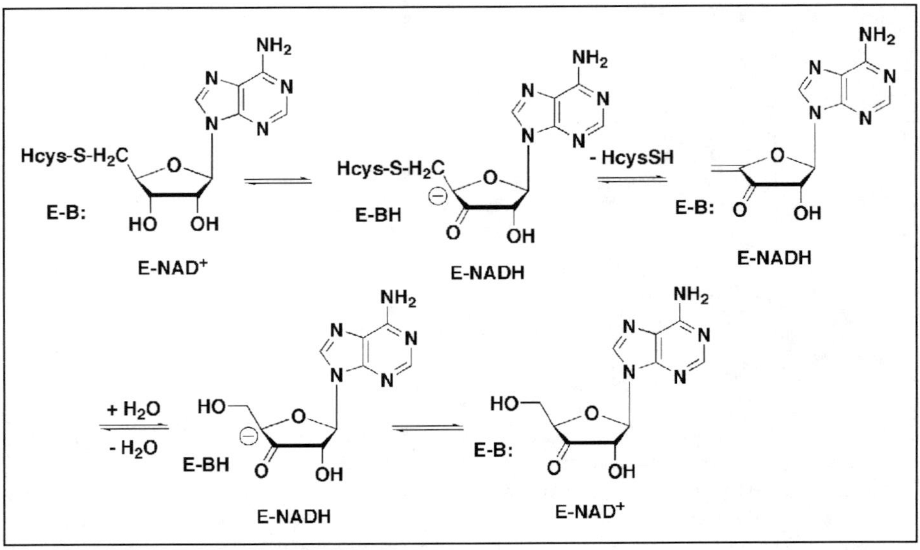

Figure 3. The mechanism of action for SAHase adapted from Palmer and Abeles.

recently published, and this will greatly assist in the design of new SAHase inhibitors. The reaction mechanism proposed by Palmer and Abeles involves a cycle of oxidation-reduction of the substrate and the enzyme-bound NADH cofactor, which remains tightly bound to the enzyme (Fig. 3).[19] It has been shown that the substrate-bound SAHase exists in a different form than that of the substrate-free SAHase.[22] A flexible hinge element between the catalytic and NAD+ binding domains has been suggested, which allows for large differences in the spatial arrangements. The two domains form a deep active site cleft containing the cofactor and the bound substrate.

Nucleoside Inhibitors

Unlike the pyrimidine nucleosides which inhibit DNA MeTase directly, the purine analogues inhibit methylations indirectly, by inhibiting SAHase. The most potent substrate inhibitors of SAHase are those that are oxidized by the enzyme-bound NAD+ to give the inactive NADH.[20,22] These are classified as Type I mechanism-based inhibitors (Type II mechanism-based inhibitors are those that are activated by the enzyme in the first step of the mechanism and irreversibly inactivate the enzyme once they become covalently bound).[13,23,24] Originally there was speculation that the closed form might result from this oxidation of the substrate rather than the binding of the substrate, but recently an SAHase inhibitor was co-crystallized with the human form and a series of elegant studies carried out on this structure provided proof that the substrate-bound active site is closed upon substrate binding, not upon substrate oxidation.[22] There is a structural difference of 17 degree rigid body movement of the catalytic domain upon substrate binding.[25]

One of the initial purines to be investigated was 3-deazaadenosine (3-deazaAdo), where the N-3 nitrogen has been replaced by a methyne group (Fig. 4). 3-DeazaAdo has been the focus of many investigations too numerous to list completely, but the biological effects of 3-deazaAdo vary widely from antiviral to antiparasitic and antitumor.[13,18,26-31] The general mode of action has been attributed to the inhibition of SAHase via the aforementioned depletion of the NADH cofactor. In addition, 3-deazaAdo can also act as a substrate of SAHase to yield 3-deazaadenosylhomocysteine, which has also been shown to inhibit methylations in vivo by cellular accumulation with SAH.[27-29,32-34] 3-DeazaAdo also inhibits the induction of murine erythroleukemia cell differentiation,[35,36] but induces 3T3-L1 fibroblast differentiation[37] hence the keen interest in incorporating this structural modification into potential inhibitors. One

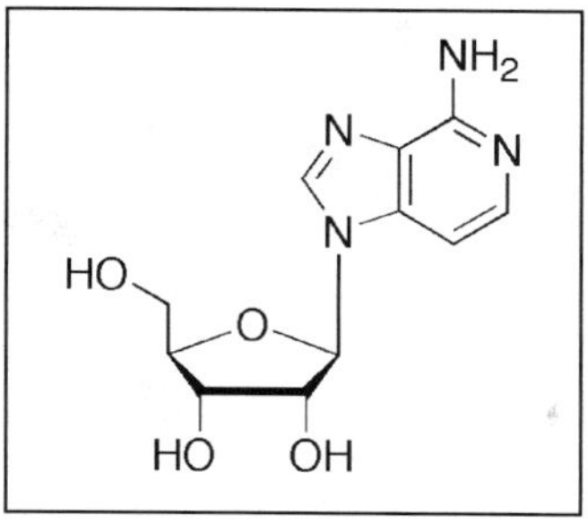

Figure 4. 3-Deazaadenosine.

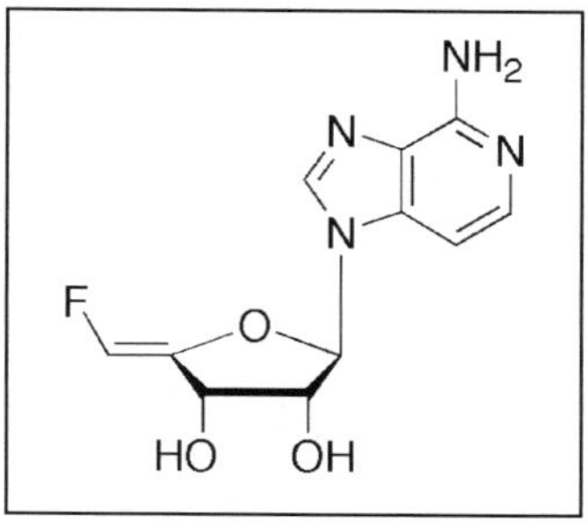

Figure 5. MDL 28,842

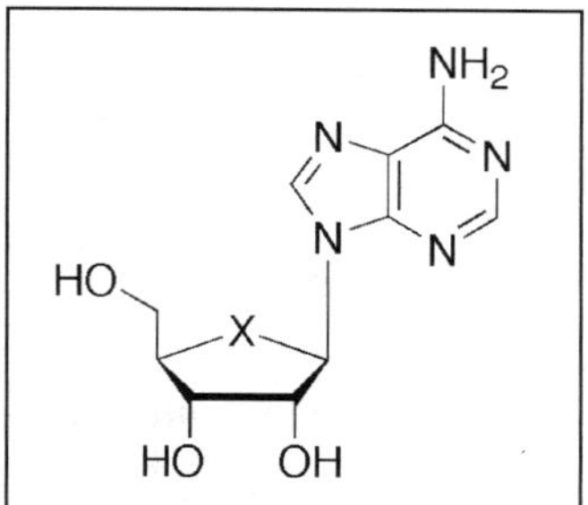

X=O, Adenosine
X=CH$_2$, Aristeromycin

Figure 6.

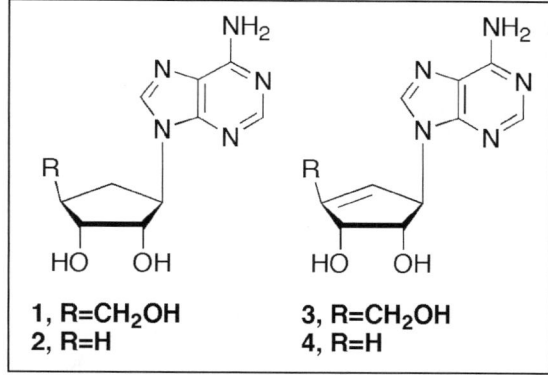

1, R=CH$_2$OH
2, R=H

3, R=CH$_2$OH
4, R=H

Figure 7. Carbocylic analogues.

notable property inherent in the 3-deaza nucleosides is that they are immune to phosphorylation by cellular enzymes. They also do not undergo deamination, and although neither of these phenomena has yet to be explained, it does endow the nucleosides with intriguing possibilities. It also results in much lower levels of toxicity as compared to the parent compounds.

Another modification that was investigated included fluorinated analogues such as Z-4',5'-didehydro-5'-deoxy-5'-fluoroadenosine (MDL 28,842) shown in Figure 5, since several fluorinated pyrimidines have shown potent anticancer activity.[38] MDL 28,842 is a potent inhibitor of SAHase, and inhibits the proliferation of cutaneaous squamous carcinoma cell lines, but in contrast, induces tumor cell differentiation in keratinocytes, although not to a great extent.

Carbocyclic Nucleoside

Another class of structurally modified nucleosides that have been shown to be exceptional inhibitors of SAHase, and indirectly inhibitors of DNA MeTase, are the carbocyclic nucleosides. Replacement of the furanose oxygen of the sugar moiety with a methylene group as represented in Figure 6 for adenosine and aristeromycin, imparts an increased level of stability due to the transformation of the unstable hemiaminal glycosidic linkage to a stable tertiary amine. This bestows upon the nucleosides the ability to resist cleavage by phosphorylases, as well as to increase their overall lipophilicity. This increased stability has led to significant biological activity against numerous viruses and parasites,[13,26,30,39] as well as against colon carcinoma,[40] leukemia cells,[35,36,41] among other cancers.

Two of the most widely recognized carbocyclic nucleosides are Ari[42,43] (**1**, Fig. 7), the carbocyclic analogue of adenosine, and its unsaturated analogue, neplanocin A (NpcA)[44, 45] (**3**,

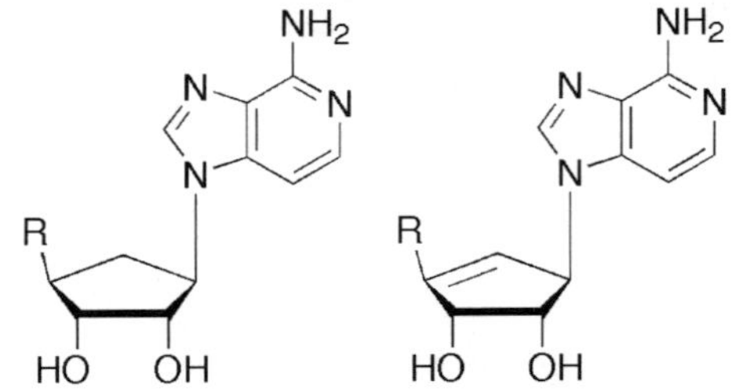

Figure 8. Carbocyclic 3-deaza analogues.

Fig. 7). These two nucleosides have provided the lead for a plethora of investigations.[39, 46-52] In addition to inhibition of SAHase and MeTase, NpcA is also a substrate for adenosine deaminase and adenosine kinase, two other enzymes involved in the regulatory cycles of many diseases. NpcA induces differentiation in human promyelocytic leukemia HL-60 cells, while Ari induces differentiation in erythroid leukemia K562 cells.

Despite the significant activity shown by Ari and NpcA, they also exhibit high levels of toxicity as a result of conversion to their triphosphate forms by phosphorylating enzymes. This observation has led to the pursuit of further structural modifications in an effort to increase the biological activity while reducing the levels of cellular toxicity. One of the most fruitful modifications to emerge out of those efforts was to eliminate the 4'-hydroxymethyl group of Ari and NpcA, to afford the 4',5'-tetrahydro and 4',5'-enyl analogues (2 and 4, respectively, Fig. 7).[46,50] These derivatives proved to be potent inhibitors of SAHase as well, but interestingly, were not found to be substrates of adenosine deaminase and adenosine kinase.[53-62]

Removal of the N-3 from Ari, NpcA and their 4'-derivatives proved extremely fortuitous; the 3-deaza analogues of Ari and NpcA, and their 4'-derivatives (5-8, respectively, Fig. 8) have shown extremely potent levels of biological activity.[40,60,61,63,64] And, as was seen with Ari and NpcA, the 3-deaza congeners, 3-deazaaristeromycin and 3-deazaneplanocin and their 4'-derivatives are also potent inhibitors of SAHase and MeTase.[14,40,46,55,57,60,63,65-68] In contrast to Ari and NpcA however, the 3-deaza analogues of Ari and NpcA and their 4'-derivatives, like 3-deazaadenosine and the parent 4'-derivatives, are also not recognized by deaminating or phosphorylating enzymes, and therefore not converted to their mono-, di- and triphosphate forms.[40,64] As mentioned previously, although this phenomenon has yet to be explained fully for the 3-deaza analogues (it is understandably obvious that the 4'-derivatives cannot undergo phosphorylation due to the lack of the 4'-hydroxymethyl group), it nonetheless endows the 3-deaza carbocyclic series with significant chemotherapeutic properties without the accompanying toxicity levels as were seen with Ari and NpcA.

Enzyme Inhibition and Cell Differentiation

The 3-deaza nucleosides, including 3-deazaAdo have all exhibited the ability to either induce or to inhibit cell differentiation in a variety of cell lines. 3-DeazaAdo induces differentiation in 3T3-L1 fibroblasts,[18,37] but inhibits differentiation in murine leukemia cells.[35] 3-DeazaAri and 3-deazaNpc both induce differentiation in human promyelocytic leukemia HL-60 cells.[4,69] All three 3-deaza analogues activate collagen IV gene expression in F9 Tetratocarcinoma cells.[27]

Another connection between methylation, cell differentiation and purine SAHase inhibitors has recently been reported that suggests that induction of erythroid cell differentiation of MEL cells is associated with changes in methylation of poly(A)$^+$ RNA and the subsequent stability of RNA transcripts.[70] Studies with 3'-deoxyadenosine and N^6-methyladenosine (Fig. 9), both inhibitors of polyadenylation and methylation of RNA, have provided new leads and further studies are presently underway. Given the strong evidence to date connecting DNA

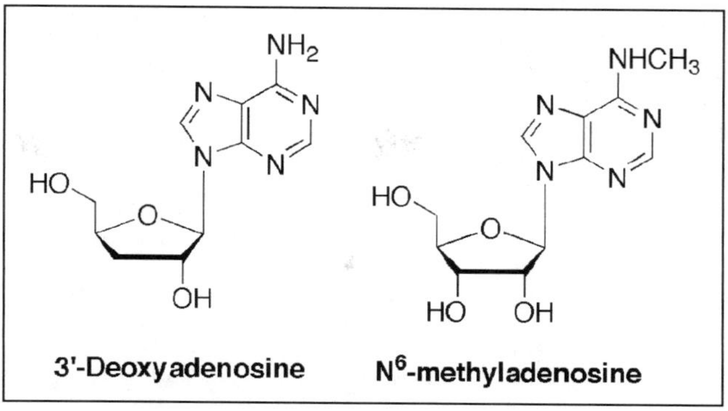

Figure 9. 3'-Deoxyadenosine and N^6-methyladenosine are inhibitors of polyadenylation and methylation of RNA.

methylation with cell differentiation, it will be interesting to see how this new evidence implicating RNA develops.

Future Directions

While incredible strides have been made towards understanding the mechanistic implications of DNA methylation and tumor cell differentiation, and indeed, many of the leads discussed herein are promising, none have provided the ultimate answer sought in cancer chemotherapy. As a result, there remains an urgent need to continue the search for new and more potent chemotherapeutic agents. The ever increasing availability of crystal structures for many newly identified methyltransferases and hydrolases, combined with the advances in bioinformatics and computer modeling techniques will serve to provide scientists with a better understanding of the complexities of these highly significant targets. This will hopefully provide the guidance necessary to cure one of mankind's most deadly maladies.

References

1. Chu CK, Baker DC. Nucleosides and Nucleotides as Antitumor and Antiviral Agents. New York: Plenum Press; 1993.
2. Montgomery JA. Studies on the biologic activity of purine and pyrimidine analogs. Med Res Rev 1982; 2(3):271-308.
3. Suhadolnik RJ. Nucleosides as Biological Probes. New York: Wiley Interscience; 1979.
4. Hatse S, De Clercq E, Balzarini J. Role of antimetabolites of purine and pyrimidine nucleotide metabolism in tumor cell differentiation. Biochem Pharmacol 1999; 58:539-55.
5. Montgomery JA, Bennett LL. Inhibitors of purine biosynthesis. In: Sandler M, Smith HJ, eds. Design of Enzyme Inhibitors as Drugs. Oxford: Oxford University Press; 1989.
6. Borchardt RT, Creveling CR, Ueland PM. Biological methylation and drug design. Clifton: Humana Press; 1986.
7. Szyf M. The DNA methylation machinery as a target for anticancer therapy. Pharmacol Ther 1996; 70(1):1-37.
8. Nishimune Y, Koscik D, Nishina Y et al. Inhibition of DNA synthesis causes stem cell differentiation: induction of teratocarcinoma F9 cell differentiation with nucleoside analogues of DNA synthesis inhibitors and their inducing abilities counterbalanced specifically by normal nucleosides. Biochem Biophys Res Commun 1989; 163:1290-97.
9. Marks PA, Rifkind RA. Differentiating factors. In: De Vita VT Jr, Hellman S, Rosenberg SA, eds. Biological Therapy of Cancer. Philadelphia: Lippincott; 1991:754-62.
10. Szyf M. DNA methylation properties: consequences for pharmacology. Trends Pharmacol Sci 1994; 15:233-38.
11. Banerjee AK. 5'-Terminal cap structure in eucaryotic messenger ribonucleic acids. Microbiol Rev 1980; 44(2):175-205.
12. Szyf M. Targeting DNA methyltransferase in cancer. Cancer Metastasis Rev 1998; 17:219-31.
13. Wolfe MS, Borchardt RT. S-Adenosyl-L-homocysteine hydrolase as a target for antiviral chemotherapy. J Med Chem 1991; 34(5):1521-30.
14. Liu S, Wolfe MS, Borchardt RT. Rational approaches to the design of antiviral agents based on S-Adenosyl-L-homocysteine hydrolase as a molecular target. Antiviral Res 1992; 19:247-65.

15. Jones PA, Taylor SM. Cellular differentiation, cytidine analogs and DNA methylation. Cell 1980; 20:85-93.

16. Juttermann R, Li E, Jaenisch R. Toxicity of 5-Aza-2'-deoxycytidine to mammalian cells is mediated primarily by covalent trapping of DNA methyltransferase rather than DNA demethylation. Proc Natl Acad Sci USA 1994; 91:11797-801.

17. Cantoni G. The Centrality of S-Adenosylhomocysteinase in the regulation of the biological utilization of S-adenosylmethionine. In: Borchardt RT, Creveling CR, Ueland PM, eds. Biological Methylation and Drug Design. Clifton: Humana Press; 1986:227-38.

18. Chiang PK. Biological effects of inhibitors of S-adenosylhomocysteine hydrolase. Pharmacol Ther 1998; 77(2):115-34.

19. Palmer JL, Abeles RH. The mechanism of action of S-adenosylhomocysteine. J Biol Chem 1979; 254(4):1217-26.

20. Turner MA, Yuan C-S, Borchardt RT et al. Structure determination of selenomethionyl S-adenosylhomocysteine hydrolase using data at a single wavelength. Nat Struct Biol 1998; 5(5):369-76.

21. Hu Y, Komoto J, Huang Y et al. Crystal structure of S-adenosylhomocysteine hydrolase from rat liver. Biochemistry 1999; 38:8323-33.

22. Yin D, Yang X, Hu Y et al. Substrate binding stabilizes S-adenosylhomocysteine hydrolase in a closed conformation. Biochemistry 2000; 39:9811-18.

23. Yuan C-S, Liu S, Wnuk SF et al. Rational approaches to the design of mechanism-base inhibitors of S-adenosylhomocysteine hydrolase. Nucleosides & Nucleotides 1995; 14:439-47.

24. Yuan C-S, Liu S, Wnuk SF et al. Design and synthesis of S-adenosylhomocysteine hydrolase inhibitors as broad-spectrum antiviral agents. In: De Clercq E, ed. Advances in Antiviral Drug Design. Greenwich: JAI Press, Inc., 1996:41-88.

25. Turner MA, Yang X, Yin D et al. Structure and function of S-adenosylhomocysteine hydrolase. Cell Biochem Biophy 2000; 33:101-25.

26. De Clercq E. Antiviral activity spectrum and target of action of different classes of nucleoside analogues. Nucleosides & Nucleotides 1994; 13(6 & 7):1271-95.

27. Chiang PK, Burbelo PD, Brugh SA et al. Activation of collogen IV gene expression in F9 teratocarcinoma cells by 3-deazaadenosine analogs. J Biol Chem 1992; 267:4988-91.

28. Chiang PK, Cantoni GL. Perturbation of biochemical transmethylations by 3-deazaadenosine in vivo. Biochem Pharmacol 1979; 28:1897-902.

29. Bader JP, Brown NR, Chiang PK et al. 3-deazaadenosine, an inhibitor of adenosylhomocysteine hydrolase, inhibits reproduction of rous sarcoma virus and transformation of chick embryo cells. Virology 1978; 89:494-505.

30. De Clercq E. S-adenosylhomocysteine hydrolase inhibitors as broad-spectrum antiviral agents. Biochem Pharmacol 1987; 36(16):2567-75.

31. Pugh CSG, Borchardt RT. Effects of S-adenosylhomocysteine analogues on vaccinia viral messenger ribonucleic acid synthesis and methylation. Biochemistry 1982; 21:1535-41.

32. Svardal A, Djurhuus R, Ueland PM. Disposition of homocysteine and S-3-deazaadenosylhomocysteine in cells exposed to 3-deazaadenosine. Mol Pharmacol 1986; 30:154-58.

33. Bennett LL Jr, Brockman RW, Allan PW et al. Alterations in nucleotide pools induced by 3-deazaadenosine and related compounds. Biochem Pharmacol 1988; 37:1233-44.

34. Chiang PK, Im YS, Cantoni GL. Phospholipids biosynthesis by methylations and choline incorporation: effect of 3-deazaadenosine. Biochem Biophys Res Commun 1980; 94:174-81.

35. Sherman ML, Shafman TD, Spriggs DR et al. Inhibition of murine erthroleukemia cell differentiation by 3-Deadeazaadenosi Cancer Res 1985; 45:5830-34.

36. Mizutani Y, Masuoka S, Imoto M et al. Induction of erthroid differentiation in leukaemic K562 cells by an S-adenosylhomocysteine hydrolase inhibitor, aristeromycin. Biochem Biophy Res Commun 1995; 207:69-74.

37. Chiang PK. Conversion of 3T3-L1 fibroblasts to fat cells by an inhibitor of methylation: effect of 3-deazaadenosine. Science 1981; 211:1164-66.

38. Paller AS, Arnsmeier SL, Clark SH et al. Z-4',5'-Didehydro-5'-deoxy-5'-fluoroadenosine (MDL 28,842), and irreversible inhibitor of s-adenosylhomocysteine hydrolase, suppresses proliferation of cultured keratinocytes and squamous carcinoma cell lines. Cancer Res 1993; 53:6058-60.

39. De Clercq E. Carbocyclic adenosine analogues as s-adenosylhomocysteine hydrolase inhibitors and antiviral agents: recent advances. Nucleosides & Nucleotides 1998; 17(1-3):625-34.

40. Glazer RI, Knode MC, Tseng CKH et al. 3-Deazaneplanocin A: a new inhibitor of S-adenosylhomocysteine synthesis and its effects in human colon carcinoma cells. Biochem Pharmacol 1986; 35(24):4523-27.

41. Linevsky J, Cohen MB, Hartman KD et al. Effect of neplanocin A on differentiation, nucleic acid methylation, and c-myc mRNA expression in human promyelocytic leukemia cells. Mol Pharmacol 1985; 28:45-50.

42. Kishi T, Muroi M, Kusaka T et al. The structure of aristeromycin. Chem Pharm Bull 1972; 20:940-46.

43. Kusaka T, Yamamoto H, Shibata M et al. Streptomyces citricolor Nov. Sp. and a vew antibiotic, aristeromycin. J Antibiot 1968; 21(4):255-63.

44. Yaginuma S, Muto N, Tsujino M et al. Studies on neplanocin A, new antitumor antibiotic. I. producing organism, isolation and characterization. J Antibiot 1981; 34(4):359-66.

45. Hayashi M, Yaginuma S, Yoshioka H et al. Studies on neplanocin A, new antitumor antibiotic. II. structure determination. J Antibiot 1981; 34(6):675-80.

46. Wolfe MS, Lee Y, Bartlett WJ et al. 4'-modified analogues of aristeromycin and neplanocin A: synthesis and inhibitory activity toward S-adenosyl-L-homocysteine hydrolase. J Med Chem 1992; 35:1782-91.

47. Borchardt RT, Keller BT, Patel-Thombre U. Neplanocin A. A potent inhibitor of S-adenosylhomocysteine hydrolase and of vaccinia virus multiplication in mouse L929 Cells. J Biol Chem 1984; 259(7):4353-58.

48. Glazer RI, Knode MC. Neplanocin A. A cyclopentyl analog of adenosine with specificity for inhibiting RNA methylation. J Biol Chem 1984; 259:12964-69.

49. Saunders PP, Tan M-T, Robins RK. Metabolism and action of neplanocin A in chinese hamster ovary cells. Biochem Pharmacol 1985; 34(15):2749-54.

50. Borcherding DR, Scholtz SA, Borchardt RT. Synthesis of analogues of neplanocin A: utilization of optically active dihydroxycyclopentenones derived from carbohydrates. J Organic Chem 1987; 52:5457-61.

51. Bennett LL Jr, Allan PW, Rose LM et al. Differences in the metabolism and metabolic effects of the carbocyclic adenosine analogs, neplanocin A and aristeromycin. Mol Pharmacol 1986; 29:383-90.

52. De Clercq E. Antiviral and antimetabolic activities of neplanocins. Antimicrob Agents Chemother 1985; 28(1):84-89.

53. Narayanan SR, Keller BT, Borcherding DR et al. 9-(trans-2', trans-3'-Dihydroxycyclopent-4'-enyl) Derivatives of adenine and 3-deazaadenine: potent inhibitors of bovine liver S-adenosylhomocysteine hydrolase. J Med Chem 1988; 31:500-03.

54. Ault-Riche DB, Lee Y, Yuan C-S et al. Effects of 4'-Modified analogs of aristeromycin on the metabolism of S-adenosyl-L-homocysteine in murine L929 cells. Mol Pharmacol 1993; 43:989-97.

55. Hasobe M, Liang H, Ault-Riche DB et al. (1'R, 2'S, 3'R)-9-(2',3'-Dihydroxycyclopentan-1'-yl)-adenine and -3-Deaza-adenine: analogues of aristeromycin which exhibit potent antiviral activity with reduced toxicity. Antiviral Chem Chemother 1993; 4(4):245-48.

56. Hasobe M, McKee JG, Borcherding DR et al. 9-(trans-2', trans-3'-Dihydroxycyclopent-4'-enyl)-adenine and -3-Deazaadenine: analogs of neplanocin A which retain potent antiviral activity but exhibit reduced cytotoxicity. Antimicrob Agents Chemother 1987; 31:1849-51.

57. Hasobe M, McKee JG, Borcherding DR et al. Effects of 9-(trans-2', trans-3'-Dihydroxycyclopent-4'-enyl)-adenine and -3-Deazaadenine on the metabolism of S-adenosylhomocysteine in mouse L929 cells. Mol Pharmacol 1988; 33:713-20.

58. Hasobe M, McKee JG, Ishii H et al. Elucidation of the mechanism by which homocysteine potentiates the anti-vaccinia virus effects of the S-adenosylhomocysteine hydrolase inhibitor 9-(trans-2', trans-3'-Dihydroxycyclopent-4'-enyl)-adenine. Mol Pharmacol 1989; 36:490-96.

59. Paisley SD, Hasobe M, Borchardt RT. Elucidation of the mechanism by which 9-(trans-2', trans-3'-Dihydroxycyclopent-4'-enyl)-adenine inactivates S-adenosylhomocysteine hydrolase and elevates cellular levels of S-adenosylhomocysteine. Nucleosides & Nucleotides 1989; 8:689-98.

60. Cools M, De Clercq E. Correlation between the antiviral activity of acyclic and carbocyclic adenosine analogues in murine l929 cells and their inhibitory effect on L929 cell S-adenosylhomocysteine hydrolase. Biochemical Pharmacology 1989; 38(7):1061-67.

61. De Clercq E, Cools M, Balzarini J et al. Broad-spectrum antiviral activities of neplanocin A, 3-deazaneplanocin A, and their 5'-Nor derivatives. Antimicrob Agents Chemother 1989; 33:1291-97.

62. Villalon MDG, Gil-Fernandez C, De Clercq E. Activity of several S-adenosylhomocysteine hydrolase inhibitors against african swine fever virus replication in vero cells. Antiviral Res 1993; 20:131-44.

63. Tseng CKH, Marquez VE, Fuller RW et al. Synthesis of 3-deazaneplanocin A, a powerful inhibitor of S-adenosylhomocysteine hydrolase with potent and selective in vitro and in vivo antiviral activities. J Med Chem 1989; 32:1442-46.

64. Glazer RI, Hartman KO, Knode MC et al. 3-deazaneplanocin: a new and potent inhibitor of S-adenosylhomocysteine hydrolase and its effects on human promyelocytic leukemia cell line HL-60. Biochem Biophys Res Commun 1986; 135:688-94.
65. Bray M, Driscoll J, Huggins JW. Treatment of lethal ebola virus infection in mice with a single dose of an S-adenosyl-L-homocysteine hydrolase inhibitor. Antiviral Res 2000; 45:135-47.
66. Secrist JA, III, Comber RN, Gray RJ et al. Synthesis of 5'-substituted analogues of carbocyclic 3-deazaadenosine as potential antivirals. J Medl Chem 1993; 36:2102-06.
67. Cosstick R, Li X, Tuli DK. Molecular recognition in the minor groove of the DNA helix. studies on the synthesis of oligonucleotides and polynucleotides containing 3-deaza-2'-deoxyadenosine. Nucleic Acids Res 1990; 18:4771-78.
68. Houston DM, Dolence EK, Keller BT et al. Potential inhibitors of S-adenosylmethionine-dependent methyltransferases. 8. Molecular dissections of carbocyclic 3-deazaadenosine as inhibitors of S-adenosylhomocysteine hydrolase. J Med Chem 1985; 28:467-71.
69. Aarbakke J, Miura GA, Prytz PS et al. Induction of HL-60 cell differentiation by 3-deaza-(±)-aristeromycin, an Inhibitor of S-adenosylhomocysteine hydrolase. Cancer Res 1986; 46:5469-72.
70. Vizirianakis IS, Tsiftsoglou AS. Induction of murine erythroleukemia cell differentiation is associated with methylation and differential stability of poly(A)$^=$ RNA transcripts. Biochimica et Biophysica Acta 1996; 1312:8-20.

CHAPTER 14

DNA Methyltransferase Inhibitors:
Paving the Way for Epigenetic Cancer Therapeutics

Gregory K. Reid and A. Robert MacLeod

Introduction

Our increased understanding of the molecular pathophysiology of cancer is beginning to impact our ability to effectively treat this disease. The recent success of the tyrosine kinase inhibitor Gleevec™ is a prime example of this. In this case, the molecular etiology of the disease (chronic myelogenous leukemia (CML) and gastrointestinal stromal tumors (GIST)) was understood and targeted with a highly selective pharmaceutical agent, the result being dramatic clinical benefit and amazingly few side effects.[1] While only the beginning, this success story provides renewed hope that the ultimate goal of conquering cancer is attainable. However, CML and GIST are quite rare, and in genetic terms, very simple forms of cancer. The BCR-ABL translocation is the principal, if not the sole genetic alteration leading to this disease. This is in contrast to the situation that exists for more prevalent tumors such as those of the lung, breast, colon and others, where dozens of genetic abnormalities (translocations, chromosome duplication, and deletions) are seen. Understandably, these have proven more difficult to treat; however, further insights into the molecular events at work in these more complex tumors will certainly yield improved therapeutic strategies.

In the last several years it has become clear that aberrant DNA methylation (with associated tumor suppressor gene silencing) is a universal feature of cancer cells. Of particular interest is the possibility that targeting the cellular activities controlling aberrant DNA methylation in cancer cells may be a means of simultaneously affecting multiple epi-genetic events underlying their pathophysiology.

Here we review the prevalence of DNA hypermethylation in human cancer, evaluate DNA methyltransferase enzymes as the first class of epigenetic cancer targets, and discuss potential future therapeutic strategies.

DNA Methylation: Discovery of the First Epigenetic Modifier

All the cells that comprise the human body contain within their nucleus exactly the same genetic information.[a] It is the epigenetic information, however, that provides the instructions for orchestrating the expression of that genetic information, giving rise to cells expressing unique sets of genes and thus capable of specialized functions. DNA methylation, or the covalent addition of a methyl group to the 5' position of cytosines within the CpG dinucleotide, is a major determinant of epigenetic information. Methylation of DNA has been shown to play roles in many cellular processes including regulation of chromatin structure,[2] genomic imprinting,[3] somatic X-chromosome inactivation in females,[4] and the timing of DNA replication.[5] However, it is the role of DNA methylation in transcriptional regulation and tissue-specific gene expression that has been most extensively studied. An inverse relationship exists between the extent of DNA methylation within a given gene and its transcriptional activity,[6] such that actively transcribed genes have low level methylation of CpGs within their promoter regions

DNA Methylation and Cancer Therapy, edited by Moshe Szyf. ©2005 Eurekah.com and Kluwer Academic/Plenum Publishers.

while transcriptionally silent genes are usually heavily methylated in these 5' regulatory regions. DNA methylation within promoter regions of genes can inhibit their expression by several mechanisms, including physically blocking access to the cellular transcription machinery or by recruiting other gene expression regulators such as histone deacetylase enzymes (HDACs), histone methyltransferases (HMETs), SWI/SNF proteins, the activities of which can alter chromatin states to regulate gene expression (for a review see ref. 7). Inhibition of both DNA methylation and HDAC activity can synergistically reactivate silenced tumor suppressor genes, although, in most cases it appears that silencing by methylation of DNA is dominant to that of deacetylation of histones.[8] It will be of great interest in the future to investigate the relationship between DNA methylation and other epigenetic modifications including histone acetylation-deacetylation, histone methylation and histone phosphorylation.

While epigenetic mechanisms allow cells to express unique sets of genes, mis-regulation can lead to disease by the inappropriate silencing of important genes. Inappropriate inactivation of genes involved in tumor suppressor functions by these mechanisms would clearly provide a growth advantage to cancer cells. Most importantly for therapeutic intervention, epigenetic mutations in contrast to genetic modifications are more readily reversible, and therefore drugs acting on these pathways may have therapeutic value in the treatment of cancer.

DNA Methylation and Cancer: A Correlation

The first link between DNA methylation and cancer came from studies showing that DNA from tumor cells had significantly lower 5-methyl cytosine (5M-Cyt) levels than those of normal cells.[9] This loss of DNA methylation in cancer was consistent with the development of hepatocellular cancer in mice fed diets severely deficient in sources of methyl groups.[10,11] Moreover, the loss of DNA methylation was found to be an early event in tumorigenesis, occurring even in pre-neoplastic colonic epithelium of individuals with familial polyposis coli.[12-14] These findings, coupled with the then recently defined role for oncogenes in cancer,[15] crystallized the hypothesis that activation of oncogenes through loss of DNA methylation was involved in the genesis of human cancer. The hypothesis was supported when hypomethylation of the H-ras and MYC oncogenes found in a variety of human tumors,[16,17] although not associated with over expression of these genes.

Studies with various demethylating agents on cultured cells, however, did not support the above hypothesis. Demethylation by these pharmacologic agents was associated with antiproliferative effects and the induction of differentiation in a variety of human and murine cancer cell lines.[18-20] This inconsistency was generally believed to be due to the well known pleotropic effect of the demethylating agents used. The nucleoside analogue 5-aza-dC, for example, forms stable complexes with several nuclear proteins, is mutagenic, causes DNA damage in fission yeast [21] and in Escherichia coli,[22] and has been shown to alter cellular differentiation even in organisms that do not carry methylated bases in their genomes.[23] The demonstration by Constantinides et al 77 and Jones et al 1990 that 5-aza-dC caused CH310T1/2 cells to differentiate into muscle, chondrocytes and adipocytes by demethylation and reactivation of the CpG island of the MYOD1 gene (a master regulatory gene for muscle differentiation), together, provided the first suggestion that gene reactivation by demethylation was potentially involved in cellular transformation. The discovery of cancer cell CpG island methylation in this case was itself very important. CpG islands are sequences rich in the CpG dinucleotide and are generally found within promoter regions of ubiquitously expressed housekeeping genes, and are not normally methylated. The discovery of aberrant hypermethylation in cancer cells with associated transcriptional inactivation of an important regulator of differentiation suggested for the first time that hypermethylation of critical DNA sequences may play a role in oncogenesis. These findings prompted a flurry of studies aimed at addressing four key questions regarding the importance of DNA methylation in cancer and its prospect as a therapeutic target.

First, was the CpG island and gene silencing observed in cultured cells also observed in primary human tissue? Secondly, did the nature of methylation-silenced genes support a causal association between their inactivation and the cancer phenotype? Thirdly, what were the cellular

Table 1. Epigenetic events [hypermethylation] are as fundamental to cancer as mutation

Gene	Locus	Tumors with Methylation	Hypermethylation in Human Tumors Incidence
Tumor Suppressor Genes Associated with Promoter Region			
Documented tumor suppressor genes			
Rb	13q14.2	Retinoblastoma	10%
VHL	3p25	Renal carcinoma	20-33%
p16^{NK4A}	9p21	Most solid tumors and lymphomas	30-40% (colon), 31% (breast), 25% (NSCLC), 67% (head and neck), 10% (melanoma), 18% (pancreatic), 40% (esophageal)
p15^{NK4B}	9p21	Primary acute leukemias and Burkitt lymphoma	71-94% in AML, 24% CML
p14^{ARF}	9p21	Colorectal	28% (carcinomas), 32% (adenomas)
E-cadherin	16q22.1	Bladder, breast, colon, liver, pancreatic	7% (pancreatic), 83% (papillary thyroid)
H-cadherin	16q	Ovarian, NSCLC	45% (NSCLC)
hMLH1	3p21	Colon, endometrial, gastric	91% (endometrial MSI), 95-100% (gastric MSI)
PTEN/MMAC1	10q23	Prostate	50% (recently derived xenografts)
p73	1p36.2	Leukemias	30%
BRCA1	17q11	Breast: Medullary Mucinous	13% 67% 55%
Probable tumor suppressor genes			
ER	6q25	Breast, colon, lung, leukemia, prostate	95% (grade III-IV prostate)
HIC1	17p13.3	Brain, breast, colon, renal, leukemia	83% (AML), 67% (breast)
AR	Xq13	Prostate	13%
RAR beta	3p24	Breast, pancreatic, lung	20% (pancreatic), 62% (SCLC), 43% (SCLC)
APC	5q21	Colorectal	18%
GSTPi	11q13	Prostate, breast	70-91% (prostate), 30% (breast), 20% (renal), 85% (heptocellular carcinoma)
TIMP-3	22q13.1	Pancreatic	11%
RASSF1	3p21.3	Lung	40%

activities responsible for the aberrant CpG island methylation in cancer cells? Fourth, if identified, could these activities be targeted effectively and would this reverse the aberrant methylation to alter the phenotype of cancer cells. Over the course of the next several years it became clear that the answer to the first two questions was a resounding yes. CpG island methylation and silencing of well characterized tumor suppressor genes such as RB1, p16ink4a, VHL, and many more, were observed in many types of primary human tumors (for a review see ref. 24). The list of genes displaying hypermethylation in human tumors has since grown long (see Table 1). Included are genes controlling such diverse cellular functions as cycle progression, apoptosis control, cell adhesion, hormone signaling and immune surveillance. These findings suggested a model where tumor suppressor silencing by promoter gene methylation was involved in the initiation of oncogenesis and perhaps the maintenance of the cancer phenotype (see Fig. 1). For many of these genes, inactivation by methylation appears to play an analogous role to inactivation by mutation. In fact, methylation and mutation can work independently to inactivate both individual alleles of given tumor suppressor genes. In an elegant study by Baylin et al, it was

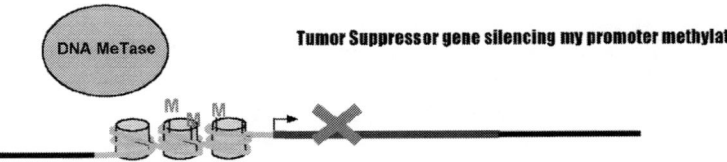

Cancer-specific DNA hypermethylation emerges as a novel Cancer target

DNA MeTase

Tumor Suppressor gene silencing my promoter methylation

Findings:

•*Inactivation of genes important in suppressing cancer (P16ink4a, ER, hMLH1) is associated with Promoter CpG methylation*

•*Decreased expression of murine Dnmt reverses transformed phenotype of cancer cells*

•*Dnmt -\+ knock outs crossed with Min- mice have decreased intestinal neoplasia*

Proposed Model:
•Epigenetic modification by DNA methylation can functionally replace mutation in cancers by transcriptional inactivation of important growth regulating genes.

Figure 1. DNA Methylation and Transcriptional Silencing. Over-expression of DNMT1 as in cancer cells can lead to hypermethylation and inactivation of tumor suppressor and growth regulator genes. Loss of these functions can result in cancer.

shown that HCT116 colon cancer cells contain a wild type allele of the p16ink4aink4a that has been selectively silenced by methylation whereas the other unmethylated allele contains a frame shift mutation, coding for a truncated non-functional protein.[25] This demonstrated for the first time that epigenetic modifications in the form of DNA methylation could effectively replace mutational inactivation in achieving a "second hit" to eliminate cellular tumor suppressor function. The third and fourth questions reflected the need to identify the cellular activities controlling aberrant methylation in human cancer cells in order to develop strategies to effectively target them therapeutically. These will be discussed in subsequent sections.

The DNA Methyltransferase Family of Enzymes

To date there are three known active mammalian DNA methyltransferase enzymes, DNMT1, DNMT3a and DNMT3b. These enzymes catalyze the transfer of a methyl group from the cofactor S-adenosylmethionine (SAM) to the C-5 position of cytosine within CpG dinucleotide sequences of DNA.[26] DNMT1 was the first methyltransferase to be discovered[27,28] and is the most abundant DNMT in somatic cells. DNMT1 localizes to replication foci and interacts with proliferating cell nuclear antigen (PCNA). DNMT1 has a 10-40 fold preference for hemi-methylated DNA over non-methylated DNA, and is often referred to as the 'maintenance' methyltransferase, as it is believed to be the enzyme responsible for copying methylation patterns from parent to newly synthesized DNA molecules during replication. The DNMT3 family of methyltransferases was discovered more recently.[29] These enzymes are required to establish methylation patterns during embryogenesis and have therefore been called the de novo methyltransferases.

Both the maintenance (DNMT1) and de novo enzymes (DNMT3a and DNMT3b) are required for proper embryonic development, as mice with targeted disruption of both alleles (-/-, homozygous null) for any of the DNMTs do not survive.[30,31] In contrast, mice with targeted disruption of only one of the alleles (+/-, heterozygous null) are viable and are indistinguishable from normal animals. Each of the three DNMT enzymes have been shown to be expressed at higher levels in cancer cells compared to normal cells,[32] and are all likely to be involved in the oncogenic process. More recently it has been found that, in humans, mutations in the catalytic domain of DNMT3b gives rise to the ICF syndrome (immunodeficiency, centromeric instability, facial anomalies).[33] This is a rare disorder in which individuals exhibit profound loss of DNA methylation from satellite 2 and 3 sequences adjacent to centromeres of chromosomes 1, 9 and 16, resulting in instability of these chromosomes. ICF is a developmental disease and, therefore, therapeutic inhibition of DNMT3b in an adult will likely not give rise to symptoms of ICF. Recent results suggest that DNMT3b may also be an interesting cancer target.[34]

DNA Methylation an Active Player in Oncogenesis: Validation of DNMT1 As a Therapeutic Cancer Target

The correlation between DNA methylation and transcriptional silencing of tumor suppressor genes in human tumors formed the basis of a cogent argument for DNA methylation playing a causal role in the genesis and maintenance of human cancer rather than being a consequence of the transformed phenotype. However, results from studies aimed at testing the causal relationship were met with skepticism because experimental inhibition of DNA methylation relied for the most part on the nucleoside analogs 5-azacytidine (5-aza-C) and 5-azadeoxycytidine (also called 5-aza-CdR, or decitabine). As a class of molecules, nucleoside analogues are known to affect many cellular processes, particularly those of DNA replication.[35] 5-aza-C and 5-aza-CdR treatment alter cellular differentiation in organisms that do not bear methylated bases in their genomes. To exert their biological effects, 5-aza-C and 5-aza-CdR must be incorporated into the DNA where they irreversibly trap the DNA methyltransferase enzymes onto DNA in a covalent manner. Thus, it is possible that, in addition to targeting DNMT enzymes, other proteins associated with DNMTs may also become trapped or stalled at replication forks along with DNMT1-5-aza-CdR complexes (see Fig. 2). It is the covalent trapping of DNA methyltransferase enzymes bound to DNA, and not the loss of DNA methylation itself, that causes the mutagenic and cytotoxic effects of 5-aza-CdR.[36] Despite the vast amount of experimental evidence supporting the role of DNA methylation in tissue-specific gene expression and gene silencing its importance is still actively debated.[37]

Nonetheless, mounting evidence continued to implicate DNA methylation and DNMT enzymes in oncogenesis. A study of the control elements regulating DNMT1 expression revealed that DNMT1 was regulated by the RAS signal transduction pathway through activation of the AP-1 transcription factor complex.[38] Moreover, attenuation of this oncogenic signal transduction pathway by the endogenous inhibitor GAP reversed the aberrant methylation in cancer cells.[39] Taken together, these studies provided the first mechanistic link between establishment of aberrant DNA methylation and cancer pathways.

The first direct evidence demonstrating that DNA methylation mediated by DNMT1 was required by mouse cancer cells to maintain their tumorigenic phenotype soon followed.[40] In this case specific depletion of DNMT1 in Y1 mouse adrenocortical tumor cells was achieved by over-expression of an antisense cDNA. This study was extended to in vivo models of murine tumors.[41] Together, they demonstrated for the first time that inhibition of DNMT1 could both reverse DNA methylation and the cancer phenotype. These findings were subsequently supported by genetic models employing DNMT1 knock out mice. When DNMT1 +/- mice were crossed with mice genetically predisposed to high rates of intestinal neoplasia (Min- mice), the progeny had dramatically reduced rates of neoplasia.[42] This finding demonstrated that DNMT1 inhibition could prevent the development of neoplasia and that there was a clear therapeutic window for doing so. Taken together, the results from these murine studies suggested that therapeutic agents that selectively inhibit DNMT1 and do not rely on incorporation into cellular DNA and covalent trapping of the DNMT enzymes may have utility as human cancer therapeutics.

DNMT1 was the only mammalian DNA MeTase enzyme known at the time of these initial findings. Since then the DNMT3a and DNMT3b enzymes have been discovered.

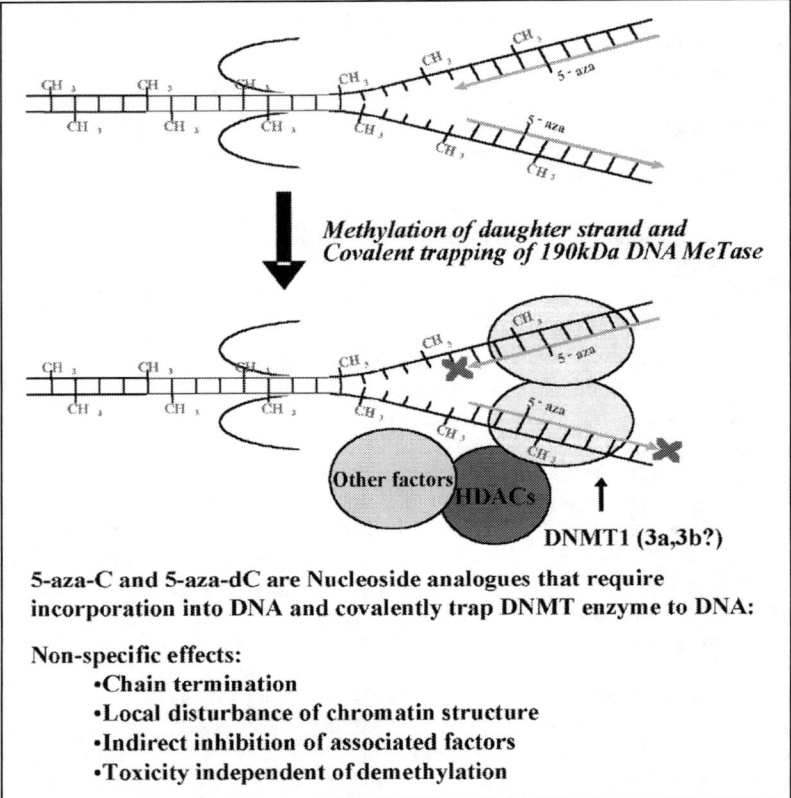

Figure 2. Non-specific effects of Nucleoside analogue inhibitors of DNA methylation. 5-aza-C and 5-aza-CdR are incorporated into genomic DNA during replication. Transfer of methyl groups to these bases leads to covalent trapping of the DNMT enzyme to DNA at the replication fork. DNMT-associated proteins (HDACs and others) may also remain trapped at these sites thus depleting their activities at other regions. Known non-specific effects of these nucleoside analogues are listed.

DNMT3a and DNMT3b clearly play a role in establishment of DNA methylation during development and are therefore considered to be de novo methyltransferases.[29] Given the understanding of the DNMT enzymes, the "maintenance" methyltransferase activity of DNMT1 represented the most reasonable target to therapeutically modulate the aberrant DNA methylation already established within tumor cells.

However, isolation of the role of DNMT1 itself in gene expression and dissociation of it from effects of the newly identified DNMT family members and on chromatin structure requires specific DNMT1 inhibitors that are not incorporated into genomic DNA. To this end, potent antisense inhibitors capable of specifically reducing cellular DNMT1 mRNA and protein levels have been employed to study the response of cancer cells to reduction of cellular DNA MeTase levels.

Isotype-Selective Inhibition of DNMTs: Antisense to SiRNA

An antisense inhibitor is an oligonucleotide analog designed to bind to a given region of a target gene's mRNA. The "antisense" oligonucleotide is designed to have a complementary sequence to the "sense" strand of the target mRNA and therefore binds to its target by Watson-Crick base pairing. The high affinity binding of the antisense oligonucleotide to its

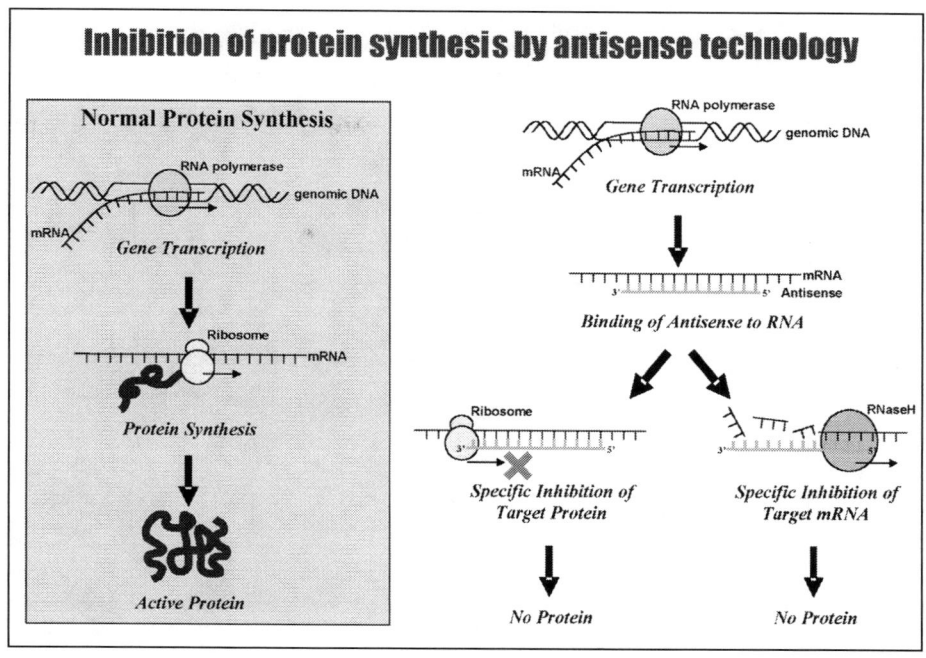

Figure 3. Antisense mechanism of action. Antisense inhibitors are chemically stabilized DNA analogues that bind to a specific region of their target mRNA. Once bound, the DNA-RNA duplex formed is recognized by cellular RNAse H that selectively cleaves on the mRNA strand. Antisense inhibitors may also inhibit translation by sterically blocking progression of the ribosome.

target mRNA can inhibit synthesis of the target protein by physically blocking the ribosome from progressing along the mRNA thus inhibiting synthesis of the target protein. In addition, the target mRNA itself may be rapidly degraded by the enzyme RNAse H that is recruited to and cleaves the RNA strand of the DNA/RNA duplex (see Fig. 3). As antisense inhibitors can theoretically be designed to inhibit any target mRNA regardless of its cellular function it is ideally suited to functional genomic studies where only sequence information is known or to validation of molecular targets for which no specific small molecule inhibitors are available.

As addressed above, we hypothesized that targeting the DNMT1 enzyme would result in clinical benefit by reversing the aberrant methylation of cancer cells and reactivating the inappropriately silenced genes. The role of each of the DNMT enzymes in gene expression and in cancer was examined by employing potent isotype-specific antisense inhibitors (2'-O-methyl-modified phosphorothioates) capable of specifically reducing cellular DNMT1, DNMT3a and DNMT3b levels. To identify antisense oligodeoxynucleotides (ODNs) capable of specific and potent inhibition of DNMT gene expression in human tumor cells, we screened a large number of phosphorothioate ODNs (20 bases in length) targeted to DNMT1, DNMT3a or DNMT3b mRNA sequences. We first assessed antisense activity by the ability of ODNs to reduce cellular DNMT mRNA after 24 hour treatments. From this screen we identified highly potent DNMT antisense inhibitors capable of specifically inhibiting cellular DNMT1 mRNA in a dose-dependent manner (Fig. 4A). These inhibitors were then shown to specifically deplete cellular levels of DNMT1 in a dose-dependent manner (Fig. 4B). Several other human tumor cell lines, including breast, lung, and colon cancer cell lines, were used to evaluate the activities of DNMT antisense inhibitors with essentially identical results. For the purposes of this review we will focus on the characterization and clinical development of DNMT1 antisense inhibitors. In addition to the antisense inhibitors described we developed DNMT1 specific siRNA inhibitors (Fig. 4C).

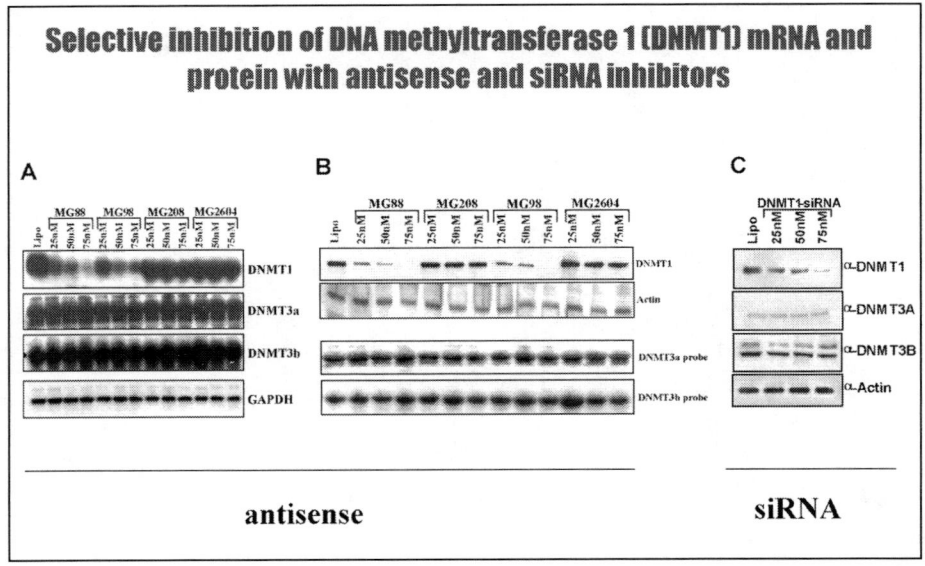

Figure 4. Identification of DNMT isoform-specific antisense inhibitors. A) Northern blot analysis showing specific dose-dependent inhibition of DNMT1 mRNA in A459 cells treated with increasing concentrations of inhibitors for 24 hours. B) Western blot analysis showing specific dose-dependent inhibition of DNMT1 protein in A459 cells treated with increasing concentrations of inhibitors for 48 hours. C) DNMT1 siRNA: Western blot analysis A549 cells treated for 48 hours with DNMT1 siRNA shows selective depletion of DNMT1 protein levels.

Three highly potent DNMT1 inhibitors (MG88, MG98 and siDNMT1) were evaluated. While these three inhibitors target entirely different regions of the DNMT1 mRNA, they produce identical results in terms of DNMT1 mRNA and protein depletion, tumor suppressor reactivation and subsequent effects on cancer cells.

The cyclin-dependent kinase inhibitor (CDKI) p16(*INK4a*) regulates cell proliferation by controlling the transition from G_1 (growth phase) to S-phases (DNA synthesis phase) of the cell cycle. Inactivation of the p16ink4a gene is one of the most frequently observed abnormalities in human cancer.[43] Genetic alterations in p16(*INK4a*), including point mutations and, to a greater extent homozygous deletion, are often found in tumors. Transcriptional inactivation with associated hypermethylation of the p16(*INK4a*) gene have also been observed in virtually all types of cancer.[44]

To determine whether specifically reducing cellular DNMT1 levels would induce the activation of a silenced p16(*INK4a*) gene, we treated human cancer cells that contain a hypermethylated and silenced p16(*INK4a*) gene (either T24 bladder cancer or HCT116 colon cancer cells) with the human DNMT1 inhibitors MG88, MG98 or siRNA-DNMT1). As a result of DNMT1 depletion, re-expression of p16(*INK4a*) protein was detected after 5 days of treatment with either inhibitor (Fig. 5A).

To determine whether re-methylation and inactivation of the activated p16(*INK4a*) gene occurs when DNMT1 returns to control levels, treatments were stopped. As DNMT1 protein levels increased in the absence of MG88 and returned to control levels, p16(INK4a) protein expression decreased steadily over the post treatment period until it was barely detectable (Fig. 5B). This finding suggests that methylation and inactivation of p16(*INK4a*) occurred when DNMT1 levels recovered and that DNMT1 contributes to the inactivation of this tumor suppressor gene observed in human cancer.

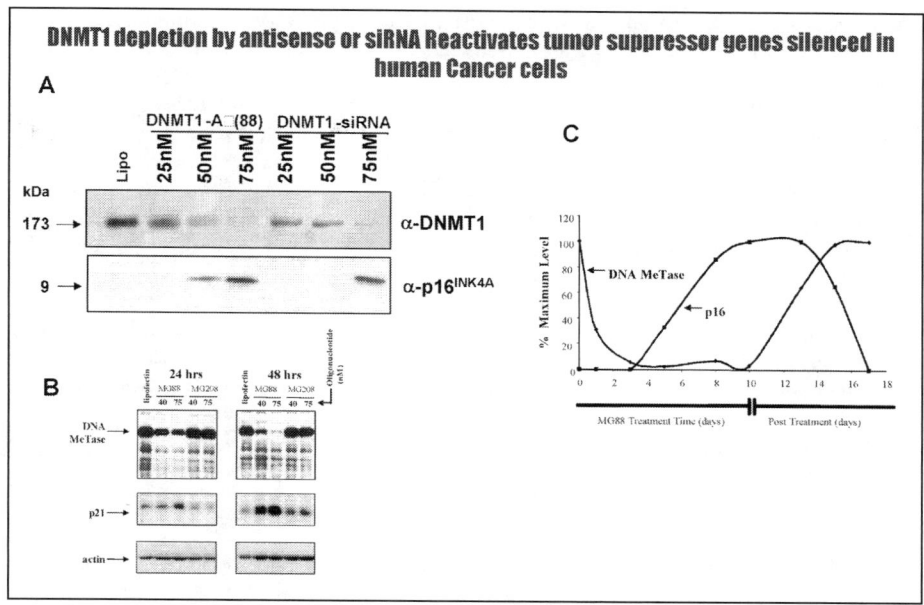

Figure 5. Specific inhibition of the DNMT1 induces tumor suppressor proteins p16ink4a and p21WAF1/Cip1 expression in cancer cells. A) Western blot showing expression of p16ink4a protein in DNMT1 depleted cells (antisense or siRNA). B) Western blot showing induction of p21WAF1/CIP1 protein in DNMT1 depleted cells. C) Schematic shows the rise in p16ink4a protein levels as DNMT1 levels decrease.

Treatment of human cancer cells with DNMT1 antisense inhibitors causes a rapid inhibition of cancer cell growth prior to the reactivation of the tumor suppressor protein p16(INK4a). Therefore, other regulators of cell growth must also be induced in response to DNMT1 inhibition. Further investigation demonstrated that the cell cycle regulator protein p21[WAF1/CIP1] was induced in a rapid fashion in response to DNMT1 inhibition (Fig. 5C). Induction of p21[WAF1/CIP1] was not due to transcriptional reactivation and may be regulated by post-translational mechanisms. These results demonstrated for the first time that DNMT1 is required by cancer cells to maintain silencing of tumor suppressor genes.[45] Studies on DNMT1-/- HCT116 clones isolated by multiple rounds of homologous recombination suggest that DNMT1 depletion alone is not sufficient to alter cancer cell methylation.[46,47] However, isotype-selective inhibition of DNMT1 alone by pharmacologic means (antisense or siRNA) that are not biased due to negative selection for growth arrested cells, demonstrate that DNMT1 is the principal methyltransferase required to maintain cancer cell CpG methylation.[48] Since these initial findings, we have found that MG98-mediated depletion of DNMT1 in many different cancer cell types leads to reactivation of multiple other genes regulating cell growth, apoptosis and resistance to cancer therapeutics (e.g., *RASSF1A* reactivation in renal cancer cells, *hMLH1* reactivation in colon cancer cells). Further in vivo evaluation of these inhibitors, including human tumor xenograft studies and preclinical toxicology, led to the selection of MG98 for clinical development.

Medicinal Chemistry of Oligonucleotides: Towards Antisense Drugs

Natural single stranded oligonucleotides are unacceptable as therapeutics because they are rapidly degraded. This degradation occurs principally through the action of phosphodiesterases that cleave the phosphodiester "backbone" of the oligonucleotide. A major advance came when backbone modifications such as phosphorothioates and methylphosphonates were introduced.

Both these modifications successfully increased the stability of oligonucleotides. The methylphosphonate modification also increased the lipophilicity relative to the parent phosphodiester molecules. Importantly, neither of these modifications interferes with the Watson-Crick base pairing that is essential for antisense mechanism of action. However, of these two first generation chemical modifications, phosphorothioates have proven to be the most useful. Having overcome the issue of stability, modifications designed to increase the affinity of the antisense molecule for its target mRNA were tested. Hundreds of chemical modifications including modifications to the sugar component, the bases and additional backbone modifications have been tested by researchers in recent years. The most promising of these seems to be modification of the sugar moiety such as 2'-O-methyl or 2'-O-propyl modifications. Introduction of 2'-O-methyl nucleotides into an oligonucleotide significantly increases the affinity of the antisense for its target (for review see ref. 49). This modification also leads to additional stability, increased half-life and a reduction of non-specific in vivo side effects of oligonucleotides. Thus, the significant advances in chemistries have provided us with antisense molecules that are powerful tools for identifying gene function as well as promising therapeutic molecules themselves.

DNA Methyltransferase Inhibitors: Cancer Specificity and Potential Therapeutic Window

An important point to be noted is that methylation-mediated inactivation of tumor suppressor genes in cancer cells is distinctly different from methylation-regulated gene expression in normal cells. The aberrant methylation in cancer blocks expression of a gene that would otherwise be expressed in that cell type and, hence, demethylation leads to regained access by transcription factors and re-expression. In contrast, demethylation of genes not programmed for expression in that cell (i.e., without appropriate transcription machinery) will not lead to re-expression. Thus, it can be viewed that a degree of specificity towards cancer cells and a 'molecular therapeutic window' may exist for demethylation therapies in cancer. This model is supported by experimental results, in particular the work of Bender CM et al where demethylating drugs selectively inhibited the growth of cancer cells over normal cells.[50] In this case the specificity was achieved by reactivation of the tumor suppressor gene p16(*INK4a*), whose inactivation is found only in cancer cells. In addition to the changes in DNA methylation observed in cancer cells, many cancer cells also over-express the DNMT enzymes. Although it is not clear what causes increased DNMT levels in cancer cells, regulation by cell signaling pathways seems to play a role. One of the most common mutations in human cancer occurs in the signal transduction protein RAS—a central point in many pathways leading to tumorigenesis. DNMT1, a downstream target of this signal transduction pathway, may, in fact, be required for transformation by RAS.[38,39,51] Further evidence suggests that other important oncogenes on the ras pathway, such as *fos*, require DNMT1 for cellular transformation [51] such that, when activated by oncogenic signaling pathways or other stimuli, DNMT1 induces tumorogenesis-causing alterations in gene expression. Very recently, it has been shown that the leukemia specific fusion protein, PML-RAR, mediates its oncogenic effects by recruiting DNMT proteins to target gene promoters, thus inappropriately inactivating their transcription.[52] Taken together, these finding demonstrate that many oncogenic pathways can deregulate DNA methylation in cancer cells, resulting in silencing of critical genes that would otherwise prevent development of cancer.

Methylation-Independent Mechanisms of DNMT1 Depletion

The maintenance methyltransferase DNMT1 is clearly required to maintain the aberrant methylation found in cancer cells. It is likely that this function requires the catalytic activity of DNMT1 enzyme. However, DNMT1 is a large (190 kD) protein that contains multiple functional domains of which the catalytic domain represents only 35 kD of the carboxy terminus of the molecule. The remainder of the DNMT1 protein performs various functions through multiple protein-protein interactions.

A potentially central interaction involving DNMT1 is the one with proliferating cell nuclear antigen (PCNA), an essential replication protein.[53] Recently it has been found that the interaction of DNMT1 with PCNA increases its activity towards hemi-methylated DNA substrates[54] suggesting that this protein-protein interaction can regulate DNMT1 enzymatic activity at the point in the cell cycle when it is required the most, at the time of DNA replication. Interestingly, the tumor suppressor protein p21waf1/cip1, a key negative regulator of entry into the DNA synthesis phase of the cell cycle, competes with PCNA for the same binding site on DNMT1.[53] In addition, the DNMT enzymes can recruit various members of the chromatin remodeling HDAC family of enzymes[55] and thus can induce repressed chromatin structures and gene silencing independent of DNA methylation. DNMT1 is also found in complexes with Rb protein. The interaction of DNMT1 with Rb protein inhibits its methyltransferase activity by disruption of DNMT1-DNA complexes.[56] DNMT1 complexed with RB and E2F can also recruit HDAC enzymatic activity to represses transcription from E2F responsive promoters whose expression is associated with cell cycle progression.[57] The maintenance methyltransferase DNMT1 and the de novo enzymes DNMT3a and DNMT3b also interact and recent results suggest that the two processes are co-regulated.[58] The unexpected extent to which mammalian cells have evolved levels of control over the DNA methylation machinery highlights its position as a nodal point for both positive and negative regulation of cell cycle control and as central regulator of gene expression, cellular physiology and cancer pathophysiology. Thus, agents that deplete DNMT1 protein levels such as antisense or siRNA, would not only induce demethylation by loss of DNMT1 catalytic activity, but would also alter cell biology mediated by multiple DNMT1-protein interactions.

Implications of Tumor DNA Methylation in Clinical Oncology

The potential of DNA methyltransferase as front-line cancer therapy capable of re-establishing "normal" gene expression programs by reactivation of silenced tumor suppressor genes in cancer cells is evident. Equally or perhaps more important from a clinical perspective is the prospect that 1) DNMT inhibitors may in some cases sensitize tumor to existing chemotherapeutic agents and 2) that evaluation of patient tumor DNA methylation profiles may allow clinicians to "tailor" the most appropriate treatment for that particular tumor.

DNA Methylation: Diagnostics and Rationally Designed Combination Therapy

Analysis of DNA methylation in biological samples including plasma, serum, sputum, urine and primary tumor tissue is being performed as a means of early detection, prognosis determination and ultimately guiding clinical treatment. Aberrant DNA methylation is an early event in the development of many tumor types and is often present in pre-neoplastic lesions.[59] In these settings DNA methylation analysis may be used to define an "at risk" population to be followed closely for early signs of disease onset or progression. DNA methylation-mediated inactivation of tumor suppressor genes has been associated with poor prognosis for survival and in such cases a more aggressive or entirely modified course of therapy may be indicated. For example, it has been shown in pre-clinical models that methylation-mediated inactivation of the mismatch repair gene hMLH1 renders tumor cells resistant to alkylating chemotherapeutic drugs such as carboplatin, temozolomide and epirubicin and that this resistance can be overcome by reactivation of hMLH1 expression by treatment with demethylating agents.[60]

Analogously, clinical resistance to interferons and retinoids may be overcome with demethylating drugs where methylation-mediated inactivation of components of the JAK/STAT signal transduction pathway[61] or retinoic acid receptors, respectively,[62] are known to be present. More generally, it is possible that methylation-mediated inactivation of various components of the apoptotic machinery or molecules involved in immune surveillance may impart resistance to numerous, commonly used chemotherapeutics. Inactivation of apoptosis effectors such as caspases 8 and 10,[63] DAPK (death associated kinase)[64] and RASSF1[65] by methylation is a common event in many tumor types and is often associated with poor response to chemotherapy and thus poor prognosis for survival.[66] Methylation of genes critical for mitigating proper activation of immune response towards tumor cells is another mechanism by which cancer cells employ gene silencing by DNA methylation to ensure their survival. Re-expression of silenced HLA class I antigens by demethylating agents restores an antigen-specific cytotoxic T-cell response against melanoma cells,[67] suggesting that, in addition to direct effects on cancer cell proliferation and apoptosis, demethylation within tumor cells may be targeted by the host (cancer patient's) immune system.

DNMT Inhibitors: The First of Many Epigenetic Therapeutics

As discussed previously, DNA methylation and HDACs can co-operate to inactivate tumor suppressor genes in cancer cells. HDAC inhibitors are attractive anticancer compounds in their own right (for review see ref. 68). However, an exciting clinical avenue is one that explores the combination of inhibitors regulating these two epigenetic pathways. The combination of 5-aza-dC with the HDAC inhibitor trichostatin (TSA) has shown promising results on cancer cells growing in culture.[8] Human clinical trials combining DNA methylation inhibitors and HDAC inhibitors are currently under way. Perhaps even more exciting is the identification of many new players involved in regulation of what has been termed "the histone code" that may coordinate processes as diverse as epigenetic regulation of gene expression, cell cycle progression, chromosome segregation, and cellular memory (see Fig. 6 and for review ref. 69). It remains to be seen which of these represent the most suitable therapeutic targets and which ones will result in effective cancer treatments.

Clinical Experience with Demethylating Agents for Cancer Therapy

In vitro experiments with agents capable of inhibiting DNA methylation have proven extremely useful in elucidating the biology of this important epigenetic regulator, particularly its central role in oncogenesis. Five inhibitors have entered human clinical testing. Four of these agents (5-azacytidine, 5-aza-2'-deoxycytidine (decitabine), 1-beta-D-arabinofuranosyl-5-aza-cytosine and dihydro-5-aza-cytidine) are analogs of the nucleoside deoxycytidine. The fifth agent, MG98, is a DNMT1 selective second generation antisense molecule.

While Phase I and II clinical trials of decitabine have revealed only mild activity against solid tumors [70-73], this agent has demonstrated more efficacy in certain hematologic malignancies, including different phases of myelodysplastic syndromes (MDS), acute myeloid leukemia (AML) and myeloid blast crisis of chronic myeloid leukemia (CML).[74-76] Phase I and II trials of both 5-aza-C and decitabine have reported response rates ranging from 9-89%, depending on the disease and agent examined. Although side effects of these agents have complicated their application, these results are consistent with the high incidence of tumor suppressor gene methylation in these diseases[77-79] and suggest a useful role for pharmacologic demethylation of DNA in the treatment of hematologic malignancies.

Antisense Cancer Therapy Targeting DNMT1

Antisense oligonucleotides were first brought into the clinic against several targets using first generation chemistry (either partially or fully phosphorothioated). These targets included the mRNA for protein kinase C-alpha, the C-raf-1 oncogene, and bcl-2. Antitumor activity demonstrated using antisense oligonucleotides targeting the mRNA of protein kinase C-alpha[80] (ISIS Pharmaceuticals) or bcl-2[81] (Genta Corporation) gave impetus for the development of novel antisense oligonucleotides against other potentially important cancer targets, including DNA methyltransferase.

While the small molecules 5-aza-C and decitabine have shown some activity in hematologic malignancies, the side effect profiles of these nucleoside analogues are not radically different

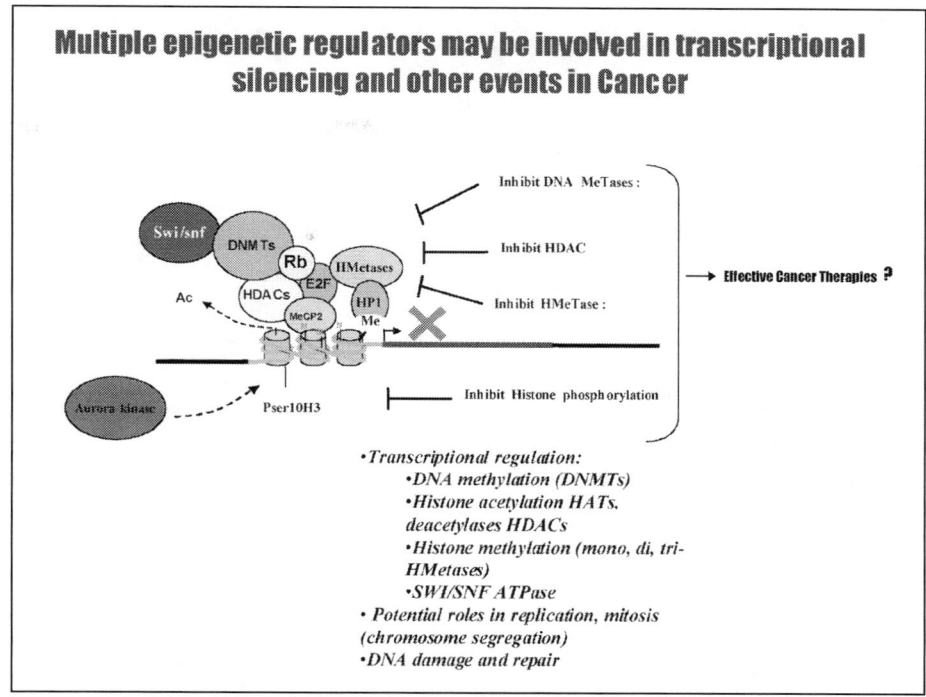

Figure 6. Multiple cellular activities are involved in the regulation of epigenetic information. Coordinated regulation of multiple histone modifications (methylation, acetylation and phosphorylation), DNA methylation and other chromatin modifying activities such as Swi/Snf ATPases are involved in transcriptional control and other processes. Single agents or combinations of agents regulating these may prove to be effective therapies for cancer or other diseases.

from traditional cytotoxic drugs. Their postulated mechanism of action, based on indirect inhibition of methylation via incorporation into DNA, as well as their presumed lack of specificity for the DNMT enzymes may contribute to their toxicity profile. In the light of mounting evidence favoring DNA methyltransferase as a potential target for oncology therapeutics, it became of interest to attempt the design of a specific inhibitor of DNMT1 using novel, second generation antisense technology.

Clinical Development of MG98

As the first antisense compound specifically directed to the DNMT1 target, MG98 was entered into clinical development in mid-1999. Several Phase I and Phase II safety, dose-optimizing and efficacy trials were conducted in patients using intermittent and continuous infusion schedules.[82,83]

As a key component of the clinical evaluation of this novel cancer agent we have studied surrogate pharmacodynamic markers of drug activity. These studies are especially useful when optimizing dose and treatment regimes. In particular, as DNMT1 mRNA is the direct target of MG98, we have monitored DNMT1 mRNA levels in peripheral blood mononuclear cells (PBMCs) and tumor samples where possible in patients prior to and following exposure to MG98. While DNMT1 mRNA levels in PBMCs were significantly suppressed (>35% compared to baseline) in a good proportion of patients, stabilization or regression of disease was only seen in a subset of those patients. A patient who experienced a sustained partial response after 6 months of treatment showed an 80-90% suppression of DNMT1 mRNA in his PBMCs,

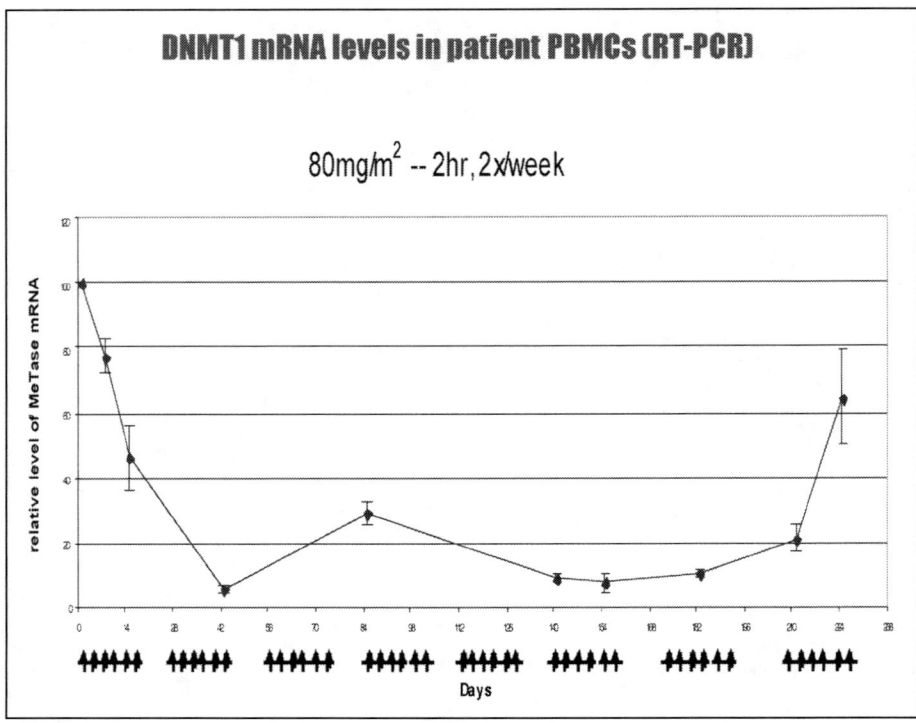

Figure 7. RT-PCR analysis of DNMT1 mRNA levels in PBMCs from MG98 treated patient. Peripheral blood mononuclear cells (PBMCs) were isolated from MG98 treated patients whole blood at the indicated time points. RNA from PBMCs was used to determine DNMT1 mRNA levels by RT-PCR. Such phamacodynamic analysis is useful in optimizing dose and scheduling regimes.

first seen after 6 weeks of treatment and lasting for more than an additional 6 months (see Fig. 7). This time course suggests that there may be a lag between suppression of DNMT1 message and the consequent demethylation and re-expression of tumor suppressor promoters. The ultimate pharmacodynamic marker of demethylating drugs for cancer therapy is presumably the demethylation of DNA within tumor cells. While analysis of DNA methylation from tissue samples is technically challenging, recent technological advances are making this valuable analysis a reality. Among other technologies is the methylation chip-based arrays developed by Epigenomics AG. In collaboration with Epigenomics we have evaluated the methylation status of numerous tumor suppressor genes in tumor DNA samples for baseline and MG98 treated patients. Results suggest MG98 induces demethylation of previously hypermethylated sequences in tumor DNA (Fig. 8).

Side effects seen under either regimen were similar to what had been observed in preclinical testing, and were similar to those seen previously with first generation chemistry antisense molecules, suggesting that these effects are likely to be class- (i.e., oligonucleotide) rather than target-related.

Clinical benefit was seen at doses far below the maximum tolerated dose, suggesting that the traditional cytotoxic chemotherapy paradigm of "more is better" may not hold for this cytostatic agent. Indeed, the fact that regression of tumor was first observed in the patient with a partial response only after 4 cycles of treatment may suggest that the ability of specific DNMT1 inhibitors to affect tumors is predicated on initial control of disease such that demethylation

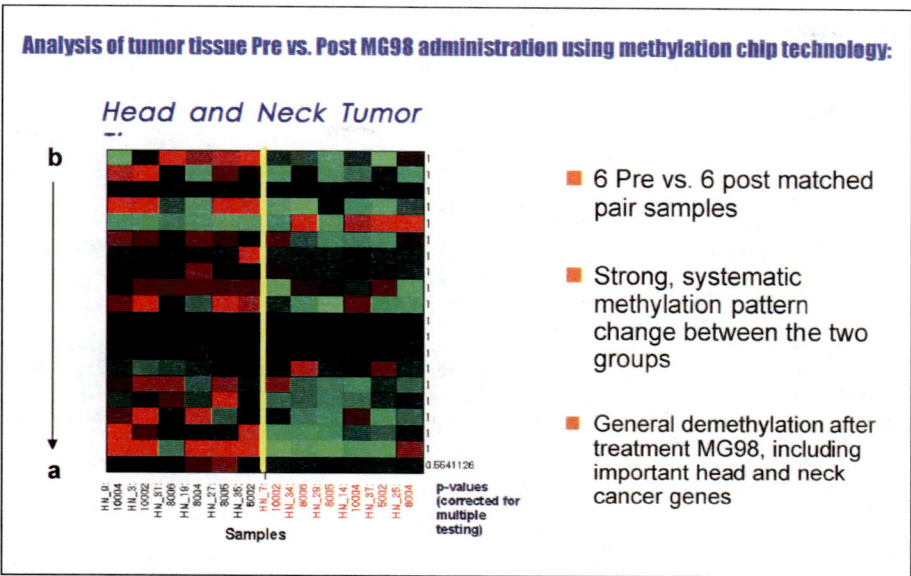

Figure 8. Methylation analysis of tumor tissue before and after MG98 treatment. Tumor samples pre vs. post MG98 administration. Ranked matrix on matched pair samples, discrimination power of CpG positions increases from b to a. Only the top scoring 20 CpGs shown (out of 256 analysed). On the left side 6 samples taken from the pre administration group are displayed with black colored identifiers, on the right 6 post administration samples are displayed with red colored identifiers. The vertical yellow bar highlights the grouping structure. Labels on the right: p-values adjusted for multiple testing using the single step Bonferroni method obtained from Wilcoxon rank statistic.

and reactivation of tumor suppressor genes by such molecules can occur before progression of disease. In addition to the ongoing DNMT1 mRNA evaluations in PBMCs and tumor samples, the effects of MG98 on the methylation status of many important growth regulation genes, especially tumor suppressor genes, before and after exposure to MG98 has been examined in tumor biopsy samples. Using a chip array of 64 genes identified as important in growth control, including many known tumor suppressor genes, DNA from patient samples have been tested for their methylation status pre- and post-exposure to MG98. Analysis of the results is ongoing. It is hoped that epigenetic methylation profiling of this sort may allow for the pharmacogenetic identification of patients most likely to respond to MG98 therapy.

Conclusions and Future Perspectives

DNA methyltransferases have been shown to be potentially important targets in the treatment of cancer. MG98, a second generation antisense oligonucleotide to DNMT1 mRNA, has shown that it can be given safely and that it may have antitumor activity in humans when administered on its own. Further development will focus on optimizing dose and schedule, and will likely include evaluation of combination therapy with currently available drugs. One goal may be to optimize treatment such that MG98 is given sufficient time to exert its effects, in the light of accumulated clinical experience and published preclinical evidence that antitumor activity of demethylating agents may be subject to a latency period during which demethylation and reactivation of silenced suppressor genes occurs prior to demonstration of clinical benefit.

References

1. Cohen P. Protein kinases—the major drug targets of the twenty-first century? Nat Rev Drug Discov 2002; 1(4):309-15.
2. Keshet I, Lieman-Hurwitz J, Cedar H. DNA methylation affects the formation of active chromatin. Cell 1986; 44:535-43.
3. Li E, Beard C, Jaenisch R. Role for DNA methylation in genomic imprinting. Nature 1993; 366:362-65.
4. Shemer R et al. Dynamic methylation adjustment and counting as part of imprinting mechanisms. Proc Natl Acad Sci USA 1996; 93:6371-76.
5. Selig S, Ariel M, Goitein R et al. Regulation of mouse satellite DNA replication time. EMBO J 1988; 7:419-26.
6. Yeivin A, Razin A. Gene methylation patterns and expression. EXS 1993; 64:523.
7. Robertson KD. DNA methylation and chromatin unraveling the tangled web. Oncogene 2002; 21(35):5361-79.
8. Cameron EE, Bachman KE, Myohanen S et al. Synergy of demethylation and histone deacetylase inhibition in the re-expression of genes silenced in cancer. Nat Genet 1999; 21:103-107.
9. Lapeyre JN, Becker FF. 5-Methylcytosine content of nuclear DNA during chemical hepatocarcinogenesis and in carcinomas which result. Biochem Biophys Res Commun 1979; 87(3):698-705.
10. Ghoshal AK, Farber E. The induction of liver cancer by dietary deficiency of choline and methionine without added carcinogens.
11. Lombardi B, Shinozuka H. Enhancement of 2-acetylaminofluorene liver carcinogenesis in rats fed a choline-devoid diet. Int J Cancer 1979; 23(4):565-70.
12. Feinberg AP, Gehrke CW, Kuo KC et al. Reduced genomic 5-methylcytosine content in human colonic neoplasia. Cancer Res 1988; 48(5):1159-61.
13. Feinberg AP, Vogelstein B. Alterations in DNA methylation in human colon neoplasia. Semin Surg Oncol 1987; 3(3):149-51.
14. Goelz SE, Vogelstein B, Hamilton SR et al. Hypomethylation of DNA from benign and malignant human colon neoplasms. Science 1985; 228(4696):187-90.
15. Land H, Parada LF, Weinberg RA. Cellular oncogenes and multistep carcinogenesis. Science 1983; 222(4625):771-8.
16. Feinberg AP, Vogelstein B. Hypomethylation of ras oncogenes in primary human cancers. Biochem Biophys Res Commun 1983; 111(1):47-54.
17. Feinberg AP, Vogelstein B. Hypomethylation distinguishes genes of some human cancers from their normal counterparts. Nature 1983; 301(5895):89-92.
18. Christman JK, Price P, Pedrinan L et al. Correlation between hypomethylation of DNA and expression of globin genes in Friend erythroleukemia cells. Eur J Biochem 1977; 81(1):53-61.
19. Constantinides PG, Jones PA et al. Functional striated muscle cells from non-myoblast precursors following 5-azacytidine treatment. Nature 1977; 267(5609):364-6.
20. Jones PA, Taylor SM. Cellular differentiation, cytidine analogs and DNA methylation. Cell 1980; 20(1):85-93.
21. Taylor EM, McFarlane RJ, Price C. Mol Gen Genet 1996; 253.
22. Lal D, Som S, Friedman S. Mutat Res 1988; 193.
23. Tamame M, Antequera F, Villanueva JR et al. Mol Cell Biol 1983; 3:2287.
24. Esteller M. CpG island hypermethylation and tumor suppressor genes: a booming present, a brighter future. Oncogene 2002; 21(35):5427-40.
25. Myohanen SK, Baylin SB, Herman JG. Hypermethylation can selectively silence individual p16ink4aink4A alleles in neoplasia. Cancer Res 1998; 58(4):591-3.
26. Adams RL, McKay EL, Craig LM et al. Mouse DNA methylase: methylation of native DNA. Biochim Biophys Acta 1979; 561:345-57.
27. Bestor T, Laudano A, Mattaliano R et al. Cloning and sequencing of a cDNA encoding DNA methyltransferase of mouse cells. The carboxyl-terminal domain of the mammalian enzymes is related to bacterial restriction methyltransferases. J Mol Biol 1988; 203(4):971-83.
28. Yoder JA, Soman NS, Verdine GL et al. DNA (cytosine-5)-methyltransferases in mouse cells and tissues. Studies with a mechanism-based probe. J Mol Biol 1997; 270:385-95.
29. Okano M, Xie S, Li E. Cloning and characterization of a family of novel mammalian DNA (cytosine-5) methyltransferases. Nat Genet 1998; 19:219-20.
30. Li E, Bestor TH, Jaenisch R. Targeted mutation of the DNA methyltransferase gene results in embryonic lethality. Cell 1992; 69:915-26.
31. Okano M, Bell DW, Haber DA et al. DNA methyltransferases Dnmt3a and Dnmt3b are essential for de novo methylation and mammalian development. Cell 1999; 99:247-57.

32. Robertson KD, Uzvolgyi E, Liang G et al. The human DNA methyltransferase (DNMTs) 1, 3a and 3b: coordinate mRNA expression in normal tissues and overexpression in tumors. Nucleic Acids Res 1999; 27:2291-98.

33. Hansen RS, Wijmenga C, Luo P et al. The DNMT3B DNA methyltransferase gene is mutated in the ICF immunodeficiency syndrome. Proc Natl Acad Sci USA 1999; 96:14412-7.

34. Beaulieu N, Morin S, Chute IC et al. An essential role for DNA methyltransferase DNMT3B in cancer cell survival. J Biol Chem 2002; 277(31):28176-81

35. Galmarini CM, Mackey JR, Dumontet C. Nucleoside analogues and nucleobases in cancer treatment. Lancet Oncol 2002; 3(7):415-24.

36. Juttermann R, Li E, Jaenisch R. Toxicity of 5-aza-2'-deoxycytidine to mammalian cells is mediated primarily by covalent trapping of DNA methyltransferase rather than DNA demethylation. Proc Natl Acad Sci USA 1999; 91:1797-801.

37. Baylin S, Bestor TH. Altered methylation patterns in cancer cell genomes: cause or consequence? Cancer Cell 2002; 1(4):299-305.

38. Rouleau J, MacLeod AR, Szyf M. Regulation of the DNA methyltransferase by the Ras-AP-1 signaling pathway. J Biol Chem 1995; 270(4):1595-601

39. MacLeod RA, Rouleau J, Szyf M. Regulation of DNA methylation by the Ras signaling. J Biol Chem 1995; 270:11327-337.

40. MacLeod RA, Szyf M. Expression of antisense to DNA methyltransferase mRNA induces DNA demethylation and inhibits tumorigenesis. J Biol Chem 1995; 270:8037-43.

41. Ramchandani S, MacLeod AR, Pinard M et al. Inhibition of tumorigenesis by a cytosine-DNA, methyltransferase, antisense oligodeoxynucleotide. Proc Natl Acad Sci USA 1997; 94(2):684-9

42. Laird PW, Jackson-Grusby L, Fazeli A et al. Suppression of intestinal neoplasia by DNA hypomethylation. Cell 1995; 81:197-205.

43. Hussussian CJ et al. Germline p16ink4a mutations in familial melanoma. Nat Genet 1994; 8:15-21.

44. Gonzalez-Zulueta M et al. Methylation of the 5' CpG island of the p16ink4a/CDKN2 tumor suppressor gene in normal and transformed human tissues correlates with gene silencing. Cancer Res 1995; 55:4531-5.

45. Fournel M, Sapieha P, Beaulieu N et al. Down-regulation of human DNA-(cytosine-5) methyltransferase induces cell cycle regulators p16ink4a(ink4A) and p21(WAF/Cip1) by distinct mechanisms. J Bio Chem 1999; 274:24250-56.

46. Rhee I, Jair KW, Yen RW et al. CpG methylation is maintained in human cancer cells lacking DNMT1. Nature 2000; 404(6781):1003-7.

47. Rhee I et al. DNMT1 and DNMT3b cooperate to silence genes in human cancer cells. Nature 2002; 416(6880):552-6.

48. Robert MF, Morin S, Beaulieu N et al. DNMT1 is required to maintain CpG methylation and aberrant gene silencing in human cancer cells. Nat Genet 2003; 33(1):61-5.

49. Opalinska JB, Gewirtz AM. Nucleic-acid therapeutics: basic principles and recent applications. Nat Rev Drug Discov 2002; 1(7):503-14.

50. Bender CM, Pao MM, Jones PA. Inhibition of DNA methylation by 5-aza-2'-deoxycytidine suppresses the growth of human tumor cell lines. Cancer Res 1998; 58(1):95-101.

51. Bakin AV, Curran T. Role of DNA 5-methylcytosine transferase in cell transformation by fos. Science 1999; 283:387-90.

52. Di Croce L, Raker VA, Corsaro M et al. Methyltransferase recruitment and DNA hypermethylation of target promoters by an oncogenic transcription factor. Science 2002; 295:1079-82.

53. Chuang LS, Ian HI, Koh TW et al. Human DNA-(cytosine-5) methyltransferase-PCNA complex as a target for p21WAF1. Science 1997; 277(5334):1996-2000.

54. Iida T, Suetake I, Tajima S et al. PCNA clamp facilitates action of DNA cytosine methyltransferase 1 on hemimethylated DNA. Genes Cells 2002; 10:997-1007.

55. Rountree MR, Bachman KE, Baylin SB. DNMT1 binds HDAC2 and a new co-repressor, DMAP1, to form a complex at replication foci. Nat Genet 2000; 25(3):269-77.

56. Pradhan S, Kim GD. The retinoblastoma gene product interacts with maintenance human DNA (cytosine-5) methyltransferase and modulates its activity. EMBO J 2002; 21(4):779-88.

57. Robertson KD, Ait-Si-Ali S, Yokochi T et al. DNMT1 forms a complex with Rb, E2F1 and HDAC1 and represses transcription from E2F-responsive promoters. Nat Genet 2000; 25(3):338-42.

58. Fatemi M, Hermann A, Gowher H et al. Dnmt3a and Dnmt1 functionally cooperate during de novo methylation of DNA. Eur J Biochem 2002; 269(20):4981-4.

59. Belinsky SA, Palmisano WA, Gilliland FD et al. Aberrant promoter methylation in bronchial epithelium and sputum from current and former smokers. Cancer Res 2002; 62(8):2370-7.

60. Plumb JA, Strathdee G, Sludden J et al. Reversal of drug resistance in human tumor xenografts by 2'-deoxy-5-azacytidine-induced demethylation of the hMLH1 gene promoter. Cancer Res 2000; 60(21):6039-44.
61. Yoshikawa H, Matsubara K, Qian GS et al. SOCS-1, a negative regulator of the JAK/STAT pathway, is silenced by methylation in human hepatocellular carcinoma and shows growth-suppression activity. Nat Genet 2001; 28(1):29-35.
62. Niitsu N, Hayashi Y, Sugita K et al. Sensitization by 5-aza-2'-deoxycytidine of leukaemia cells with MLL abnormalities to induction of differentiation by all-trans retinoic acid and 1alpha,25-dihydroxyvitamin D3. Br J Haematol 2001; 112(2):315-26.
63. Harada K, Toyooka S, Shivapurkar N et al. Deregulation of caspase 8 and 10 expression in pediatric tumors and cell lines. Cancer Res 2002; 62(20):5897-901.
64. Esteller M, Corn PG, Baylin SB et al. A gene hypermethylation profile of human cancer. Cancer Res 2001; 61(8):3225-9.
65. Dammann R, Li C, Yoon JH et al. Epigenetic inactivation of a RAS association domain family protein from the lung tumour suppressor locus 3p21.3. Nat Genet 2000; 25(3):315-9.
66. Burbee DG, Forgacs E, Zochbauer-Muller S et al. Epigenetic inactivation of RASSF1A in lung and breast cancers and malignant phenotype suppression. J Natl Cancer Inst 2001; 93(9):691-9.
67. Serrano A, Tanzarella S, Lionello I et al. Rexpression of HLA class I antigens and restoration of antigen-specific CTL response in melanoma cells following 5-aza-2'-deoxycytidine treatment. Int J Cancer 2001; 94(2):243-51.
68. Johnstone RW. Histone-deacetylase inhibitors: novel drugs for the treatment of cancer. Nat Rev Drug Discov 2002; 1(4):287-99.
69. Jenuwein T, Allis CD. Translating the histone code. Science 2001; 293(5532):1074-80.
70. Abele R, Clavel M, Dodion P et al. The EORTC Early Clinical Trials Cooperative Group experience with 5-aza-2'-deoxycytidine (NSC 127716) in patients with colo-rectal, head and neck, renal carcinomas and malignant melanomas. Eur J Cancer Clin Oncol 1987; 12:1921-4.
71. Momparler RL, Bouffard DY, Momparler LF et al. Pilot phase I-II study on 5-aza-2'-deoxycytidine (Decitabine) in patients with metastatic lung cancer. Anticancer Drugs 1997; 4:358-68.
72. Thibault A, Figg WD, Bergan RC et al. A phase II study of 5-aza-2'deoxycytidine (decitabine) in hormone independent metastatic (D2) prostate cancer. Tumori 1998; 1:87-9.
73. Schwartsmann G, Schunemann H, Gorini CN et al. A phase I trial of cisplatin plus decitabine, a new DNA-hypomethylating agent, in patients with advanced solid tumors and a follow-up early phase II evaluation in patients with inoperable non-small cell lung cancer. Invest New Drugs 2000; 1:83-91.
74. Silverman LR, Demakos EP, Peterson B et al. The CALGB, Chicago, IL. A randomized controlled trial of subcutaneous azacitidine (AZA C) in patients with the myelodysplastic syndrome (MDS). A study of the cancer and leukemia GROUP B (CALGB). Proc ASCO 1998.
75. Santini V, Kantarjian HM, Issa JP. Changes in DNA Methylation in Neoplasia: Pathophysiology and Therapeutic Implications. Ann Intern Med 2001; 134:573-86.
76. Wijermans P, Lubbert M, Verhoef G et al. Low-dose 5-aza-2'-deoxycytidine, a DNA hypomethylating agent, for the treatment of high-risk myelodysplastic syndrome: a multicenter phase II study in elderly patients. J Clin Oncol 2000; 5:956-62.
77. Aoki E et al. Methylation status of the p15[INK4B] gene in hematopoietic progenitors and peripheral blood cells in myelodysplastic syndromes. Leukemia 2000; 14:586-93.
78. Quesnel B, Fenaux P: P15[INK4b] gene methylation and myelodysplastic syndromes. Leuk Lymphoma 1999; 35:437-43.
79. Pinto A, Zagonel V. 5-Aza-2'-deoxycytidine (Decitabine) and 5-azacytidine in the treatment of acute myeloid leukemias and myelodysplastic syndromes: past, present and future trends. Leukemia 1993; 7(Suppl 1):51-60.
80. Yuen A, Halsey J, Fisher G et al. Phase I/II Trial of ISIS 3521, an Antisense Inhibitor of PKC-Alpha, with Carboplatin and Paclitaxel in Non-Small Cell Lung Cancer. Proc ASCO 2001; 20:1234.
81. Morris MJ et al. Clin Phase I trial of BCL-2 antisense oligonucleotide (G3139) administered by continuous intravenous infusion in patients with advanced cancer. Cancer Res 2002; 8(3):679-83.
82. Stewart DJ, Donehower R C, Eisenhauer EA et al. Phase I study of the DNA methyltransferase 1 inhibitor MG98 administered twice weekly as a two hour intravenous infusion. Ann Onc 2003; 14:766-774.
83. Davis AJ, Gelmon KA, Siu LL et al. Phase I and Pharmacologic Study of the Human DNA Methyltransferase Antisense Oligodeoxynucleotide MG98 Given as a 21-Day Continuous Infusion Every 4 Weeks. Invest New Drugs 2003; 21:85-97.

CHAPTER 15

Preclinical and Clinical Studies on 5-Aza-2'-Deoxycytidine, a Potent Inhibitor of DNA Methylation, in Cancer Therapy

Richard L. Momparler

Abstract

The preclinical and clinical investigations by the author on the antineoplastic activity of 5-Aza-2'-deoxycytidine (5AZA), a potent inhibitor of DNA methylation are reviewed. These include studies on the molecular, cellular and animal pharmacology of 5AZA. These preclinical studies indicated that 5AZA has enormous potential in cancer therapy. This potential is supported by reports in the literature indicate that 5AZA can reactivate many different types of genes that suppress tumorigenesis and were silenced by aberrant DNA methylation. However, the potential still remains to be demonstrated in clinical investigations where the author observed interesting responses in both patients with leukemia and lung cancer. Several suggestions are made concerning the design of the optimal dose-schedule for 5AZA in cancer therapy. Preclinical studies show that 5AZA in combination with inhibitors of histone deacetylase (HDI) show a synergistic interaction against neoplastic cells. 5AZA plus HDI may have the potential to be a very effective chemotherapeutic regimen in patients with cancer.

Historical Perspective

My interest in the potential use of 5-Aza-2'-deoxycytidine (5AZA) in cancer therapy was a result from my investigations on the pharmacology of the related analog, cytosine arabinoside (ARA-C). These latter studies on ARA-C included investigations on drug resistance,[1] activation of the prodrug by deoxycytidine kinase[2] and the mechanism of action as a chain terminator.[3] These studies lead me to propose the use in high dose ARA-C therapy of leukemia which was tested for the first time in a patient with relapsed leukemia in Montreal.[4] In 1973 I joined the Department of Hematology at Children's Hospital of Los Angeles. The head of Hematology at this time was Dr Myron Karon, a brilliant clinician who also had experience working in the laboratory on the pharmacology of antineoplastic agents. He had the intuition that important advances could be made doing translational research, taking discoveries from the bench to the clinic. At this time Dr Karon had just completed a clinical trial on 5-azacytidine in children with relapsed leukemia.[6] He encouraged me to perform preclinical investigations on 5-azacytidine. From my studies on ARA-C I realized that analogs of deoxycytidine had much more potential in cancer therapy than the cytidine analogs, which were much more toxic due to their incorporation into RNA and they lacked S phase specificity. The group at the Czechoslovak Academy of Science in Prague reported in the literature that the related deoxycytidine analog, 5AZA, showed significant activity in the mouse model of leukemia.[5] I was fortunate to obtain some 5AZA from Dr J. Vesely during a visit to Prague. When I tested 5AZA in the mouse model with L1210 leukemia, it was much more potent than ARA-C.[7]

DNA Methylation and Cancer Therapy, edited by Moshe Szyf. ©2005 Eurekah.com and Kluwer Academic/Plenum Publishers.

The initial investigations on the antineoplastic activity of 5AZA that showed so much promise motivated me to make every effort to bring this agent into clinical trial in patients with leukemia. I performed toxicology studies in animals to prepare a new drug application.[8,9] Realizing that it would be very difficult to obtain permission from the FDA in the USA for a clinical trial on an orphan drug, like 5AZA, I moved back to Canada to join the Research Centre at Hôpital Sainte Justine and the Department of Pharmacology, Unviersité de Montréal. With the help of my clinical colleague, Dr Georges E. Rivard, we succeeded in obtaining permission from the health authorities in Ottawa to perform a clinical trial in children with leukemia. An excellent chemist in Prague, Dr A. Piskala, performed the chemical synthesis of 5AZA for our clinical trial. In patients with very advanced leukemia we were successful in obtaining some very interesting responses, including complete remissions.[10] In collaboration with Dr Martin Gyger of Hôpital Maisonneuve-Rosemont in Montreal we also tested 5AZA in adult patients with adult leukemia and also obtained some complete remissions.[11] The durations of the remissions using a very conservative dose-schedule of 5AZA were short. The research grant to support our study was not renewed and the clinical studies were terminated.

Fortunately, interest in 5AZA was growing in Europe. Dr Dick deVos of Pharmachemie BV in Holland was interested in the potential of 5AZA in cancer therapy and provided funds to continue our clinical trials. This company also supported clinical trials on cancer in Europe by other groups. The data from my in vitro and in vivo animal studies on tumors indicated that 5AZA also had potential for tumor therapy. With funds provided by Pharmachemie I initiated a clinical trial on patients with metastatic nonsmall cell lung cancer (NSCLC) in collaboration with Dr Joseph Ayoub at Hôpital Notre Dame in Montreal. We performed a pilot study on these patients with 5AZA.[12,13] Using a conservative dose-schedule of 5AZA we obtained some interesting responses including one patient with NSCLC survived over 6 years, which is remarkable for this disease where the historical survival rate is less than 2 years. Grant applications to continue this study were not funded and so the study was terminated. Recently, Pharmachemie BV sold the rights of 5AZA to SuperGen, USA who have provided us with funds to continue our clinical trial on 5AZA in patients with NSCLC. The recent reports that 5AZA can activate so many different classes of genes that suppress neoplastic progression and are silenced by aberrant methylation has revived the interest in the potential of this analog in cancer therapy.

Pharmacology of 5-Aza-2'-Deoxycytidine (5-AZA)[14]

Chemistry

The chemical synthesis of 5AZA (Fig. 1) was first reported by Pliml and Sorm.[15] 5AZA is chemically unstable, especially in an alkaline environment in which an opening of the 5-azacytosine ring occurs between positions 1 and 6 followed by decomposition.[16] At pH 7 and 37°C the decomposition half-life of 5AZA is about 12 hr. The stability of 5AZA is markedly increased by lowing the temperature. For administration to patients, 5AZA should be prepared in a solution at pH ~7 and kept at about 5-10°C. Under these conditions there is minimal decomposition during an 8 hour infusion.

Metabolism

5AZA is a prodrug that is activated by phosphorylation by deoxycytidine kinase.[17] Other kinases in the cell rapidly convert the monophosphate (5AZA-dCMP) to its triphosphate form (5AZA-dCTP). This latter nucleotide analog has a similar Km for DNA polymerase as the natural substrate, dCTP, and is rapidly incorporated into DNA.[18] Deamination of 5AZA by cytidine deaminase[19] or 5AZA-dCMP by dCMP deaminase[20] results in a loss of antineoplastic activity.

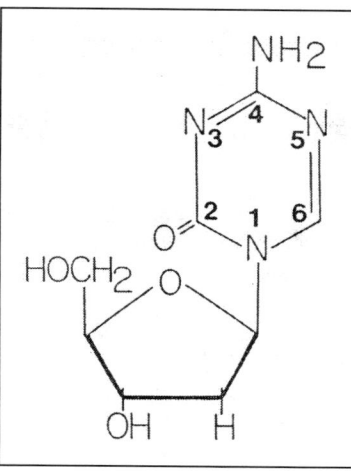

Figure 1. Chemical structure of 5-aza-2'-deoxycytidine (5AZA). The numbers in the 5-azacytosine ring indicate the positions of the different atoms. The only difference between 2'-deoxycytidine and 5AZA is the presence of a nitrogen atom at position 5. The 5 position of cytosine is the site of methylation in genomic DNA.

Molecular Pharmacology

The incorporation of 5AZA into CpG sites during the replication of the daughter strand of DNA results in a potent inhibition of DNA methylation[18] as a result of the inactivation of DNA methylase 1 (Fig. 2) due to a formation of a covalent bond between this enzyme and the 5-azacytosine ring.[21] Depletion of DNA methylase activity in the cell results in global hypomethylation of the genome. During a second cell division in the absence of 5AZA most of the hypomethylated CpG sites are remethylated, except the CpG sites that contain 5-azacytosine.

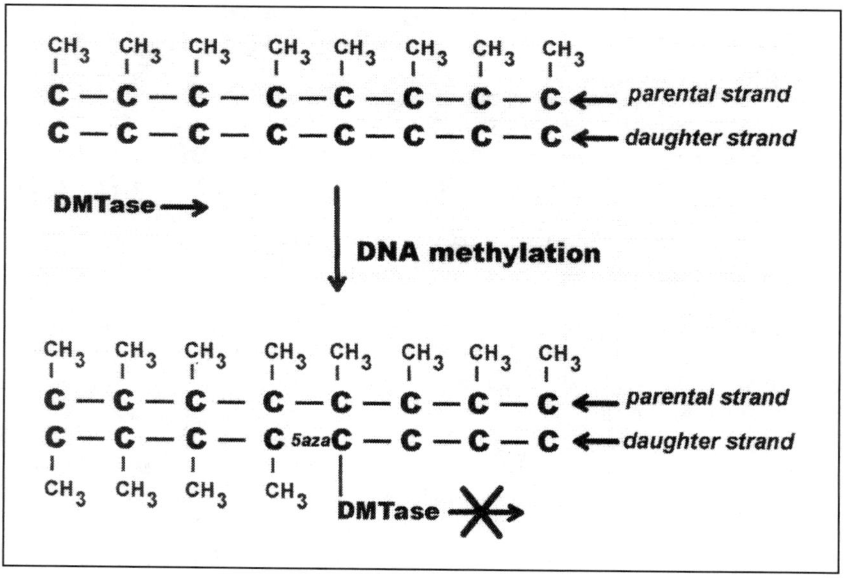

Figure 2. Inhibition of DNA methylation by 5AZA. After DNA replication DNA methylase 1 (DMTase) methylates the cytosines in the daughter strand that contain a complementary 5-methylcytosine in the parental strand. The incorporation of 5AZA into designated methylation sites results in the covalent linking of DMTase to 5AZA, which result in enzyme inactivation. Due to the inactivation of DMTase the downstream cytosines in the daughter strand are not methylated.

The demethylation action of 5AZA on genes that suppress tumorigenesis and that were previously silenced by aberrant methylation can result in their reactivation (see related chapters for more details). Although not fully understood, this reactivation of tumor suppressor genes plays an important role with respect to the antineoplastic action of 5AZA.

Cellular Pharmacology

5AZA is a S phase specific agent. This means that 5AZA is only pharmacologically active in cells during the S phase of the cell cycle.[22] In addition, 5AZA does not block the progression of cells from G1 phase into S phase. In many tumor cell lines there is a delay in the onset of growth and DNA synthesis inhibition during 5AZA treatment. In some cell lines this inhibition becomes apparent only after 2 or 3 cell divisions. For most human leukemic and tumor cell lines the IC50 (concentration that produces 50% response) for the loss of clonogenicity are in the range of 10 to 50 ng/ml for a 48 hr drug exposure.[23,24] The longer the drug exposure the greater the antineoplastic activity of 5AZA. In mammalian cell assays 5AZA showed no significant mutagenic activity.[22]

Drug Resistance

Cells that lack deoxycytidine kinase, the enzyme that activates 5AZA, are completely resistant to this analog.[25] Cells with increased activity of cytidine deaminase, the enzyme that inactivates 5AZA, show signs of drug resistance 5AZA.[26] Another possible mechanism of resistance to 5AZA is an increased intracellular pool of dCTP.[27] The enhanced level of dCTP will compete with 5AZA-dCTP for incorporation into DNA and reduce the phosphorylation of 5AZA by deoxycytidine kinase since dCTP is a feedback inhibitor of this enzyme.[2] Since cells contain two alleles for the deoxycytidine kinase gene, both alleles have to be inactivated for the cells to become completely resistant to 5AZA, a very rare event.

Evaluation of Antineoplastic Activity in Animal Models

Leukemia

In order to have an animal model that was as close as possible to the clinical disease in patients, we used an iv injection of L1210 leukemia cells and administration of 5AZA by continuous iv infusion.[28] Pathological analysis of the mice with L1210 leukemia showed a histology in the different organs that was identical to the human disease. Using the L1210 leukemia model we observed that 5AZA was more potent than ARA-C and very much more potent than its related riboside analog, 5-azacytidine.[7] In some experiments depending on the stage of the L1210 leukemia we observed that the maximal tolerated dose 5AZA, but not ARA-C could cure the mice. The differences in the antileukemic activity of these two deoxycytidine analogs could not be due to differences in metabolism since both analogs are metabolized by the same enzymes. The only difference between the two analogs is their mechanism of action. ARA-C is a potent inhibitor of DNA replication whereas 5AZA is a potent inhibitor of DNA methylation.

The antileukemic activity of 5AZA in the L1210 leukemia model increased with the dose and duration of the iv infusion. We observed a very significant correlation between the antileukemic activity at different dose levels of 5AZA and the extent of inhibition of global DNA methylation in the leukemic cells.[29] A dose of 21.4 mg/kg administered as a 15 hour iv infusion showed curative potential in this leukemia model. The estimated steady state plasma level of 5AZA in this experiment was 1.1 μg/ml. This observation supports the hypothesis that the antileukemic action of 5AZA is related to the extent of its inhibition of DNA methylation.

Tumors

Using an in vitro clonogenic assay we observed that all the human tumor cell lines that we tested, regardless of their phenotype, were sensitive to the antineoplastic action of 5AZA. Since silencing of tumor suppressor genes by aberrant methylation has been reported for all tumor types, it may explain why all the tumor cells lines showed drug sensitivity to 5AZA. We also evaluated the antitumor activity of 5AZA in the mouse model with EMT6 mammary tumor.[30] The differentiation action of 5AZA has the potential to convert a malignant tumor to a "benign" tumor with minimal reduction in tumor size. For this reason we designed an in vivo-in vitro model to obtain a more precise evaluation of the antitumor activity produced by 5AZA. Following 5AZA treatment of mice with EMT6 tumor, the tumor was excised, the tumor cells disaggregated and plated in petri dishes to measure survival by a colony assay. With this type of assay we observed that the antitumor activity of 5AZA increased with the dose and duration of treatment. The "curative" dose-schedule in this tumor model was 30 mg/kg for 18-hour infusion. The estimated steady state plasma level of 5AZA in this experiment was 1.3 μg/ml.

Toxicity[8,9]

The major toxicity produced by 5AZA in animal models is leukopenia and intestinal ulceration. This side effect is expected since both these organ systems show very high incidence of rapid cell turnover and proliferation. Due to the S phase specificity of 5AZA rapidly proliferating tissue is very sensitive to the inhibitory action of this analog. Apparently, 5AZA induces the proliferating stem hematopoietic cell to undergo terminal differentiation losing their capacity for self-renewal. However, the resting hematopoietic stem cells escape the toxic effects of 5AZA due to their presence in a Go or G1 phase. The onset of leukopenia stimulates the recruitment of these resting hematopoietic stem cells to enter the cell cycle leading to full recovery of the hematopoietic system 3 to 4 weeks after treatment with 5AZA. A similar chain of events following 5AZA treatment occurs for the intestinal stem cells located in the crypts of the small intestine. In the mouse model the LD50 of 5AZA for an 8-hour infusion is in the range of 54 mg/kg. For the schedule of daily iv injections of 5AZA for 5 days the estimated LD50 was 15.5 mg/kg. Dogs administered a 12 h iv infusion of 5AZA at a dose of 3 mg/kg showed a marked leukopenia and thrombocytopenia with complete recovery by day 20.

Clinical Trials in Leukemia[10,11]

Our preclinical data on the antileukemic activity of 5AZA in mice and the toxicology in mice and dogs was sufficient to prepare a new drug application for a clinical trial on this analog in children with relapsed leukemia. Since a 12 hour iv infusion was used in most of our mouse experiments we started a phase I study using the same infusion time and in a stepwise manner increased the infusion time to 18, 24, 30, 36 hours, etc. The first remission was obtained with an infusion time of 36 hours with a total dose of 36 mg/kg. This clinical trial was also extended to adults with relapsed leukemia where the infusion time was extended to 60 hours with a total dose of 67 mg/kg. An example of a response to the 5AZA therapy in a patient with acute myeloid leukemia is shown in Figure 3. In some patients who went into remission, bone marrow analysis at 21 days after 5AZA showed extensive leukemic blast cells indicating a delayed action of this analog as we observed in our in vitro studies. In comparison, when high dose ARA-C induces a complete remission in patients, the bone marrow at day 21 is usually completely cleared of leukemic blasts. This illustrates the cell kinetic differences in the action between these two deoxycytidine analogs.

The remission rate in these studies which included both acute lymphoblastic leukemia and acute myeloid leukemia was in the range of 28 %. The remissions were of short duration. This response rate was encouraging taking into consideration that most of the patients prior to entering the 5AZA study received extensive chemotherapy with a variety of different antineoplastic agents. In some patients we were capable of demonstrating that the 5AZA therapy produced a significant inhibition of DNA methylation in the blood leukemic blast cells.[31]

One possible explanation for the short remissions is that the duration of the 5AZA therapy was too short to permit all of the leukemic blasts to enter the S phase. Another explanation is that the prior treatments of the patients with ARA-C lead to the survival of leukemic cells resistant to deoxycytidine analogs.[27] In order to test this possibility we performed in vitro drug sensitivity tests before and after treatment with 5AZA. Indeed we detected in two patients signs

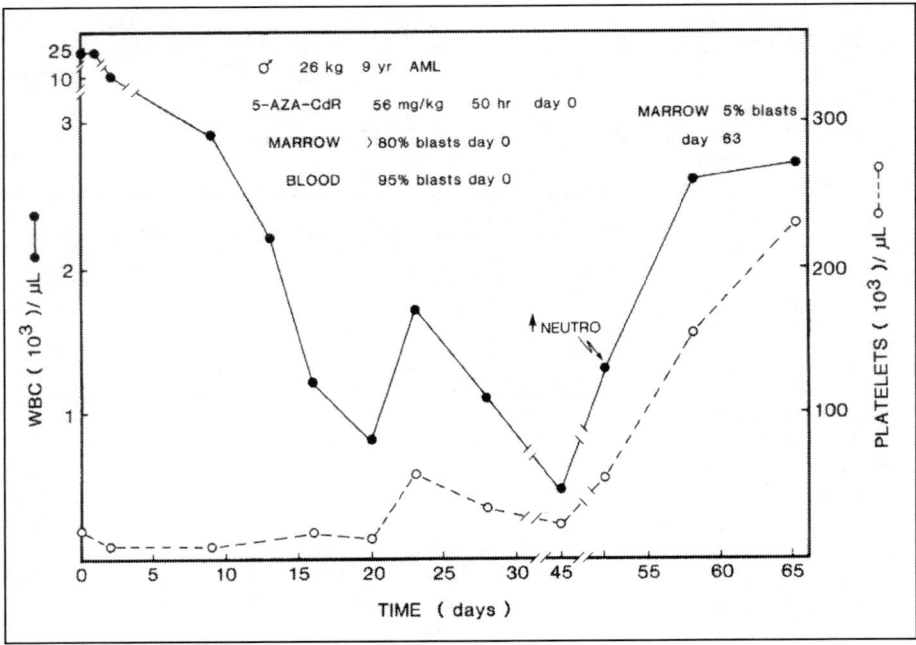

Figure 3. Hemogram of a pediatric male patient (age 9 yr) with acute myeloid leukemia (AML) after treatment with a continuous 50 h iv infusion of 5AZA at a total dose of 56 mg/kg. The blood and the bone marrow before start of treatment contained 95% and >80% leukemic blasts, respectively. At day 63 after treatment the bond marrow contained 5% leukemia blasts. The neutrophhil count showed a significant increase at day 48 after treatment. Before treatment the platelet count was <30,000/µl and increased to >200,000/µl by day 65 after treatment. This figure illustrates the kinetics of reduction of leukemic blasts and the recovery of the normal hematopoiesis after 5AZA treatment. It is possible that in this phase I study the duration of the 5AZA infusion (50 h) was probably too short to permit all the leukemic blasts to enter S phase so as to produce a prolonged remission (ref. 11).

of drug resistance due to a significant reduction in deoxycytidine kinase activity and an increase in cytidine deaminase activity.[32]

Clinical Trials in Tumors

Our preclinical studies on tumor cell lines and in mice indicated that 5AZA has the potential to be an effective chemotherapeutic agent for the treatment of malignant disease. Since the antineoplastic action of 5AZA is related to its activation of tumor suppressor genes silenced by aberrant methylation, prior treatment of tumors with DNA-damaging anti-cancer drugs has the potential to mutate these target genes. If the silent tumor suppressor gene is mutated there is no possibility for its reactivation by 5AZA. After careful consideration of this possibility I concluded that the evaluation of the antitumor potential of 5AZA should be performed on cancer patients that did not receive any prior chemotherapy. Since metastatic NSCLC responds poorly to conventional chemotherapy and expected patient survival is short, we choose to investigate the antitumor activity of 5AZA in this cohort of patients. In this pilot study most patients received several cycles of 5AZA administered as an 8-hour infusion at doses of 200 to 660 mg/m2.[12,13] At this very conservative dose-schedule the survival of the patients increased with the dose. We observed some interesting responses. Remarkably, one patient that received 5 cycles of 5AZA was removed from the study due to signs of tumor progression, but subsequently survived almost 7 years. I think that the response in this patient was related to the

5AZA therapy. The delayed response is in accord with the delayed in vitro action of 5AZA on tumor cells. Other investigators have reported of a delayed action on the proliferative potential of tumor cell lines after treatment with 5AZA.[33] Perhaps in this latter patient the 5AZA therapy activated a senescence program in which the tumor stem cells could undergo a limited number of cell divisions before losing completely their proliferative potential. Future research will clarify the delayed action of 5AZA on malignant cells.

Future Perspectives on 5AZA in Cancer Therapy

The major question is does 5AZA have significant chemotherapeutic potential for the clinical therapy of cancer? The preliminary clinical trials performed on 5AZA indicate that this interesting differentiating agent has promising anti-antineoplastic activity. Before a final conclusion can be made on the role of 5AZA in cancer therapy, the optimal dose-schedule for this analog has to be tested on patients with malignant disease. In a classical study in animal models, Skipper and Schabel demonstrated that the antineoplastic activity of deoxycytidine analogs, such as ARA-C, is very dose-schedule dependent.[34] Since 5AZA is a deoxycytidine analog, its antineoplastic activity is also dose-schedule dependent. This is due in part to its S phase specificity and short in vivo half-life.

Any attempt to design the optimal dose schedule for 5AZA in cancer therapy has to take into account the pharmacology of this analog, the cell kinetics of the neoplastic cells and normal tissue and the stage of disease of the patient. Should oncologists make a serious effort to determine the optimal dose-schedule to use 5AZA in cancer therapy? If one looks at the long list of genes that suppress neoplastic transformation and which 5AZA has the potential to reactivate,[35] in my opinion the answer to this question should be "yes". In a concise manner I summarized below some of the key parameters that should be taken into account in the attempt to design the optimal dose schedule for 5AZA.[36] These parameters include dose, duration of treatment, disease stage, hematopoietic toxicity and combination chemotherapy.

Dose of 5AZA

Since the half-life of 5AZA in man is only 15 to 25 minutes, this agent should be administered as a continuous infusion. The dose should be defined in terms of the plasma level of 5AZA, which should be above the minimal pharmacologically active concentration in all anatomical compartments. In my opinion based on preclinical studies and on clinical observations in patients this level should be in the range greater than 100 ng/ml. The mechanism of action of 5AZA should be taken into account when selecting the optimal plasma concentration of this analog. The objective should be to incorporate the maximum number of molecules of 5AZA into tumor DNA to demethylate the methylated CpG sites to reactivate the genes that suppress tumorigenesis and induce terminal differentiation[37,38] or senescence. Since 5AZA-dCTP competes with dCTP for incorporation into DNA,[18] the higher the level of the nucleotide analog the greater the number of molecules that will be incorporated into the key sites. Pharmacokinetic factors are also important since neoplastic cells in the liver may escape the therapeutic action of this analog when administered at low dose, since this organ contains high levels of cytidine deaminase, the enzyme that inactivates 5AZA. The use of inhibitors of cytidine deaminase to prevent the inactivation of deoxycytidine analogs is an approach to overcome this potential problem.[39]

Duration of 5AZA Therapy

Since 5AZA is a S phase specific agent, the duration of therapy should be long enough to permit all the neoplastic cells to enter this phase. In the case of leukemia the estimated cell cycle times are in the range of 72 to 120 h or greater.[40] High dose ARA-C has been used effectively in the treatment of acute myeloid leukemia at a dose of 2 g/m2 every 12 hours for 5 to 6 days.[41] If a similar schedule is used for 5AZA, its hematopoietic toxicity should similar to ARA-C, which is also a S phase specific drug. My suggestion for patients with acute leukemia is to use a 4-day continuous infusion of 5AZA at a dose that produces steady state plasma level greater than 100 ng/ml in the initial study. If this regimen does not produce unacceptable hematopoietic toxicity, the plasma level of 5AZA or duration of therapy can be increased in a stepwise manner until the maximal tolerated dose is obtained. This type of dose-schedule may also be effective against rapidly growing lymphomas.

Since the cell kinetics for most tumors are different from leukemic cells, it may be necessary to use a different schedule for 5AZA. In general, the cells cycle for most tumors is much longer than leukemic cells.[40] Due to this fact, in advanced disease it may not be possible to produce a complete eradication of the tumor with a single course of therapy with 5AZA. An interesting approach for the therapy of tumors would be to use 5AZA infusions of 24 to 48 hr in duration for several cycles with a steady state plasma level of 5AZA greater than 100 ng/ml. In patients with minimal residual disease the duration of 5AZA therapy may be increased to completely eradicate the last surviving tumor stem cells.

Stage of Disease

The pharmacology of 5AZA suggests that this analog should be used in the early stages of the disease prior to the use of genotoxic antineoplastic drugs. Since the objective of the 5AZA therapy is to reactivate genes that suppress tumorigenesis, if a genotoxic anti-cancer drug is used in advance it may mutate or damage many of the target genes of this cytosine analog making it impossible to reactivate their expression.[36] Perhaps the ideal candidates for 5AZA therapy are leukemic patients at diagnosis or in remission. Tumor patients at diagnosis with a minimal tumor burden should also be good candidates for the 5AZA therapy. Women with a high genetic risk to develop breast cancer, as an alternative to ablative surgery, may be potential candidates for 5AZA treatment as a form of chemoprevention. In this regard it has been reported that BRCA1 gene can be silenced by hypermethylation in breast cancer.[42]

Hematopoietic Toxicity

The most serious side effect of 5AZA is its suppression of normal bone marrow function producing prolonged leukopenia and thrombocytopenia. The start of leukopenia probably triggers the production of colony stimulating growth factors (CSF) to activate the resting hematopoietic stem cells to enter the cell cycle.[43] Under these circumstances, it may be ideal to use CSF immediately after 5AZA therapy to shorten the duration of leukopenia. Another approach that can be used in patients with tumors that do not show metastasis to the bone marrow is to remove the hematopoietic stem cells prior to 5AZA therapy and to reinfuse them back into the patient after treatment. We have also been investigating the use of gene therapy for chemoprotection of the bone marrow. This latter approach involves the transduction of normal hematopoietic stem cells with the cytidine deaminase genes to confer drug resistance to 5AZA.[26]

Combination Chemotherapy

The remarkable report that 5AZA in combination with the inhibitor of histone deacetylase (HDI), Trichostatin A (TSA), produces a synergistic activation of the tumor suppressor genes in both leukemic and tumor cells[44] suggests that the combination of these two different classes may be a very effective chemotherapeutic regimen. HDI are classified as an epigenetic agent since they can activate certain silent genes in neoplastic cells. They are under current clinical investigation for their anti-cancer activity.[45] We have observed using a clonogenic assay that the combination of 5AZA with different types of HDI produces a synergistic antineoplastic effect against both human leukemic and tumor cell lines.[23,24,47,48] The mechanism behind this interesting interaction is most likely due to the attachment of 5-methylcytosine binding protein (MBP) to the methylated promoter, which in turn recruits histone deacetylase resulting in a compactation of the chromatin structure.[46] These two events act in a complementary manner to suppress transcription. When the target genes that are silenced by these epigenetic events have the function to suppress tumorigenesis, a combination of an inhibitor of DNA methylation with HDI can reactivate gene expression and produce a very interesting antineoplastic effect.

Another interesting agent that can be used in combination with 5AZA is retinoic acid (RA). Retinoids have the potential to induce leukemic cell and tumor cell differentiation. When the expression of the transcription factor, retinoic acid receptor beta (RARβ), is silenced by aberrant methylation, RA loses its antineoplastic activity. We were the first to demonstrate that 5AZA in combination with RA produces a synergistic antineoplastic effect against human DLD1 colon carcinoma cells and that the RARβ gene is silenced by methylation in this cell line.[49,50] On major advantage in using RA in combination with 5AZA is that this retinoid is relatively nontoxic.

The antineoplastic activity of 5AZA can also be increased by biochemical modulation, using agents such as 3-deazauridine (3DU) or cyclopentylcytosine (CPC), which reduce the pool size of dCTP resulting in an enhanced incorporation of 5AZA into DNA.[51,52] Inhibitors of cytidine deaminase, the enzyme that inactivates 5AZA, also have the potential to increase its antineoplastic action in cells that express high levels of this enzyme[26] and in vivo by increasing the plasma half-life of cytosine nucleoside analogs.[39] Another interesting approach would be to use 5AZA in combination with other inhibitors of DNA methylation, such as antisense oligonucleotides that target mRNA of DNA methylase.[53]

Conclusions

The discovery of so many different silent genes that suppress tumorigenesis that reactivated in vitro by treatment with 5AZA,[35] suggests that this deoxycytidine analog may have enormous potential as anti-cancer agent. This aspect is supported by my demonstration of very potent antineoplastic activity of 5AZA in animal models with both leukemia and tumors.[28-30] It is of interest to note that other investigators reported that 5AZA was a more potent antineoplastic agent in the human tumor xenograft model in mice than several of the effective anticancer drugs used in clinical therapy.[54] My pilot clinical studies on 5AZA on leukemia and lung cancer are also supportive of the chemotherapeutic potential in cancer therapy.[10-13] Also in support of the anticancer potential of 5AZA are the positive responses observed in recent clinical investigations on patients with acute and chronic myeloid leukemia and myelodysplastic syndrome, a preleukemic disease.[55-58]

Past research on deoxycytidine analogs has shown that their the antineoplastic activity is very dose-schedule dependent, primarily due to the fact that they are S phase specific agents[34] and have a short in vivo half-life. The full chemotherapeutic potential of 5AZA in cancer therapy will be only realized when its "optimal dose-schedule" is used in clinical trials.[36] The design of the optimal dose-schedule of 5AZA will require an in depth analysis of its pharmacology, tumor cell kinetics and host toxicity. Based on my preclinical and clinical investigations on 5AZA, I favor the use of intensive therapy with this interesting analog. The highest dose with acceptable toxicity of 5AZA used in a clinical trial in patients with acute leukemia was 250 mg/m2 as 6 h infusion every 12 h for 6 days for a total dose of 1500 mg/m2.[55] The intensive therapy with 5AZA should only be investigated in patients with an adequate hematological status. Due to the complex nature of clinical trials, the limited number of available patients, ethical considerations, and the long time interval to obtain results in terms of patient survival, I favor the extensive use of animal models in parallel with clinical investigations to help design the optimal dose-schedule 5AZA. Animal models can make important contributions provided we have to learn how to translate the results to patients. The design of the optimal dose schedule for 5AZA in combination with other antineoplastic agents is even more complex and may require computer modeling.

Specific or nonspecific activation of the immune system to target neoplastic cells may also increase the effectiveness of 5AZA therapy. In this regard 5AZA was reported to upregulate several tumor-associated antigens in tumor cells.[59-62] Some of these genes were demonstrated to be silenced by promoter methylation.[62] In addition to upregulating tumor associated antigens, 5AZA may possibly also activate T lymphocytes to target the tumor cells. by an immunological event that still remains to identified and may involve the activation of specific immune genes. These reports give some insight into the possible mechanisms to explain the interesting observation that an immune adjuvant increased remarkably the antineoplastic activity of 5AZA in mice with L1210 leukemia.[63]

The HDI are a very exciting class of agents to use in combination with 5AZA due to the synergistic interaction as discussed above. As single agents, HDI show very interesting antitumor activity in animal models and in preliminary clinical trials.[45] In combination with 5AZA these agents show synergistic antineoplastic activity.[23,24,47,48] Future investigation should also focus on finding the optimal schedule to use HDI in combination with 5AZA in cancer therapy. Since HDI have the potential in to inhibit the entry of G1 phase cells into S phase due to their activation of the p21 (WAF1) gene,[45] they can possibly interfere with the antineoplastic action of 5AZA on some tumor cells. In order to avoid such an event, it may be more effective to use intermittent treatment with HDI during a continuous infusion of 5AZA.

Epigenetic therapy of cancer using 5AZA has tremendous potential for the treatment of malignant disease. In order to uncover this potential, investigators should realize that the mechanism of action of 5AZA is very different from most conventional antineoplastic agents. In this regard it may be necessay to use of novel approaches with different end points in clinical trials of this interesting inhibitor of DNA methylation to make significant advances in the chemotherapy of cancer.

Addendum

In this chapter the author focused primarily on his own experimental work. The author acknowledges that many other investigators have made important contributions in this field, but due to space limitations they were too numerous to be included in the chapter.

Acknowledgements

This work was supported in part by grants from the Canadian Breast Cancer Research Initiative, Leukemia & Lymphoma Society (USA) and Canadian Institutes of Health Research. The author is very cognizant of the important contributions made by his former students, research assistants and clinical colleagues including Joseph Ayoub, Christian Beauséjour, Anne-Julie Boivin, Veronica Bovenzi, Jacques Bouchard, David Bouffard, Guy Chabot, Sylvie Côté, Benoit Doré, Nicoletta Eliopoulos, Jacynthe Gagnon, Martin Gyger, Louise F. Momparler, Melanie Primeau, Georges E. Rivard and Sepideh Shaker.

References

1. Momparler RL, Chu MY, Fischer GA. Studies on a new mechanism of resistance of L5178Y murine leukemic cells to cytosine arabinoside. Biochem Biophys Acta 1968; 161:481-493.
2. Momparler RL, Fischer GA. Mammalian deoxynucleoside kinases. 1. Deoxycytidine kinase: Purification. Properties and kinetic studies with cytosine arabinoside. J Biol Chem 1968; 243:4398-4304.
3. Momparler RL. Effect of cytosine arabinoside 5'-triphosphate on mammalian DNA polymerase. Biochem Biophys Res Commun 1968; 34:465-471.
4. Momparler RL. A model for the chemotherapy of acute leukemia with 1-β-arabinofuranosylcytosine. Cancer Res 1974; 34:1775-1787.
5. Sorm F, Vesely J. Effect of 5-aza-2 deoxycytidine against leukemic and hemopoietic tissues in AKR mice. Neoplasma 1968; 15:339.
6. Karon M, Sieger L, Leimbrock S et al. 5-Azacytidine: A new active agent for the treatment of acute leuekmia. Blood 1973; 42:359-365.
7. Momparler RL, Momparler RL, Samson J. Comparison of the antileukemic activity of 5-aza-2'-deoxycytidine, 1-β-D-arabinofuranosylcytosine and 5-azacytidine against L1210 leukemia. Leukemia Res 1984; 8:1043-1049.
8. Momparler RL, Frith CH. Toxicology in mice of the antileukemic agent 5-aza-2'-deoxycytidine. Drug Chem Toxicology 1981; 4:373-381.

9. Momparler RL, Gonzales FA, Momparler LF et al. Preclinical evaluation of hematopoietic toxicity of antileukemic agent, 5-Aza-2'-deoxycytidine. Toxicology 1989; 57:329-336.

10. Rivard GE, Momparler RL, Demers J et al. Phase I study on 5-aza-2'-deoxycytidine in children with acute leukemia. Leukemia Res 1981; 5:453-462.

11. Momparler RL, Rivard GE, Gyger M. Clinical trial on 5-aza-2-deoxycytidine in patients with acute leukemia. Pharmac Ther 1986; 30:277-286.

12. Momparler RL, Bouffard DY, Momparler LF et al. Pilot phase I-II study on 5-Aza-2'-deoxycytidine (Decitabine) in patients with metastatic lung cancer. Anticancer Drugs 1997; 8:358-368.

13. Momparler RL, Ayou J. Potential of 5-aza-2'-deoxycytidine (Decitabine) a potent inhibitor of DNA methylation for therapy of advanced nonsmall lung cancer. Lung Cancer 2001; 334:S111-S115.

14. Momparler RL. Molecular, cellular and animal pharmacology of 5-aza-2-deoxycytidine. Pharmac Ther 1986; 30:287-299.

15. Pliml J, Sorm F. Synthesis of a deoxy-D-ribofuranosyl-5-cytosine. Collection Czech Chem Commun 1964; 29:2576-2577.

16. Lin KT, Momparler RL, Rivard GE. High performance liquid chromatographic analysis of chemical stability of 5-aza-2'-deoxycytidine. J Pharm Sci 1981; 70:1228-1232.

17. Momparler RL, Derse D. Kinetics of phosphorylation of 5-aza-2'-deoxycytidine by deoxycytidine kinase. Biochem Pharmacol 1979; 28:1443-1444.

18. Bouchard J, Momparle RL. Incorporation of 5-aza-2'-deoxycytidine 5'-triphosphate into DNA interactions with mammalian DNA polymerase and DNA methylase. Mol Pharmacol 1983; 24:109-114.

19. ChaboTG, Bouchard J, Momparler RL. Kinetics of deamination of 5-aza-2'-deoxycytidine and cytosine arabinoside by human liver cytidine deaminase and its inhibition by 3-deazauridine, thymidine or uracil arabinoside. Biochem Pharmacol 1983; 32:1327-1328.

20. Momparler RL, Rossi M, Bouchard J. Kinetic interaction of 5-aza-2'-deoxycytidine 5'-monophosphate and its 5'-triphosphate with deoxycytidine deaminase. Mol Pharmacol 1984; 25:436-440.

21. Juttermann R, Li E, Jaenisch R. Toxicity of 5-aza-2'-deoxycytidine to mammalian cells is mediated by covalent trapping of DNA methyltransferase rather than DNA demethylation. Proc Natl Acad Sci USA 1994; 91:11797-11801.

22. Momparler RL, Samson J, Momparler LF et al. Cell cycle effects and cellular pharmacology of 5-aza-2'-deoxycytidine. Cancer Chemotherapy Pharmacol 1984; 13:191-194.

23. Bovenzi V, Momparler RL. Antineoplastic action of 5-aza-deoxycytidine and histone deacetylase inhibitor and thir effect on the expression of retinoic acid receptor a and estrogen receptor β genes in breast carcinoma cells. Cancer Chemother Pharmacol 2001; 71-76.

24. Primeau M, Gagnon J, Momparler RL. Synergistic antineoplastic action of DNA methylation inhibitor 5-aza-2'-deoxycytidine and histone deacetylase inhibitor depsipeptide on human breast carcinoma cells. Intl J Cancer 2003; 103:177-184.

25. Momparler RL, Momparler LF. Chemotherapy of L1210 and L1210/ARA-C leukemia with 5-aza-2'-deoxycytidine and 3-deazauridine. Cancer Chemother and Phamacol 1980; 25:51-54.

26. Eliolopoulos N, Momparler RL. Drug resistance to 5'-aza-2'-deoxycytidine, 2,2'-difluorodeoxycytidine and cytosine arabinoside conferred by retroviral-mediated transfer of human cytidine deaminase cDNA into murine cells. Cancer Chemother Pharmacol 1998; 42:373-378.

27. Momparler RL, Onetto-Pothier N. Drug resistance to cytosine arabinoside. In: Kessel D, ed. Resistance to Antineoplastic Drugs. Boca Raton: CRC Press Inc., 1988:353-367.

28. Momparler RL, Gonzales FA. Effect of intravenous infusions of 5-aza-2'-deoxycytidine on survival time of mice with L1210 leukemia. Cancer Res 1978; 38:2673-2678.

29. Wilson VL, Jones PA, Momparler RL. Inhibition of DNA methylation in L1210 leukemic cells by 5-aza-2'-deoxycytidine as a possible mechanism of chemotherapeutic action. Cancer Res 1983; 43:349-3497.

30. Chabot G. Pharmacocinétique et effets antinéoplasiques de al 5-aza-2'-desoxycytidine chez les animaux. Ph.D. thesis. Université de Montréal 1983:166-174.

31. Momparler RL, Bouchard J, Onetto N. 5-Aza-2'-deoxycytidine therapy in patients with acute leukemia inhibits DNA methylation. Leukemia Res 1984; 8:181-185.

32. Onetto N, Momparler RL, MomparleR LF et al. In vitro tests to evaluate the response to therapy of acute leukemia with cytosine arabinoside or 5-aza-deoxycytidine. Seminars Oncol 1987; 14:231-237.

33. Bender CM, Pao MM, Jones PA. Inhibition of DNA methylation by 5-aza-2'-deoxycytidine suppresses the growth of tumor cell lines. Cancer Res 1998; 58:95-101.

34. Skipper HE, Shabel FM, Willcox WS. Experimental evaluation of potential anticancer agents. XXI. Scheduling of arabinosylcytosine to take advantage of its S phase specificity. Cancer Chemother Rep 1967; 51:125-165.

35. Momparler RL, Bovenzi V. DNA Methylation and Cancer (A Review) J. Cell Physiol 2000; 183:145-154.
36. Momparler RL, Côté S, Eliopoulos N. Pharmacological approach for optimization of the dose-schedule of 5-Aza-2'-deoxycytidine (Decitabine) for the therapy of leukemia. Leukemia 1997; 11:175-180.
37. Jones PA, Taylor SM. Cellular differentiation, cytidine analogs and DNA methylation. Cell 1980; 20:85-93.
38. Pinto A, Attadia V, Fusco A et al. 5-Aza-2 deoxycytidine induces terminal differentiation of leukemic blasts from patients with acute myeloid leukemias. Blood 1984; 64:922-929.
39. Kreis W, Budman DR, Chan K et al. Therapy of refractory/relapsed leukemia with cytosine arabinoside plus tetrahydrouridine (an inhibitor of cytidine deaminase)- a pilot study. Leukemia 1991:991-998.
40. Andreeff M, Goodrich DW, Pardee AB. Cell proliferation, differentiation and apoptosis. In: Bast RC, Kufe DW, Pollock RE et al. eds. Cancer Medicine, 5th ed. London: BC Decker Inc., 2000:17-20.
41. Weick JK, Kepecky KJ, Appelbaum FR et al. A randomized investigation of high dose versus standard-dose cytosine arabinoside with daunorubicin in patients with previously untreated acute myeloid leukemia: A Southwest Oncology Group study. Blood 1996; 88:2841-2851.
42. Dobrovic A, Simplefendorfer D. Methylation of the BRAC1 gene in sporadic breast cancer. Cancer Res 1997; 57:3347-3350.
43. Ganser A, Heil G. Use of hematopoietic growth factors in the treatment of aclute myelogenous leukemia. Curr Opin Hematol 1997; 4:191-195.
44. Cameron EE, Bachman KE, Myohanen S et al. Synergy of demethylation and histone deacetylase inhibition in the reexpression of genes silenced in cancer. Nat Genet 1999; 21:103-112.
45. Marks PA, Rifkind RA, Richon VM et al. Histone deacetylases and cancer: Causes and therapies. Nature Reviews 2001; 1:194-202.
46. Jones PA, Baylin SB. The fundamental role of epigenetic events in cancer. Nature Rev 2002; 415-428.
47. Bovin A-J, Momparler LF, Hurtubise A et al. Antineoplastic action of 5-aza-2'-deoxycytidine and phenylbutyrate on human lung cancer. Anti-Cancer Drugs 2002; 13:1-6.
48. Shaker S, Bernstein M, Momparler LF et al. Preclinical evaluation of antineoplastic activity of inhibitors of DNA methylation (5-aza-2'-deoxycytidine) and histone deacetylation (trichostatin A, depsipeptide) against myeloid leukemic cells. Leukemia Res 2003 in press.
49. Côté S, Momparler RL. Antineoplastic action of all-trans retinoic acid and 5-aza-2'-deoxycytidine on human DLD-1 colon carcinoma cells. Cellular Pharmacol 1995; 2:221-228.
50. Côté S, Sinnett D, Momparler RL. Demethylation by 5'-aza-2'-deoxycytidine of specific 5-methylcytosine sites in the promoter region of the retinoic acid receptor beta gene in human colon carcinoma cells. Anti-Cancer Drugs 1998; 9:743-750.
51. Momparler RL, Vesely J, Momparler LF et al. Synergistic action of 5-aza-2'-deoxycytidine and 3-deazauridine on L1210 leukemic cells and EMT tumor cells. Cancer Res 1979; 39:3822-3827.
52. Bouffard DY, Momparler LF, Momparler RL. Enhancement of the antileukemic activity of 5-aza-2'-deoxycytidine by cyclopentenyl cytosine in HL-60 leukemic cell line. Anti-Cancer Drugs 1994; 5:223-228.
53. MacLeod AR, Szyf M. Expression of antisense to DNA methyltransferase mRNA induces DNA demethylation and inhibits tumorigenesis. J Biol Chem 1995; 270:8037-8043.
54. Braakhius BJM, van Dongen GAMS, van Walsum M et al. Preclinical antitumor activity of 5-aza-2'-deoxycytidine against human head and neck cancer xenografts. Invest New Drugs 1988; 6:299-304.
55. Richel DJ, Colly LP, Kluin-Nelemans JC et al. The antileukaemic activity of 5-aza-2 deoxycytidine (Aza-dC) in patients with relapsed and resistant leukaemia. Br J Cancer 1991; 64:144-149.
56. Pinto A, Zagonel V. 5-Aza-2 deoxycytidine (Decitabine) and 5-azacytidine in the treatment of acute myeloid leukemias and myelodysplastic syndrome: Past, present and future trends. Leukemia 1993; 7 (suppl 1):51-60.
57. Kantarjian HM, O'Brien SM, Keating M et al. Results of decitabine therapy in the accelerated and blastic phases of chronic myelogenous leukemia. Leukemia 1997; 11:1617-1620.
58. Daskalakis M, Nguyen TT, Nguyen C et al. Demethylation of a hypermethylated P15/INK4B gene in patients with myelodysplastic syndrome by 5-aza-2 deoxycytidine (Decitabine) treatment. Blood 2002; 100:2957-2964.
59. Weber J, Salgaller M, Samid D et al. Expression of the MAGE-1 tumor antigen is up-regulated by the demethylating agent 5-aza-2 deoxycytidine. Cancer Res 1994; 54:1766-1771.

60. Coral S, Sigalotti L, Gasparollo A et al. Prolonged upregultation of the expression of HLA Class I antigens and costimulatory molecules on melanoma cells treated with 5-aza-2 deoxycytidine (5-AZA-CdR). J Immunother 1999; 22:16-24.
61. Weiser TS, Sheng Guo Z, Ohnmacht GA et al. Sequential 5-aza-2 deoxycytidine-depsipeptide FR901228 treatment induces apoptosis preferentially in cancer cells and facilitates their recognition by cytolytic T lymphocytes specific for NY-ES0-1. J Immunother 2001; 24:151-161.
62. Sigalotte L, Coral S, Nardi G et al. Promoter methylation controls the expression of MAGE2,3, and 4 genes in human cutaneous melanoma. J Immumother 2002; 28:16-26.
63. Zaharko DS, Covey JM, Muneses CC. Experimental chemotherapy (L1210) with 5-aza-2 deoxycytidine in combination with pyran copolymer (MVE-4) as immune adjuvant. J Natl Cancer Inst 1985; 74:1319-1324.

CHAPTER 16

Anticancer Gene Therapy by in Vivo DNA Electrotransfer of MBD2 Antisense

Pascal Bigey and Daniel Scherman

Abstract

Harnessing the full therapeutic potential of DNA methylation machinery proteins would require efficient techniques of introducing either anti sense, iRNA or expression vectors into tumors in vivo. Efficient techniques for introducing DNA in vivo are also required for target validation. This chapter discusses the electrotransfer of introducing DNA in vivo and its use in validation of MBD2 as anticancer target as well as its potential as an anticancer gene therapy.

Introduction

Efficiently transferring DNA to eukaryotic cells is a requirement for several purposes, such as study of gene function, or gene therapy. Particularly, the sequencing of the human genome will result in the cloning and the functional characterization of numerous new proteins, which will require both in vitro and in vivo functional studies. Although in vitro gene transfer is reasonably solved by means of cationic lipid transfection (commercial kits), calcium phosphate precipitation or electroporation for example, it is much more difficult to achieve efficient in vivo gene transfer, since there is no ideal vehicle system. A wide range of methods has been recently developed, generally falling into two categories: viral and nonviral.[1-2] Viral techniques include the use of different viral vector types: adenovirus, AAV (adeno-associated virus), retrovirus, lentivirus, herpes simplex virus (for a review see ref. 1). All of them have advantages but their use is limited by safety concerns, such as immune response, possible mutagenesis and carcinogenesis, and high production costs. Nonviral vector techniques, using plasmid DNA, do not reach the efficiency of viral ones as far as expression levels are concerned. However, they are attractive for different reasons: they are less toxic, much easier and cheaper to produce, safer, tissue-specific in some cases and showing no DNA insert size limitations. Major disadvantages are low gene transfer efficiency and immunostimulatory properties of plasmid DNA.[3-4] A wide range of techniques is available, for example the use of lipoplexes, electroporation, direct DNA injection, particle bombardment, receptor-mediated gene transfer (see ref. 1 and references therein). Among these different nonviral strategies currently under study, in vivo electroporation has proven to be one of the most efficient and simple methods,[5-9] which could be applied in gene therapy and as a laboratory tool to study gene function. It also has the advantage of allowing drug delivery to the targeted cells, like the anticancer drug bleomycine for example,[10] or other foreign molecules like proteins or oligonucleotides.[10]

This chapter will give a short description of the in vivo electroporation technique and of its possible applications in gene therapy targeted at proteins of the DNA methylation machinery and cancer therapy.

DNA Methylation and Cancer Therapy, edited by Moshe Szyf. ©2005 Eurekah.com and Kluwer Academic/Plenum Publishers.

Delivery Principle

Drug Delivery and Cancer

Since the initial reports by Wong and Neumann,[11-12] the use of electricity to mediate the delivery of molecules to cells in vitro is now a routine technique. DNA and other molecules can be introduced in a variety of living cells: bacteria, yeast, animal or plant cells.[13] The development of squarewave electric pulse power supplies allowed membrane permeabilization without loss of cell viability,[14-15] which is necessary for an in vivo use. Application of a controlled electric field induces a transmembrane potential. When this potential exceeds the dielectrical strength of the cell membrane, one or more reversible pores will develop, allowing any molecule to pass through them and have direct access to the cytosol (for reviews see ref. 16-17). The first in vivo relevant application of electropermeabilization was demonstrated by the cellular uptake in tumors of the antineoplastic drug bleomycin, and was called electrochemotherapy (see refs. 16 and 18 for reviews). Bleomycin is an antibiotic chemotherapeutic agent that is used to treat a variety of cancers. It causes single-strand and double-strand breaks in DNA.[19] Its efficiency is dependent on the intracellular concentration, but it poorly enters cells. It was shown that electropermeabilization enhanced its cytotoxicity by allowing a better penetration into cells.[20] Electrochemotherapy is now a well-established technique. Important advances have been achieved and results of clinical trials are published since 1998 on small nodes of head neck squamous cell carcinoma, melanoma, basal cell carcinoma, adenocarcinoma and primary skin cancer, showing some complete regression (for a review see refs. 16, 21 and 22). Electrochemotherapy has also been applied with another anticancer drug, cisplatin, which forms DNA adducts, both in clinical trials on malignant melanoma skin metastases[23] and in veterinary use on horses.[24]

Electrochemotherapy is a promising technique to locally treat small accessible tumors. It is found tolerable by patients, and the dose of anticancer drug is much lower than the one used in classical chemotherapy protocols. It is interesting to note that the currently reported clinical trials are very encouraging. Besides electrochemotherapy, which is the first historical use of in vivo electropermeablization, the last few years have seen electric pulses mediated gene transfer as a rapidly emerging technique, under the name of electrotransfer.

DNA Delivery and Toxicity

In vivo DNA electrotransfer is a simple physical technique for gene delivery in various mammalian tissues, which consists of injecting plasmid DNA to a targeted tissue and applying a series of electric pulses.

Practically, a plasmid solution in isotonic saline (NaCl 150 mM) is injected into the targeted tissue, and electric pulses are then delivered by means of two electrodes placed on each part of the injection site (electrodes can be either needles or plates). Different types of electric pulses (voltage versus time) can be easily delivered by commercial electropulsators. Generally, square wave electric pulses are preferred for in vivo experiments. It should be noted that the electric field pattern varies depending on the electrode type, resulting in varying effective field intensity (in V/cm) in the treated area. The field is more homogeneous when using plate electrodes.[25]

Efficient conditions for plasmid DNA electrotransfer into cells depends on the tissue. Several efficient conditions can be also used for a same tissue, as it was shown for the skeletal muscle.[26] Optimal conditions result from a compromise between efficient plasmid transfer and minimal toxicity of the electric field. This toxicity may involve different parameters: permeabilization is a main factor of toxicity, since the external media diffuses into cells and modify their internal media composition. Internal medium may also leak out of the cell. This is reduced when the duration and the level of permeabilization are minimal. Another toxic effect is an oxidative stress due to the generation of free radicals induced near the membrane by electropermeabilization.[27] Furthermore, it was shown on a muscle model that electrotransfer induces plasmid-dependent muscle lesions containing necrotic myofibers, although electrotransferred muscles were indistinguishable from nontreated controls at day 56.[28] Finally, in vivo delivery of electric pulses to tissues can induce vascular effects.[29] But little data is available regarding in vivo electrotransfer toxicity, and further studies are needed to fully understand its exact mechanism and consequences.

Mechanism

The mechanism of DNA electrotransfer is not totally clarified, but it is generally accepted that efficient DNA transfer is associated with two main parameters: cell membrane permeabilization (which creates "holes" in the plasma membrane) and DNA electrophoresis (which allows the polyanionic molecule of DNA to enter the cell). This electrophoretic "force" has different possible effects as for example to favor the insertion of DNA in a membrane destabilized by an electric field. This association was further studied to determine the respective contribution of cell permeabilization and DNA electrophoresis for in vivo DNA transfer into muscle fibers. Noteworthy it was experimentally demonstrated that electrophoresis of DNA allows a very efficient transfection only if the membrane was previously destabilized by a permeabilizing pulse.[30-31] In vitro studies of the electropermeabilization/electrophoresis association by time-resolved fluorescence microscopy showed that DNA accumulates at the cathode side of the cells,[32] and that DNA was entrapped within the membrane, forming localized spots.

NMR imaging studies have shown that when muscle was submitted to a series of electrical pulses efficient for electrotransfer, the zone of permeabilization to the Gadolinium complex Gd-DTPA (a NMR contrasting agent) was similar to the zone of expression of an electrotransferred plasmid coding for *β-galactosidase* reporter gene.[33]

We do not currently know if additional active mechanisms contribute to DNA translocation through muscle fibers membrane or membrane of other tissues. In vitro evidence is in favor of such a mechanism for DNA electrotransfer,[34] showing that some domains of the membrane are probably involved in the transfection of naked DNA,[35] but this does not seem to occur with in vivo electrotransfer. We have also observed that electrotransferable plasmid DNA remains available for a long time after injection: the level of transfection did not significantly vary when the lag time between DNA injection and electric pulses delivery varied between few seconds to 4 hours.[35]

In summary, the exact mechanism of DNA electrotransfer is not yet fully understood, although significant results point to the involvement of both a permeabilization and electrophoretic effect. The best electric conditions result from a compromise between efficient DNA transfer and toxicity. These conditions have to be set up according to the targeted tissue. Also, if clinical applications are envisioned it will be necessary to further investigate the possible toxic effects of electrotransfer and how to prevent them. As a laboratory tool, this technique is undoubtedly one of the simplest ways to transfer DNA into living animals, and a variety of tissues can be targeted, as we will see later. It is of considerable value for in vivo functional studies of proteins.

In Vivo DNA Electrotransfer: Targeted Tissues

Muscle and Other Tissues

Three teams showed independently DNA electrotransfer in the skeletal muscle.[8,30,36] Skeletal muscle is the most widely targeted tissue because of its advantages: it is easily accessible, and muscle fibers have a long lifespan, which allows a long-term (more than a year) expression in transfected cells.[1,25] It is also able to produce secreted proteins. By following luciferase expression with a CCCD (conductively connect charge-coupled device) camera (an imaging technique that allows an in vivo kinetic study without sacrificing the animals), it was observed that gene expression increased with time in the first few days, and then stayed at comparable levels for at least 70 days,[37] and up to a year (see ref. 39 and Fig. 1), raising hope in the gene therapy

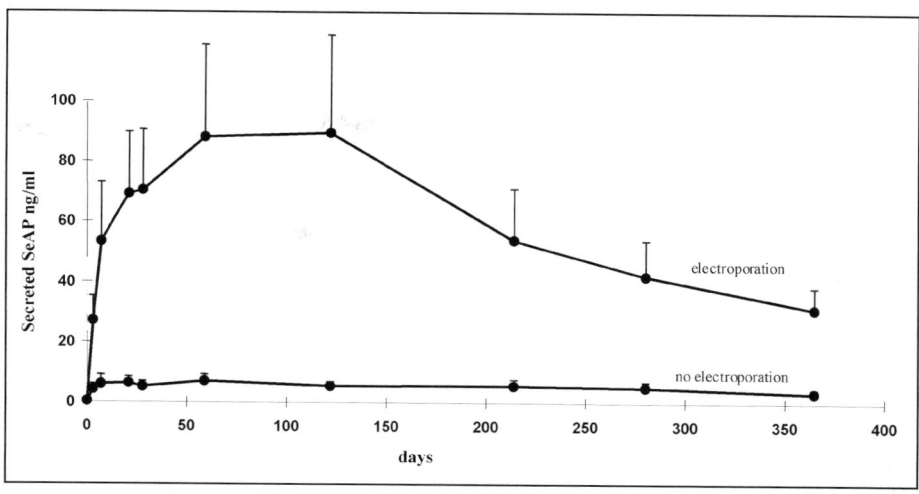

Figure 1. Long-term expression of the human secreted alkaline phosphatase (*SeAP*) reporter gene. A plasmid DNA solution is injected with a syringe into the tibial-cranialis muscle of SCID mice, and electric pulses are applied by means of two plate electrode linked to a BTX 830 power supply (8 pulses, 20 ms, 1 Hz, 200 V/cm).

field. A variety of genes have been introduced in skeletal muscles. Some examples are listed in Table 1, all of them (except for reporter genes) for a gene therapy purpose.

As therapeutic levels of secreted proteins can be reached, it is a good candidate as an endocrine tissue for expression of cytokines, growth factors or coagulation factors for example.[38-39] Two groups have recently shown that *Epo* (erythropoietin) secretion after muscle electrotransfer results in improved erythropoiesis, improvement in red cells half-life and phenotype, and high hematocrit value for several months in a β-thalassemic mouse model.[40-41] This could be developed as a treatment for β-thalassemia, or anemia as shown by Maruyama et al.[42] Skeletal muscle was also used for cytokines production that resulted in an improved survival in a mouse viral myocarditis model[43] or in rat induced myocarditis.[44] Anti-inflammatory cytokine *IL-10* showed

Table 1. Examples of cDNA containing plasmids introduced into skeletal muscle by electrotransfer

Reporter Gene	Ref.	Cytokines Family	Ref.	Other Proteins	Ref.
luciferase	30	IL-5	36	hepatocyte growth factor	46
SeAP	39	IL-10	44-45	clotting factor IX	39
β-Gal	30	vIL-10-Fc	43	endostatin	50
		IL-12	49	FGF1	30
		EPO	38	glial cell line derived growth factor	51
				laminin α2	48
				dystrophin	48

SeAP: secreted alkaline phosphatase; *IL*: interleukin; *FGF1*: fibroblast growth factor 1; *b-gal*: beta galactosidase; *EPO*: erythropoietin.

Table 2. Examples of targeted tissues by electrotransfer

Tissues	Species	Gene(s)	Reference
skin	rat	EPO	55
testis	mouse	CAT, Luc, LacZ	58
liver	rat	GFP	59
muscle			see table1
kidney	rat	LacZ, Luc	60
brain	mouse	GFP	57, 61
carotid artery	rabbit	Luc	52
retinal ganglion cells	rat	GFP	54
cornea	mouse	IL-6	53
spinal cord	rat	GFP	56

EPO: erytrhopoietin; *CAT*: Chloramphenicol acetyl-transferase; *luc*: luciferase; *LacZ*:beta-galactosidase; *GFP*: green fluorescent protein; *IL-6*: interleukin 6.

interesting properties in an atherosclerosis model.[45] *HGF* muscle secretion recently showed cytoprotective activity in mice with acute liver injury.[46] Another promising result was recently obtained by Prud'homme et al[47] who showed protection against autoimmune diabetes by muscle secretion of a ligand of *CTLA-4* (cytotoxic T lymphocyte antigen 4), a negative regulator of T cell activity.

Finally, expression of *laminin α2* chain in dystrophic mouse muscles was obtained without extended muscle damage,[48] although a loss of expression was observed with time, due to degeneration-regeneration of muscle. It is encouraging to see that gene transfer by electroporation is also possible in fragile muscles like dystrophic muscles, where no satisfactory gene transfer method is yet available.

In addition to its potential use in gene therapy, we think that DNA electrotransfer is a powerful laboratory complementary tool to study in vivo gene expression in any given tissue. Besides skeletal muscles, a number of other tissues have been shown to express reporter genes after electrotransfer, including tumors (which will be discussed below). Some are listed in Table 2, among which: rabbit carotid artery,[52] cornea,[53] retinal ganglion cells,[54] skin,[55] spinal cord[56] or brain (see ref. 57 for a short review). Each tissue requires specific electroporation parameters that have to be empirically studied. This provides a tool to study gene expression and function, in a spatially and temporally restricted manner. This is illustrated by the use of this technique in developmental biology.[57] In an excellent study, Saito and Nakatsuji performed embryonic mouse brain electroporation both in utero and exo utero.[61] They showed *GFP* expression in different targeted regions of the brain, and visualized neuronal morphologies. It was also possible to cotransfect three different plasmids in the same cells. Electroporation was also performed on zebrafish for gene invalidation by a dominant-negative in a fin regeneration study.[62] In vitro and in vivo electroporation tools were also used to decipher the transcriptional regulation of human skeletal muscle myosin heavy chain in muscle development and differenciation.[63] Finally, in vivo electroporation proved to be a valuable tool for the study of gene regulation systems, such as the tetracycline system, which requires at least cotransfection of two plasmids in the same cell. As this is allowed by electroporation, Lamartina et al studied the activity of novel doxycycline transactivators in a gene switch system,[64] and we studied a system based on hypoxia-responsive element and tetracycline transactivators.[65]

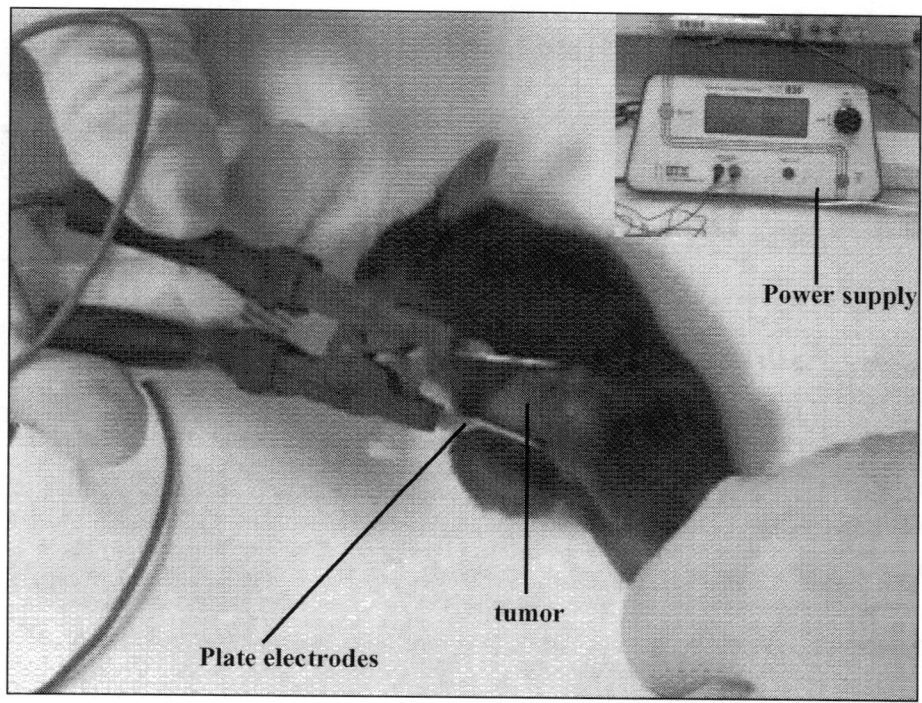

Figure 2. An example of plasmid electrotransfer into solid tumors. 3LL tumors were grafted subcutaneously in the flank of C57Bl6 mice. A plasmid DNA solution is intratumorally injected with a syringe, and electric pulses are applied by means of two plate electrode linked to a BTX 830 power supply.

Electrotransfer in Cancer Gene Therapy

As cancer is a disease of the genes, gene therapy seems to be an exciting area of research. However, effective and safe gene delivery methods to target cancer cell are still lacking. Viruses are the most effective as far as transfection is concerned, but they may elicit an immune response and raise some safety concerns. Nonviral vectors suffer from a lack of transfection efficiency. Out of 403 clinical trials in the field of cancer gene therapy, 10 only reached phase II/III or phase III: nine using viral delivery vectors, and one only using a nonviral vector.[66] If an efficient nonviral gene transfer method were to be developed, it would certainly allow great hopes for cancer gene therapy. A large range of ideas and technologies can be used, among which suicides genes, genetic enhancement of anti-tumor immune response, antiangiogenesis, tumor suppressors genes, drug resistance gene therapy.[67-68] Some of these strategies have been recently applied by in vivo electrotransfer in tumor tissues, with encouraging results, showing the feasibility of this approach. Gene electrotransfer in accessible solid tumors is easy and rapid to perform, and we showed that is can be efficient.[69] Although the transfection efficiency is low compared to viral vectors, it is a safe technique which can be repeated as much as necessary, resulting in an accumulation in the number of transfected cells. One way of approaching this is to inject DNA into tumors with a syringe, and then to apply a series of electric pulses with two external electrodes. An example is shown on Figure 2. For example, suicide gene therapy using *HSVtk*/ganciclovir technology suppressed the growth and metastasis of subcutaneously grafted mammary tumors in mice, although no complete regression was noted,[70] and considerably slowed the growth of a CT26 solid tumor in an animal model.[71] Electrotransfer of cytokines into tumors is also widely used: *IFN-α,*[72] *IL-12* or *IL-18*[73-74] have recently been shown to reduce tumor growth and increase survival times in different tumor models. In the case of *Il-12*, tumor eradication was observed in 40% of the mice, which survived for a year. It is suggested that *IL-12* induces an increase in *IFN-γ, Mig* and *IP-10*, which trigger both the immune response and antiangiogenic response.[75] Human *IL-2* or murine *GM-CSF* electroporation into a model of human esophageal T.Tn tumors grafted in nude mice suppressed the growth of these tumors and prolonged the survival.[76] Significant inhibition of tumors growth was also obtained by intratumoral electrotransfer of *TRAIL/Apo2* ligand, an apoptosis inducer,[77] and by skeletal muscle electrotransfer of a metalloproteinase-4 inhibitor.[78] In the latter case, the skeletal muscle would serve as a site of synthesis and secretion in the blood of a protein having antitumoral effect. The electrotransfer of an angionesis inhibitor, endostatin, into primary tumors and muscle tissues showed encouraging results.[50] Another encouraging result was obtained by electroporation into the liver of a liposome-encapsulated plasmid encoding the pro-apoptotic gene *bcl-xs* (member of the *bcl-2* family): the occurrence and growth of a rat hepatocellular carcinoma induced by N-nitrosomorpholine was inhibited.[79] Recently, significant antitumor immunity was achieved by skeletal muscle electroporation of a plasmid encoding a specific tumor antigen in the B16 murine melanoma model.[80]

All these promising results show the potential of in vivo electrotransfer for cancer gene therapy, which could be used for surgically inaccessible tumors, such as head and neck tumors for example. As the number of transfected cells is probably not sufficient, it is unlikely that electrotransfer of tumors will lead by itself to a cure of cancer. Furthermore, the efficiency of gene transfer depends on the tumor tissue.[81] However, electrotransfer should be used in combination with other strategies such as chemotherapy. As chemotherapy and gene therapy follow different mechanisms to kill cancer cells, a synergy between them can reasonably be expected, in addition to a different toxicity profile.

MBD2 Antisense Electrotransfer

The recent identification of the nucleotide sequence of new target genes involved in carcinogenesis and tumor growth raised hope for the development of new drugs with less toxic side effects than those of the conventional chemotherapeutic agents. Among them, the antisense strategy is particularly studied. Most of these studies use short antisense oligonucleotides, which start to give encouraging results. Some of these oligonucleotides have already reached clinical trials (see ref. 82 for a review). Although they seem promising, oligonucleotides suffer from poor cell penetration, lack of specificity, and undesirable nonantisense toxic effects.[82]

The methylated DNA binding protein *MBD2* is a member of the DNA methylation machinery that has recently been proposed to act as a suppressor of expression of methylated genes.[83] It has also been characterized to have an active demethylase activity.[84] Although some earlier data failed to confirm the demethylase activity of *MBD2*, recent work showed that *MBD2* demethylates ectopically methylated DNA in a promoter specific manner.[85] DNA methylation is an important component of the epigenome, which plays a critical role in programming gene expression.[86] A long list of data has implicated aberrations in the epigenome in cancer.[87] Both putative actions of *MBD2* are potentially important for maintaining the aberrant epigenome of transformed cells.

We have shown that intratumoral electrotransfer of an antisense of *MBD2* results in a inhibition of tumor growth in a human tumor model grafted in nude mice,[88] suggesting that *MBD2* is required for tumorigenesis. In this experiment, a single H1299 tumor was sectioned and fragments were implanted subcutaneously in the right flank of nude mice. Treatment of tumors began when they reached a size of approximately 30-100 mm^3 approximately 1 -2 weeks later. Intratumoral injections were performed 3 or 5 times every 2 to 5 days with either 50 μg plasmid (control or antisense) in 40μl, or 150mM NaCl using a Hamilton syringe and a 26 G needle. Both sides of the tumors were covered with conductive gel and placed between two flat parallel stainless steel electrodes 0.45 cm apart, as shown in Figure 2. Twenty to 30 sec after DNA injection, each tumor was subjected to 8 pulses of 20 ms duration at a voltage to

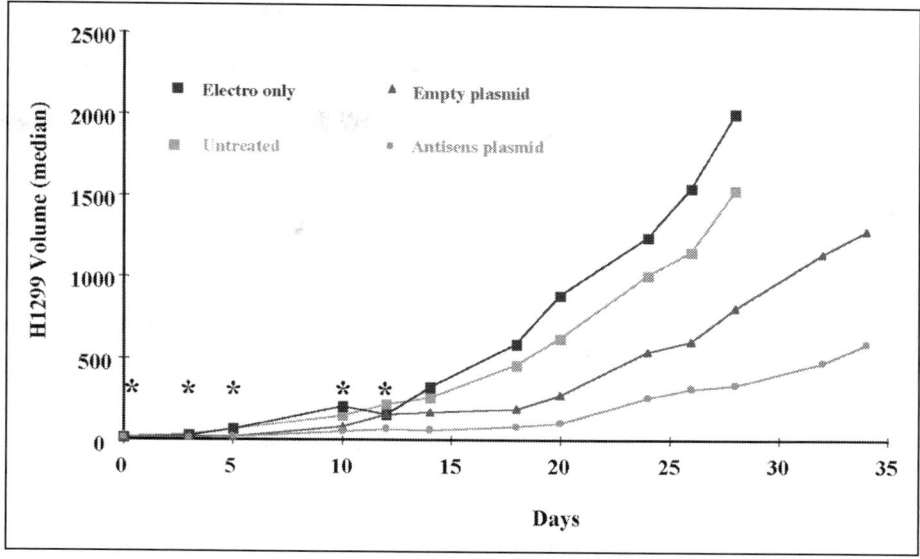

Figure 3. Antisense expression of MBD2 results in a reduction in H1299 tumor growth in vivo. indicates the treatment of H1299 xenografted tumors in Nude mice by 5 intratumoral injections (50 μg) of a MBD2 antisense coding plasmid or an empty plasmid followed by application of electric pulses.

distance ratio of 500 V/cm, delivered at the frequency of 1 Hertz, using an electropulsator PS 15 (Jouan, St Herblain, France). Tumor volume was then monitored for 4-5 weeks. Antisense treatment delayed by 9 days the time to reach an average size of 1000 mm³ relative to empty vector control and by 15 to 18 days relative to untreated controls, as shown in Figure 3. It has been described that electrically mediated delivery of vector plasmid DNA elicits an antitumor effect, which could explain the difference observed between empty vector treated versus untreated tumors.[89] This antitumor effect of a *MBD2* antisense was confirmed by another experiment using an adenoviral vector expressing *Mbd2* antisense in human lung cancer carcinoma A549 tumors in nude mice.[88]

These results strongly suggest that *MBD2* plays a critical role in maintaining the transformed state of cancer cells and is a candidate target for inhibition by antisense gene therapy in cancer, although the mechanism of inhibition of tumor growth is not elucidated yet.

As it was already mentionned above, it is unlikely that electrotransfer of tumors will lead by itself to a cure of cancer because of the relatively low number of transfected cells. As chemotherapy and *MBD2* antisense strategy follow different mechanisms to kill cancer cells or inhibit tumor growth, we thought that a synergy between these two technologies could be expected. In a separate experiment, we showed that a single intravenous injection of bleomycin combined with a MBD2 antisense treatment as previously described efficiently inhibits tumor growth in our H1299 model. Some complete tumor regression was also observed, showing the validity of our approach, and suggesting a potential therapeutic use of the DNA electrotransfer technology combined with electrochemotherapy.[90]

Conclusions

In vivo electrotransfer is a nonviral technique for reasonably efficient gene transfer. It has the main advantages of being fast, easy to perform, tissue-specific, usable in a wide range of tissues and cheap. As it does not induce the immune system, it can also be repeated as often as desired. Its exact mechanism is not yet elucidated, and improvement can be expected in its understanding from further studies. Parameters and DNA biodistribution have also to be further investigated in order to optimize this technique. Still, electrotransfer appears to be a very promising technique, both in the field of gene therapy, and of functional genomics as a laboratory tool. Although no gene therapy clinical trial is currently ongoing, it can be expected that it will soon happen. Some applications could be considered using skeletal muscle as an endocrine tissue to secrete proteins at therapeutic concentrations. For example, *Epo* in β-thalassemia or anemia, cytokines for treatment of chronic inflammatory diseases, or clotting factors in hemophilia. Furthermore, the possibility to perform multiple injections of plasmid and to use high dose injection is a great advantage.

We have reviewed data in this chapter suggesting that electrotransfer is a promising anticancer gene therapy method, particularly, electrotransfer of cytokines. Using electrotransfer of a *MBD2* antisense coding plasmid, we have provided evidence that this methylated DNA binding protein is required for tumorigenesis. As tumor growth is strongly delayed as a consequence of antisense*MBD2* gene transfer, it supports the hypothesis that it is a candidate target for an anticancer therapy. Preliminary data suggests that *MBD2* antisense electrotransfer could be used effectively in combination with electrochemotherapy. Although insufficient data is currently available on the combination of electrotransfer and electrochemotherapy, we strongly believe that important progress should be achieved in the future. Electrotransfer of methylation machinery proteins and either their antisense or iRNA inhibitors offers a novel avenue for applying our understanding of the DNA methylation machinery to anticancer therapy and target validation.

References

1. Van Tendeloo VF, Van Broeckhoven C, Berneman ZN. Gene therapy: Principles and applications to hematopoietic cells. Leukemia 2001; 15(4):523-544.
2. Nishikawa M, Huang L. Nonviral vectors in the new millennium: Delivery barriers in gene transfer. Hum Gene Ther 2001; 12(8):861-870.
3. Krieg AM, Yi AK, Matson S et al. CpG motifs in bacterial DNA trigger direct B-cell activation. Nature 1995; 374:546-549.
4. Krieg AM. CpG motifs in bacterial DNA and their immune effects. Annu Rev Immunol 2002; 20:709-760.
5. Rols M, Delteil C, Golzio M et al. In vivo electrically mediated protein and gene transfer in murine melanoma. Nature Biotech 1998; 16(2):168-171.
6. Aihara H, Miyazaki J. Gene transfer into muscle by electroporation in vivo. Nature Biotech 1998; 16(9):867-870.
7. Mir L, Bureau MF, Rangara R et al. Long-term, high level in vivo gene expression after electric pulse-mediated gene transfer into skeletal muscle. Compte-Rendu de l'Académie des sciences, Sciences de la vie/Life Sciences 1998; 321:893-899.
8. Mathiesen I. Electropermeabilization of skeletal muscle enhance gene transfer in vivo. Gene Ther 1999; 6(4):508-514.
9. Vicat JM, Boisseau S, Jourde P et al. Muscle transfection by electroporation with high voltage and short-pulse currents provide high-level and long lasting gene expression. Hum Gene Ther 2000; 11(6):909-916.
10. Teissie J, Eynard N, Gabriel B et al. Electropermeabilization of cell membranes. Adv Drug Deliv Rev 1999; 35(1):3-19.
11. Wong TK, Neumann E. Electric field mediated gene transfer. Biochem Biophys Res Com 1982; 107(2):584-587.
12. Neumann E, Schaefer-Ridder M, Wang Y et al. Gene transfer into mouse lyoma cells by electroporation in high electric fields. EMBO J 1982; 1(7):841-845.
13. Potter H. Electroporation in biology: Methods, application and instrumentation. Anal Biochem 1988; 174(2):361-373.
14. Rols MP, Teissié J. Electropermeabilization of mammalian cells. Quantitative analysis of the phenomenon. Biophys J 1990; 58(5):1089-1098.
15. Mir LM, Banoun H, Paoleti C. Introduction of definite amount of nonpermeant molecules into living cells after electropermeabilization: Direct access to the cytosol. Exp Cell Res 1988; 175(1):15-25.

16. Mir LM. Therapeutic perspectives of in vivo cell electropermeabilization. Bioelectrochemistry 2000; 53(1):1-10.
17. Neumann E, Kakorin S, Toensing K. Fundamentals of electroporative delivery of drugs and genes. Bioelectrochem Bioenerg 1999; 48(1):3-16.
18. Mir LM, Orlowki S. Mechanism of electrochemotherapy. Adv Drug Deliv Rev 1999; 35(1):107-118.
19. Chabner BA, Allegra CJ, Curt GA et al. Antineoplastic agents. In: Hardman JC, Limbird LE, Goodman and Gilman's, eds. The pharmacological basis of therapeutics. 9th edition. McGraw-Hill Co., 1996:1266-1268.
20. Belehradek Jr J, Orlowski S, Ramirez LH et al. Electropermeabilization of cells in tissues assessed by the qualitative and quantitative electroloading of bleomycin. Biochim Biophys Acta 1994; 1190(1):155-163.
21. Rols MP, Bachaud JM, Giraud P et al. Electrochemotherapy of cutaneous metastases in malignant melanoma. Melanoma Res 2000; 10(5):468-474.
22. Rodriguez-Cuevas S, Barroso-Bravo S, Almanza-Estrada J et al. Electrochemotherapy in primary and metastatic skin tumors: Phase II trial using intralesional Bleomycin. Arch Med Res 2001; 32(4):273-276.
23. Sersa G, Stabuc B, Cemazar M et al. Electrochemotherapy with cisplatin: The systemic antitumour effectiveness of cisplatin can be potentiated locally by the application of electric pulses in the treatment of malignant melanoma skin metastases. Melanoma Res 2000; 10(4):381-385.
24. Rols MP, Tamzali Y, Teissié J. Electrochemotherapy of horses. A preliminary clinical report. Bioelectrochemistry 2002; 55(1-2):101-105.
25. Gehl J, SØrensen TH, Nielsen K et al. In vivo electroporation of skeletal muscle: Threshold, efficacy and relation to electric field distribution. Biochim Biophys Acta 1999; 1428(2-3):233-240.
26. Scherman D, Bureau MF. In vivo DNA electrotransfer into skeletal muscle and other tissues: Mechanism and applications. STP Pharma Sciences 2001; 11:69-74.
27. Bonnafous P, Vernhes MC, Teissié J et al. The generation of reactive-oxygen species associated with long-lasting pulse-induced electropermeabilisation of mammalian cells is based on a non destructive alteration of the plasma membrane. Biochim Biophys Acta 1999; 1461(1):123-134.
28. Hartikka J, Sukhu L, Buchner C et al. Electroporation-facilitated delivery of plasmid DNA in skeletal muscle: Plasmid dependence of muscle damage and effect of poloxamer 188. Mol Ther 2001; 4(5):407-415.
29. Gehl J, Skovsgaard T, Mir LM. Vascular reactions to in vivo electroporation: Characterization and consequences for drug and gene delivery. Biochim Biophys Acta 2002; 1569(1-3):51-58.
30. Mir LM, Bureau MF, Gehl J et al. High-efficiency gene transfer into skeletal muscle mediated by electric pulses. Proc Natl Acad Sci USA 1999; 96(8):4262-4267.
31. Satkauskas S, Bureau MF, Puc M et al. Mechanism of in vivo DNA electrotransfer: Respective contributions of cell electropermeabilization and DNA electrophoresis. Mol Ther 2002; 5(2):133-140.
32. Golzio M, Teissié J, Rols MP. Direct visualization at the single-cell level of electrically mediated gene delivery. Proc Natl Acad Sci USA 2002; 99(3):1292-1297.
33. Paturneau-Jouas M, Parzy E, Vidal G et al. Electroporation-mediated delivery of a magnetic resonance imaging contrast agent into muscle to visualize electrotransfer. Acta Myologica 2001; XX:174-178.
34. Rols MP, Delteil C, Golzio M et al. Control by ATP and ADP of voltage-induced mammalian-cell-membrane permeabilization, gene transfer and resulting expression. Eur J Biochem 1998; 254(2):382-388.
35. Satkauskas S, Bureau MF, Mahfoudi A et al. Slow accumulation of plasmid in muscle cells: Supporting evidence for a mechanism of DNA uptake by receptor mediated endocytosis. Mol Ther 2001; 4(4):317-323.
36. Aihara H, Miyazaki J. Gene transfer into muscle by electroporation in vivo. Nature Biotech 1998; 16(9):867-870.
37. Honigman A, Zeira E, Ohana P et al. Imaging transgene expression in live animals. Mol Ther 2001; 4(3):239-249.
38. Kreiss P, Bettan M, Crouzet J et al. Erythropoietin secretion and physiological effect in mouse after intramuscular plasmid DNA electrotransfer. J Gene Med 1999; 1(4):245-50.
39. Bettan M, Emmanuel F, Darteil R et al. High-level protein secretion into blood circulation after electric pulse-mediated gene transfer into skeletal muscle. Mol Ther 2000; 2(3):204-10.
40. Payen E, Bettan M, Rouyer-Fessard P et al. Improvement of mouse β-thalassemia by electrotransfer of erythropoietin cDNA. Exp Hematol 2001; 29(3):295-300.
41. Samakoglu S, Fattori E, Lamartina S et al. βminor-globin messenger RNA accumulation in reticulocytes governs improved erhytropoiesis in β thalassemic mice after erythropoïetin complementary DNA electrotransfer in muscles. Blood 2001; 97(8):2213-2220.

42. Maruyama H, Ataka K, Gejyo F et al. Long-term production of érythropoïétine after electroporation-mediated transfer of plasmid DNA into the muscles of normal and uremic rats. Gene Ther 2001; 8(6):461-468.

43. Adachi O, Nakano A, Sato O et al. Gene transfer of Fc-fusion cytokine by in vivo electroporation: Application to gene therapy for viral myocarditis. Gene Ther 2002; 9(9):577-583.

44. Watanabe K, Nakazawa M, Fuse K et al. Protection against autoimmune myocarditis by gene transfer of interleukin-10 by electroporation. Circulation 2001; 104(10):1098-1100.

45. Mallat Z, Besnard S, Duriez M et al. Protective role of interleukin-10 in atherosclerosis. Circ Res 1999; 85(8):e17-24.

46. Xue F, Takahara T, Yata Y et al. Attenuated acute liver injury in mice by naked hapatocyte growth factor gene transfer into skeletal muscle with electroporation. Gut 2002; 50(4):558-562.

47. Prud'homme GJ, Chang Y, Li X. Immunoinhibitory DNA vaccine protects against autoimmune diabetes through cDNA encoding a selective CTLA-4 (CD152) ligand. Hum Gene Ther 2002; 13(3):395-406.

48. Vilquin JT, Kennel PF, Paturneau-Jouas M et al. Electrotransfer of naked DNA in the skeletal muscles of animal models of muscular dystrophy. Gene Ther 2001; 8(14):1097-1107.

49. Lucas ML, Heller R. Immunomodulation by electrically enhanced delivery of plasmid DNA encoding IL-12 to murine skeletal muscle. Mol Ther 2001; 3(1):47-53.

50. Cichon T, Jamrozy L, Glogowska J et al. Electrotransfer of gene encoding endostatin into normal and neoplastic mouse tissues: Inhibition of primary tumor growth and metastatic spread. Cancer Gene Ther 2002; 9(9):771-777.

51. Yamamoto M, Kobayashi Y, Li M et al. In vivo gene electroporation of glial cell line-derived neurotrophic factor (GDNF) into skeletal muscle of SOD1 mutant mice. Neurochem Res 2001; 26(11):1201-1207.

52. Matsumoto T, Komori K, Shoji T et al. Successful and optimized in vivo gene transfer to rabbit carotid artery mediated by electronic pulse. Gene Ther 2001; 8(15):1174-1179.

53. Blair-Parks K, Weston BC, Dean DA. High-level gene transfer to the cornea using electroporation. J Gen Med 2002; 4(1):92-100.

54. Dezawa M, Takano M, Negishi H et al. Gene transfer into retinal ganglion cells by in vivo electroporation: A new approach. Micron 2002; 33(1):1-6.

55. Maruyama H, Ataka K, Higuchi N et al. Skin-targeted gene transfer using in vivo electroporation. Gene Ther 2001; 8(23):1808-1812.

56. Lin CR, Tai MH, Cheng JT et al. Electroporation for direct spinal gene transfer in rats. Neurosci Lett 2002; 317(1):1-4.

57. Inoue T, Krumlauf R. An impulse to the brain using in vivo electroporation. Nat Neurosci 2001; 4 suppl:1156-1158.

58. Muramatsu T, Shibata O, Ryoki S et al. Foreign gene expression in the mouse testis by localized in vivo gene transfer. Biochem Biophys Res Commun 1997; 233(1):45-49.

59. Suzuki T, Shin BC, Fujikura K et al. Direct gene transfer into rat liver cells by in vivo electroporation. FEBS Lett 1998; 425(3):436-440.

60. Tsujie M, Isaka Y, Nakamura H et al. Electroporation-mediated gene transfer that targets glomeruli. J Am Soc Nephrol 2001; 12(5):949-954.

61. Saito T, Nakatsuji N. Efficient gene transfer into the embryonic mouse brain using in vivo electroporation. Dev Biol 2001; 240(1):237-246.

62. Tawk M, Tuil D, Torrente Y et al. High-efficiency gene transfer in adult fish: A new tool to study fin regeneration. Genesis 2002; 32(1):27-31.

63. Konig S, Burkman J, Fitzgerald J et al. Modular organization of phylogenetically conserved domains controlling developmental regulation of the human skeletal myosin heavy chain gene family. J Biol Chem 2002; 277(31):27593-27605.

64. Lamartina S, Roscilli G, Rinaudo CD et al. Stringent control of gene expression in vivo by using novel doxycycline dependant trans-activators. Hum Gene Ther 2002; 13(2):199-210.

65. Payen E, Bettan M, Henri A et al. Oxygen tension and a pharmacological switch in the regulation of transgene expression for gene therapy. J Gene Med 2001; 3(5):498-504.

66. The J Gene Med website. http://www.wiley.co.uk/wileychi/genmed/clinical/. Nov 2002.

67. McCormick F. Cancer gene therapy: Fringe or cutting edge? Nature Rev Cancer 2001; 1(2):130-141.

68. Wadhwa PD, Zielske SP, Roth JC et al. Cancer gene therapy: Scientific basis. Annu Rev Med 2002; 53:437-452.

69. Bettan M, Ivanov MA, Mir LM et al. Efficient DNA transfer into tumors. Bioelectrochemistry 2000; 52(1):83-90.

70. Shibata MA, Morimoto J, Otsuki Y. Suppression of murine mammary carcinoma growth and metastasis by HSVtk/GCV gene therapy using in vivo electroporation. Cancer Gene Ther 2002; 9(1):16-27.

71. Goto T, Nishi T, Tamura T et al. Highly efficient electro-gene therapy of solid tumor by using an expression plasmid for the herpes simplex virus thymidine kinase gene. Proc Natl Acad Sci USA 2000; 97(1):354-359

72. Li S, Xia X, Zhang X et al. Regression of tumors by IFN-α electroporation gene therapy and analyss of the responsible genes by cDNA array. Gene Ther 2002; 9(6):390-397.

73. Kishida T, Asada H, Satoh E et al. In vivo electroporation-mediated transfer of interleukin-12 and interleukin-18 genes induces significant antitumor effects against melanoma in mice. Gene Ther 2001; 8(16):1234-1240.

74. Tamura T, Nishi T, Goto T et al. Intratumoral delivery of interleukin 12 expression plasmids with in vivo electroporation is effective for colon and renal cancer. Human Gene Ther 2001; 12(10):1265-1276.

75. Li S, Zhang X, Xia X. Regression of tumor growth and induction of long-term antitumor memory by interleukin 12 electro-gene therapy. J Natl Cancer Inst 2002; 94(10):762-768.

76. Matsubara H, Gunji Y, Maeda T et al. Electroporation-mediated transfer of cytokine genes into human esophageal tumors produces anti-tumor effects in mice. Anticancer Res 2001; 21(4A):2501-2503.

77. Yamashita Y, Shimada M, Tanaka S et al. Electroporation-mediated tumor necrosis factor-related apoptosis-inducing ligand (TRAIL)/Apo2L gene therapy for hepatocellular carcinoma. Human Gene Ther 2002; 13(2):275-286.

78. Celiker MY, Wang M, Atsidaftos E et al. Inhibition of Wilms' tumor growth by intramuscular administration of tissue inhibitor of metalloproreinases-4 plasmid DNA. Oncogene 2001; 20(32):4337-4343.

79. Baba M, Iishi H, Tatsuta M. Transfer of bcl-xs plasmid is effective in preventing and inhibiting rat hepatocellular carcinoma induced by N-notrosomorpholine. Gene Ther 2001; 8(15):1149-1156.

80. Kalat M, Küpcü Z, Schüeller S et al. In vivo plasmid electroporation induces tumor antigen-spectic CD8+ T-cell responses and delays tumor growth in a syngenic mouse melanoma model. Cancer Res 2002; 62(19):5489-5494.

81. Cemazar M, Sersa G, Wilson J et al. Effective gene transfer to solid tumors using different nonviral delivery techniques: Electroporation, liposomes, and integrin-targeted vectors. Cancer Gene Ther 2002; 9(4):399-406.

82. Jansen B, Zangemeister-Wittke U. Antisense therapy for cancer-the time of truth. Lancet Oncol 2002; 3(11):672-683.

83. Ng HH, Zhang Y, Hendrich B et al. MBD2 is a transcriptional repressor belonging to the MeCP1 histone deacetylase complex. Nat Genet 1999; 23(1):58-61.

84. Bhattacharya SK, Ramchandani S, Cervoni N et al. A mammalian protein with specific demethylase activity for mCpG DNA. Nature 1999; 397:579-583.

85. Detich N, Theberge J, Szyf M. Promoter-specific activation and demethylation by MBD2/ demethylase. J Biol Chem 2002; 277(39):35791-35794.

86. Razin A. CpG methylation, chromatin structure and gene silencing-a three-way connection. EMBO J 1998; 17(17):4905-4908.

87. Baylin SB, Esteller M, Rountree MR et al. Aberrant patterns of DNA methylation, chromatin formation and gene expression in cancer. Hum Mol Genet 2001; 10(7):687-92.

88. Slack A, Bovenzi V, Bigey P et al. Antisense MBD2 gene therapy inhibits tumorigenesis. J Gene Med 2002; 4(4):381-389.

89. Heller L, Coppola D. Electrically mediated delivery of vector plasmid DNA elicits an antitumor effect. Gene Ther 2002; 9(19):1321-1325.

90. Ivanov MA, Lamrihi B, Szyf M et al. Enhanced antitumor activity of a combination of MBD2-antisense electrotransfer gene therapy and bleomycin electrochemotherapy. J Gene Med 2003 10:893-899.

EPILOGUE

Moshe Szyf

Advances made in the last decade have firmly established the critical role of the epigenome in orchestrating the complex and dynamic gene expression program of multicellular organisms such as humans. The epigenome is composed of two distinctly different layers of information, chromatin and DNA methylation. While chromatin is associated with the genome and serves to package its different regions in either tight or open structures, DNA methylation is part of the chemical covalent structure of the DNA. DNA methylation is therefore believed to be a fixed component of the epigenome and to be a consistent and stable signal of gene inactivation. These two layers of information are tightly correlated. DNA methylation is characteristic of inactive regions of the genome that are packaged in tight chromatin, whereas hypomethylated DNA is found in open and active chromatin structures. Recent advances in understanding the relation between chromatin and DNA methylation have provided some insights into the mechanisms that tie these processes to each other. It is clear that DNA methylation has to be understood within its chromatin context and aberrations in DNA methylation must be understood in relation to changes in chromatin structure and in the proteins that remodel chromatin. Cancer is a disease of foiled programming of gene expression and could be therefore considered an epigenomic disease. Aberrations in either one or both chromatin structure and DNA methylation have been found by many studies to be a persistent hallmark of cancer. An understanding of DNA methylation changes and their diagnostic and therapeutic implications in cancer could only be achieved if they are analyzed in the context of the chromatin.

The chapters of this book unravel multiple meeting points between DNA methylation and cancer therapy. Each of these points has distinct implications for cancer therapy. A first example relates to the diagnostic potential of DNA methylation in cancer. Notwithstanding the causal role of DNA methylation in cancer, it is well established that distinct DNA methylation patterns characterize many tumors when compared with their noncancerous-paired tissue. Several of the changes in DNA methylation observed in cancer are easy to explain since they include hypermethylation of tumor suppressor genes, which marks these genes for inactivation. Similarly, methylation of repair genes, adhesion proteins and angiogenesis inhibitors confer a selective advantage upon cancer cells. Such changes in DNA methylation are consistent with a causal role for DNA methylation in cancer and could provide a clear mechanism. However, not all changes in methylation necessarily relate to a clear biological function. Paradoxically, in addition to DNA hypermethylation of specific genes, it is well established that global hypomethylation of repetitive sequences as well as of genes that promote metastasis is characteristic of many tumors as discussed earlier in the book. Whereas resolving this paradox is critical for our understanding of the mechanisms responsible for alterations in DNA methylation in cancer as well as their therapeutic potential, the diagnostic value of methylation changes is independent of these questions. The diagnostic value of DNA methylation markers is a function of their correlation with tumorigenic states and not their mechanism of action. Thus, it is possible to take advantage of the unique DNA methylation profiles of tumors without understanding their function.

The main issue that will hopefully be resolved in the near future is whether we could utilize specific DNA methylation profiles for early diagnosis of cancer, classification of tumor grades, and predicting their susceptibility to specific therapies. Up to recently, a small number of genes,

DNA Methylation and Cancer Therapy, edited by Moshe Szyf. ©2005 Eurekah.com and Kluwer Academic/Plenum Publishers.

which were selected for analysis by a candidate gene approach, were shown to be altered by DNA methylation. This limited and biased repertoire was insufficient for methylation profiling of a broad range of tumors and for classification of cancers. Some of the candidate genes are methylated only in a subset of tumors and cannot serve as markers for comprehensive diagnostic tests. The prediction however is that all cancers exhibit cancer-type and grade-characteristic DNA methylation profiles that would be unraveled once a broad range of methylation markers are defined. This book discusses a number of whole genome approaches for methylation profiling. These whole genome approaches will hopefully lead to a comprehensive directory of methylation markers. As a consequence, this might provide diagnostic methylation tools that will increase the precision of early diagnosis as well as result in a more accurate classification of cancers.

Although the aberrations in DNA methylation in cancer might have important diagnostic value irrespective of the mechanism causing them, it is essential to understand how these paradoxical patterns of methylation are generated in cancer and whether they have a causal role in tumorigenesis. Answering these questions has evidently important implications on any potential use of DNA methylation therapeutics in cancer. Without understanding how these changes in methylation come about, it would be impossible to truly determine their role in tumorigenesis. In the absence of a comprehensive understanding of the role of methylation changes in the mechanisms of tumor generation and progression, it is hard to take full advantage of the therapeutic potential of the DNA methylation machinery. It is imperative that future studies will be directed at these cardinal questions.

An important issue that needs to be resolved is whether aberrant DNA methyltransferase expression can stimulate tumorigenesis independent of DNA methylation. DNA methyltransferases are multifunctional proteins, which are involved in suppression of gene expression and DNA replication in addition to their DNA methylating activity. It is essential that the specific functions of DNA methyltransferases that lead to tumorigenesis be identified. This will allow us to direct therapies at these functions specifically. It is also critical to determine whether hypomethylation plays a causal role in cancer and what is the mechanism involved. As discussed earlier in this book, DNA methyltransferase inhibitors are tested in clinical trials for their anticancer activity. If hypomethylation of DNA can play a causal role in tumorigenesis by stimulating metastasis as previously proposed, hypomethylating agents should be used with extreme care. Other agents that inhibit the tumor promoting activity of DNA methyltransferase 1 in the absence of global hypomethylation should thus be used.

Recent studies discussed in this book suggest that our whole understanding of the DNA methylation machinery must be redirected in light of the putative involvement of DNA demethylases and chromatin structure in shaping DNA methylation patterns. Our traditional understanding of the DNA methylation pattern has been that the pattern is laid down during development and is then fixed and maintained by a semiconservative DNA methyltransferase throughout life, which copies the DNA methylation pattern as directed by the template state of methylation. This model fails to explain how methylation patterns change in somatic cells once they are transformed. Using this model, it is even more difficult to explain how it is possible to have both DNA hypermethylation and demethylation occurring simultaneously in the same cancer cell. It has originally been proposed that increased DNA methyltransferase results in increased DNA methylation. However, there is no strong data to suggest that regional hypermethylation of CG islands correlates with the levels of the DNA methyltransferases. In addition, if an increase in DNA methyltransferase activity is responsible for the changes in DNA methylation in cancer cells, how is it possible to have global hypomethylation in the presence of high levels of DNA methyltransferase activity? It is clear that our long-established understanding of the maintenance of DNA methylation patterns in somatic cells lacks a number of key players.

Two very recent advances might unveil a new understanding of DNA methylation patterns in general and particularly in cancer. These advances raise the prospect that the DN A methylation pattern is in a dynamic steady state in somatic cells, and that a relative change in the factors that maintain this dynamic steady state in cancer can result in alteration of DNA methylation. First, is the realization that chromatin structure might have a serious impact on DNA methylation patterns, and since chromatin structure is dynamic, DNA methylation might be dynamic as well? Second, is the discovery of demethylase enzymes that reverse DNA methylation patterns in a replication independent manner, thus introducing a novel understanding of DNA methylation pattern as a balance of two reversible reactions, DNA methylation and demethylation. The access of demethylases to methylated DNA is gated by chromatin structure as discussed in the chapter by Szyf et al. Local changes in chromatin structure that alter accessibility to demethylase can explain how regional hypermethylation is generated in the presence of high levels of DNA demethylase. On the other hand, a global increase in demethylase activity might explain global demethylation in cancer. Future research must delineate how chromatin structure fashions DNA methylation patterns in cancer and identify the key factors that alter the accessibility of DNA to either DNA methyltransferases or demethylases. These factors might unfold into important cancer drug targets.

Understanding the role of other factors in altering chromatin and DNA methylation patterns in cancer will help us address the issue of whether DNA methylation patterns *per se* play a causal role in tumorigenesis, as currently believed, or whether these changes are merely fingerprints of other important alterations. Addressing this question is obviously critical for DNA methylation based anticancer therapy. It is essential to define the goal of therapy as either reversing DNA methylation patterns or as interfering with the factors that cause these changes in DNA methylation. In the latter case, DNA methylation is a surrogate marker of other more significant events. A future understanding of DNA methylation pattern changes and their relevance to cancer will require a complete different perception of DNA methylation as a reversible and dynamic state, which is in an interactive relation with other components of chromatin.

Using pharmacological or genetic knock down of the different components of the DNA methylation machinery, it is possible to determine that a certain protein plays a causal role in cancer and is therefore an anticancer drug target. This could be accomplished in absence of a full understanding of the mechanisms involved. A long list of data from tissue culture, animal and clinical trials supports the hypothesis that DNA methyltransferase1 (DNMT1) is critical for cancer. Recent data also suggests that MBD2/demethylase is critical for cancer. However, these proteins play different roles in cancer. DNMT1 is important for cell growth and possibly initiation of DNA replication, while MBD2/demethylase is not required for normal cell growth. The fact that different proteins of the DNA methylation machinery are required for distinct processes involved in tumorigenesis raises the hope that in the future we will be able to accurately target specific cellular functions critical for cancer using these agents. Agents that affect tumorigenesis without affecting the cell cycle are especially attractive, since they should not have side effects on dividing normal tissue which is common to most anticancer drugs that target cell growth functions.

The role of hypomethylation in cancer has been neglected for some time. The chapter by Ehrlich and colleagues provides an incentive to revisit the therapeutic implications of these observations. In addition to the cautionary note raised above on the potential untoward effects of demethylating agents, it is possible that inhibitors of hypomethylation would also be anticancer and antimetastatic agents. The chapter by Szyf et al discusses the therapeutic implications of hypermethylating agents. The discovery of demethylases such as MBD2/demethylase, which is highly expressed in some tumors, raises the possibility that inhibition of hypomethylation could be accomplished by inhibiting demethylases. However, our understanding of demethylases and their regulation and deregulation in cancer is rudimentary. It is therefore important to characterize the demethylases that are highly expressed in cancers and are involved in tumorigenesis. It is also critical to delineate the specific tumorigenic factors, which are regulated by these demethylases. Understanding the molecular machineries which contain demethylases and determining the factors that guide their specificity is obviously highly significant for any attempt to design therapeutic agents targeting demethylases. Although this area of DNA methylation is in its infancy, it is conceivable that in the near future demethylases would become important targets for anticancer agents.

Another important issue that is unresolved and requires further attention is the relation between the level of methyl promoting agents such as folates in diets, DNA methylation, and cancer. We must understand how diets affect the levels of both the methyl donor S-adenosylmethionine (AdoMet), and S-adenosylhomocysteine, the product of the DNA methylation reaction, in target tissues. It is also important to determine the mechanisms through which AdoMet levels affect DNA methylation levels. AdoMet was originally proposed to stimulate the DNA methyltransferase reaction, but if the DNA methylation pattern is dynamic, then methylation-promoting and methylation-deficient diets might alter both sides of the DNA methylation equilibrium. Recent data from our laboratory suggests that AdoMet inhibits demethylase activity. It is important to determine whether AdoMet inhibits the specific demethylase activity responsible for the global hypomethylation in cancer. If hypomethylation plays a causal role in cancer progression or metastasis, and if it is possible to inhibit it by modulating dietary intake of methyl promoting agents, nutrition might emerge to play an important role in DNA methylation based anticancer therapy and prevention. The possibility that the deleterious effects of global hypomethylation could be modulated by diet is extremely attractive. In addition, pharmacological agents that mimic the activity of AdoMet and folates might then be developed to reverse global hypomethylation and its putative effects on tumor progression.

In summary, while a long list of data reviewed in this book has established many links between DNA methylation and cancer therapy and diagnostics, many questions remain to be resolved. We hope that the chapters of this book will inspire the reader to get involved in studying the remaining issues discussed here. Unraveling of these issues promises to unfold into new modalities of cancer diagnosis and cancer therapy, as well as a better understanding of the mechanisms involved in cancer and of the basic rules that guide and maintain epigenomic gene regulation in somatic cells.

Index